Reliability Engineering

Updated throughout for the second edition, *Reliability Engineering: A Life Cycle Approach* draws on the author's global industry experience to demonstrate the invaluable role reliability engineers play in the entire life cycle of a plant.

Applicable to both high-cost, cutting-edge plants and to plants operating under serious budget constraints, this textbook uses a practical approach to cover the theory of reliability engineering, alongside the design, operation, and maintenance required in a plant. This textbook has been updated to cover the modern standards of maintenance practice, most notably the ISO 55 000 standards. It also covers linear programming, failure analysis, financial management, and analysis. This textbook refers to case studies throughout.

This textbook will be of interest to students and engineers in the field of reliability, mechanical, manufacturing, and industrial engineering. It will also be relevant to automotive and aerospace engineers.

Reliability Engineering

A Life Cycle Approach

Second Edition

Dr Edgar Bradley

CRC Press is an imprint of the
Taylor & Francis Group, an **informa** business

Second edition published 2023
by CRC Press
6000 Broken Sound Parkway NW, Suite 300, Boca Raton, FL 33487–2742

and by CRC Press
4 Park Square, Milton Park, Abingdon, Oxon, OX14 4RN

CRC Press is an imprint of Taylor & Francis Group, LLC

First edition published by CRC Press 2017

ISBN: 978-1-032-35337-1 (hbk)
ISBN: 978-1-032-35349-4 (pbk)
ISBN: 978-1-003-32648-9 (ebk)

DOI: 10.1201/9781003326489

Typeset in Times
by Apex CoVantage, LLC

Contents

Foreword to the First Edition

This book is written by a mechanical engineer with decades of experience in reliability matters. What it adds to the corpus of books on the subject of reliability is a unique approach between theory and practice and between the various phases in the system life cycle. Design for reliability is covered, as is the importance of maintaining reliability and availability over the operating life cycle of the system, with the theory at all stages being reinforced with case study material.

I have known the author for many years and can vouch for his technical and teaching abilities. These attributes come to the fore in the theory he presents and in the case studies that he has assembled from his own experience and the experience of others around the world.

The book came to life during our discussions for a postgraduate programme at my university. Our clients in industry wanted a broad-based programme covering theory and management issues as well as some subjects of interest to maintenance and reliability professionals that are not always taught at either undergraduate or graduate level.

The book is for busy mechanical, electrical, and mining engineers, to round out their education as they face reliability problems in industry, giving them an understanding of reliability issues and techniques to solve them.

Having said that, I believe the book will also be useful at undergraduate level, where there are few texts that cover the necessary ground thoroughly yet succinctly.

Finally, the book will also be useful in business schools, where the case study approach is well-known but the emphasis on maintenance and reliability of plants, rather than just their operation, is often sadly lacking. This book fulfils a real need in that connection.

Dr Harry Wichers
Professor, Mechanical and Nuclear Engineering
North West University
South Africa
2016

Preface

The first edition of this book, produced in 2016, emphasised the connection between reliability on the one hand and maintenance on the other by combining both subjects in one volume. In this second edition, extra chapters have been added including one on ISO 55000, the internationally accepted standard on asset management.

Corrections have been made to the few typographical errors in the first edition.

This second edition then remains up to date with the latest principles in reliability and maintenance practice.

Author

Dr Edgar Bradley is a consultant specialising in reliability engineering and related topics.

His qualifications include a bachelor's degree in mechanical engineering from the University of the Witwatersrand, Johannesburg, South Africa, and a master's degree in mechanical engineering from the University of Akron, Ohio, USA, as well as an MBA from Cranfield University in the UK and a PhD from North West University.

He completed his compulsory military service in the Technical Services Corps of the South African Army, being discharged with the rank of major.

He was formerly employed by Eskom, the South African power utility, including 11 years as a project engineer before becoming a reliability specialist. He has also worked for mechanical contracting firms as a project and design engineer and has 50 years of industrial experience in project management, benchmarking, RAM engineering, power plant, pipework engineering, cryogenics, and plant maintenance.

For more than 40 years, he has also served as a part-time lecturer in the Engineering Faculty at the University of the Witwatersrand, lecturing and developing courses on the MSc programme on, inter alia, project management, reliability engineering, maintenance engineering, industrial marketing, and systems engineering. He has also lectured at the University of Pretoria, at Northwest University, and at the Namibian University of Science and Technology.

He is a registered professional engineer and a member of the South African Institute of Industrial Engineers. He was also twice president of the Southern African Maintenance Association (SAMA—now SAAMA).

In 2016, he authored this book, *Reliability Engineering: A Life Cycle Approach*, published by CRC/Taylor and Francis in the USA, now in its second edition.

Introduction

RELIABILITY ENGINEERING: A LIFE CYCLE APPROACH

Reliability engineering is growing fast and changing in the process. In the last century, it might have been considered as an adjunct to the design process, but it has since grown to cover areas like risk management and maintenance. In fact, in several technologies, such as petrochemicals and mineral processing, persons who used to be called maintenance engineers are now titled as reliability engineers. This is because it is now recognised that such persons must be competent in areas other than just maintenance.

Hence the title of this handbook—the subject of reliability engineering is becoming broader, and few current texts cover the entire field. With this book, we hope to correct that. Secondly, conciseness is necessary; otherwise, such a broad subject cannot be covered in one book!

Reliability in this text is an engineering parameter, as are the associated concepts of availability and maintainability. These are definable, measurable, achievable, and manageable. And in almost every technology, these parameters are compelled by market and other forces to increase with time. What was considered an acceptable level of reliability 25 years ago might be totally unacceptable today. Responding to this pressure to improve these parameters is also part of the reliability engineer's job, and hence the need to be familiar with the latest techniques which might assist in this goal.

Another important parameter is availability, which is what results from a systems reliability and maintainability. The importance of this parameter cannot be overemphasised—with increasing demand for a product, it may not be necessary to build new production facilities, if the availability of the existing facilities can be increased. *Availability is therefore a proxy for capacity, and therefore for capital.* Increasing availability can mean getting something for nothing.

What is fairly unique about this book's approach is the balance between theory, examples, both illustrative and to-be-solved, and case studies. And by *case studies* we mean business school–type cases, which have to be discussed in a group and answers given to practical problems, not all of them analytical. This is to help cover the *management* of any process that the reliability professional might encounter; the theory might be correct, but the management of the process might derail all the engineer's good intentions.

The cases are drawn from a variety of sources—some are the writer's own, but not all. Some are disguised to protect the innocent and the guilty; some are not. The cases cover a wide range of technologies, including mechanical, structural, electrical, and aeronautical engineering, as well as mining, nuclear power, and chemical processing.

Several subjects in this book are not usually found in reliability texts. One such example is the theory and practice of lubrication, a subject frequently neglected in reliability initiatives. Another is the use of linear programming in the solution of availability problems. Such topics have been included because the author has found them useful in his long career as a reliability professional.

Human reliability is also included, because in the end, *human reliability* is what it is all about. Equipment is neutral—if it is unreliable, it is because some humans designed, made, operated, or maintained it incorrectly. This text does therefore not neglect this important part of the reliability engineer's portfolio.

1 Reliability Fundamentals I
Component Reliability

I often say that when you can measure what you are speaking about, and express it in numbers, you know something about it; but when you cannot measure it, when you cannot express it in numbers, your knowledge is of a meagre and unsatisfactory kind; it may be the beginning of knowledge, but you have scarcely in your thoughts advanced to the state of *Science*, whatever the matter may be.

Lord Kelvin

INTRODUCTION

The Importance of Reliability

Reliability is sometimes ill-defined but always important in the minds of professional engineers and laymen alike. No one wants an unreliable product. Over time, products do tend to become more reliable. For example, twenty-first-century cars are demonstrably more reliable than cars of the 1940s, for example. This is because of the enormous production volume of motor vehicles that has enabled and forced manufacturers to correct defects that the market has highlighted. This is in fact the oldest reliability improvement method: try-fix-try. Other, more scientific methods have also been used, but try-fix-try has served the motor industry well. For systems other than motor vehicles, reliability improvement happens more slowly, unless special techniques are applied to force reliability improvement. Aircraft, which have much lower production runs than cars, are a case in point. So are chemical plants and other types of manufacturing plant.

HISTORY

Reliability engineering is a separate branch of engineering developed in World War II (WWII). In fact, engineering as a discipline has been splitting into ever more specific subdisciplines ever since its emergence as a profession centuries ago. It may be said that reliability engineering split off from aeronautical engineering, itself a subset of mechanical engineering. And it is also true to say that, unfortunately, as with many engineering advances, it was the hothouse atmosphere of war that spurred the development of the reliability engineering profession. A German engineer, Robert Lusser, was involved in the V1 weapon programme for the Luftwaffe. (More details of this will be given in Chapter 2.) Several of the early concepts of reliability engineering are attributable to him.

DOI: 10.1201/9781003326489-1

DEFINITIONS

Lusser was one of many German engineers and scientists who immigrated to the United States after WWII and became part of the US missile and space programmes. In 1956, at a symposium at the Convair aircraft works in San Diego, he defined *reliability* in engineering terms for the first time. Engineering reliability, he said, was defined as follows.

Reliability

The probability that a system will continue to work, for a stated period of time, given defined operating conditions. From this definition, we can deduce the following:

- Reliability engineering is involved with statistics, as the term *probability* implies.
- Engineering reliability is a time function.
- Engineering reliability depends on stated conditions. A piece of equipment designed for use in the Antarctic might not do so well in the Sahara.

The significance of reliability in an engineering sense is that it is an engineering parameter like any other, like efficiency, power, or whatever. The difference between reliability engineering parameters and more traditional parameters is that they are often statistical, or they are only measurable after quite a long period. The efficiency of an electric motor, for example, can be determined in minutes once the motor is instrumented and then put on load. The reliability of the motor can only be ascertained after a long operating period—perhaps years. Also, the answers we get are statistical and probabilistic rather than deterministic. The answers should always be given with limits of statistical error attached. Reliability is usually denoted by $R(t)$.

Common measures of reliability are the mean time between failures, or MTBF, the mean time to failure, or MTTF, and the maintenance-free operating period, or MFOP. The symbol θ is often used for MTBF.

Other definitions of importance are given in the sections that follow.

Maintainability

Maintainability is the probability that a system, having failed, will be restored in a given time, given a certain maintenance environment.

Thus, we see that *maintainability* is analogous to *reliability*. It is a statistical function, it is a function of time, and it depends on certain conditions, that is, the maintenance environment. Maintainability is usually denoted by $M(t)$.

A common measure of maintainability is the mean time to repair, or MTTR. The symbol used for MTTR is usually φ. MTTR is further elaborated on in the following text.

Availability

Availability is the percentage of time that a system is available for use, whether it is required or not.

Availability is, for our purposes, not a statistical function. It is simply a percentage. It is not a basic parameter, such as reliability or maintainability, but is in fact derived from them. It is, however, in the case of maintained systems, the most important parameter and the one usually measured. Availability is usually given the capital letter A in reliability formulae, for example, as given here:

$$A = \frac{\text{MTBF}}{\text{MTBF} + \text{MTTR}} \quad 1.1$$

or

$$A = \frac{\theta}{\theta + \phi} \quad 1.2$$

Unreliability

Unreliability is the complement of reliability and is usually denoted by a capital F. In other words, $R + F \equiv 1$. It is defined as follows:

> Unreliability is the probability that a system will fail before a certain time, given certain operating conditions.

Unavailability

This parameter is usually denoted by a capital U. It is the complement of availability.

Component

A *component* is a part of a system that is irreparable. When it fails, it is replaced.

Correlation Coefficient

Correlation coefficient, or simply correlation, is a measure of the accuracy to which data points conform to a line graph. If all the points lie on the line, the correlation is 100%. The higher the correlation, the more confidence we have that our data conform to the given equation.

System

A *system* is a set of components. A system can be restored by replacing components or by reworking or restoring subsystems.

Failure

Failure is the inability of a component or system to continue to function. Failures can be of two main types: catastrophic failures happen suddenly and completely, while degradation failures are those in which the system or component gradually loses the ability to function. Such failures are common in engineering systems. They must be defined and agreed on by all stakeholders. For example, if the output pressure of a pump drops below 80% of its value when new, we may agree that that is a failure and will necessitate a shutdown. The percentage chosen will depend on the effect the pressure reduction has on the rest of the system.

Failure Rate

Failure rate is the number of failures per unit time.

Failure Probability Density Function

This is the probability of failure at a specific time. It is usually denoted by $f(t)$, with f in lower case (*note:* not by capital letter $F(t)$, which denotes unreliability).

Sample

A *sample* is a set of items drawn from a parent population to provide information about the population.

Population

A *population* is a complete set of values of a variable. Information about a population comes by *sampling* from the population.

ACRONYMS

A list of acronyms is now included, which will be used at various points in the chapters that follow.

ABD availability block diagram
DCF discounted cash flow
FMEA failure modes and effects analysis
FMECA failure modes and effects criticality analysis
FRACAS failure reporting and corrective action system
FTA fault tree analysis
KPI key performance indicator
LCC life cycle costing
MDT mean downtime
MFOP maintenance-free operating period
MTBF mean time between failures
MTTF mean time to failure
MTTR mean time to repair
RBD reliability block diagram
RCA root cause analysis

RCM reliability-centred maintenance
RTF run-to-failure
TPM total productive maintenance

BASIC STATISTICS

It is impossible to appreciate reliability engineering without some knowledge of statistics.

PROBABILITY

The first concept to understand in statistics is probability. One dictionary defines *probability* as "a measure of how likely it is that some event will occur; a number expressing the ratio of favourable cases to the whole number of cases possible."

Thus, probability can be expressed either as a percentage between 0% and 100% or as a number between 0 and 1. Hence, if the reliability of some device is 90% at 2,000 hours, this means that there is a 90% chance of it having been in continuous operation for 2,000 hours. For some, this is hard to visualise. A better way of understanding this is to assume we had 100 of these devices. Then, for a reliability of 90% at 2,000 hours, 90 of them would still be working at 2,000 hours, and 10 would have failed before 2,000 hours.

Some find statistics a difficult subject because some of the (correct) conclusions drawn by statisticians are counter-intuitive to the layman. In other words, they seem to defy common sense.

The simplest of these problems is probably the case of tossing a coin. Assume that you have observed a coin tossed three times and each time it has come down heads. *What do you think is going to happen on the fourth throw? Heads or tails?* The answer is that the history does not matter—there is a 50% chance of a head and a 50% chance of a tail on the new throw. This introduces to us the concept of *randomness*, which we will re-encounter in reliability later on. A more difficult example of the strangeness of statistics is the Monty Hall problem, given in Figure 1.1.

As an example of the counter-intuitive nature of statistics, we present here what has become known as the Monty Hall problem. Monty Hall was a North American TV quiz show host who presented the following conundrum to the contestants: There are three closed doors on the stage. Behind two of the doors are goats, and behind one of the doors is a car. The quiz show host knows the locations of the goats and the car; the contestant does not. The host will not move the three around. They stay put.

He asks the contestant to choose a door. If the door is eventually opened and found to reveal the car, the contestant wins it. If the opened door reveals a goat, well, the goat is not an option—they don't want to eat it or have it as a pet, so they leave the studio empty-handed.

After the contestant has chosen a door, the host opens one of the other doors and reveals a goat.

So there are now two closed doors. Behind one is a goat and behind the other is the car. The host then asks the contestant if he wishes to change his choice of door or not before the second door is opened.

FIGURE 1.1 The Monty Hall problem.

Assignment

The question is, does the contestant increase his chances of winning the car by changing his choice? Justify your answer using probabilities.

The Failure Probability Density Function

In statistics, we are very often dealing with groups of things. Sometimes, a group will be referred to as a sample, sometimes as a population. This leads us to discuss the important concept of the failure probability density function, also known as the failure distribution.

This can take various forms, several of which are common in reliability work. The four most common are as follows:

- Infant mortality
- The negative exponential distribution
- The normal distribution
- The Weibull distribution

All these distributions are useful in analysing various types of failure. In this chapter, these four functions will be presented in different ways:

- As a simple linear plot of time
- As a graph of failure rate against time
- As a graph of the failure probability density function against time

The linear time plot is given here to establish understanding. It is not normally useful as the timescales used in practice are often too long for normal presentation. If a log scale is used for time, the plot is no longer valuable, as the idea is to present the frequency of failures over an undistorted time axis. The failure rate presentation is quite common, showing the numbers failing per unit time against time. Here, the time axis is often expressed logarithmically. The third method of presentation is to plot probability of failure against time. The resulting graph is called the failure probability density function, $f(t)$.

Three Common Failure Patterns

One type of failure pattern that might manifest is infant mortality, so called because statistical studies of failure were first done by medical statisticians before the concept was taken over by engineers. Medical statisticians were dealing with the life of human populations, hence the term. For this case, failures are clustered in the front of a graph, as shown at the top of Figure 1.2.

In Figure 1.2, in each case, we have four periods, but we see that in the top graph, the failures all occur very early on. This is typical infant mortality. A note of caution is necessary here. Just because infant mortality exists in engineering components and systems does not mean one should accept it. Infant mortality is always a sign of a *quality* problem, either with the supplier or with the installer. One should never accept infant mortality as inevitable. One must always find the source and eliminate it.

The central graph in Figure 1.2 describes random failure, which also goes by the name of constant failure rate. The question arises in the minds of some: "If the failure rate is constant, how can failures occur at random?" The word *constant* implies some form of regularity, does it not? The central diagram in Figure 1.2 should help clear up this misunderstanding. Let us assume we have two failures per period. (The period can be of any length.) The failures can then occur in any pattern, provided there are two failures in every period.

If we study the central diagram in Figure 1.2, we will see that the first two failures are very close together. We might expect, from this information, that the next failure will immediately follow, but it does not. There is a longer period, and then the next failure occurs. But then there is an even longer period before the next failure, but then a shorter period before the next. But the failure rate is constant at two per period. Hence, *constant failure rate* and *random failure* mean the same thing.

A very different type of failure pattern is shown in the lower graph in Figure 1.2, where there are four failures in the four periods combined, but they are all clustered in the final period. This is a *wearout* failure pattern and is also commonly encountered in reliability work. It is described mathematically by the equation for the *normal distribution*.

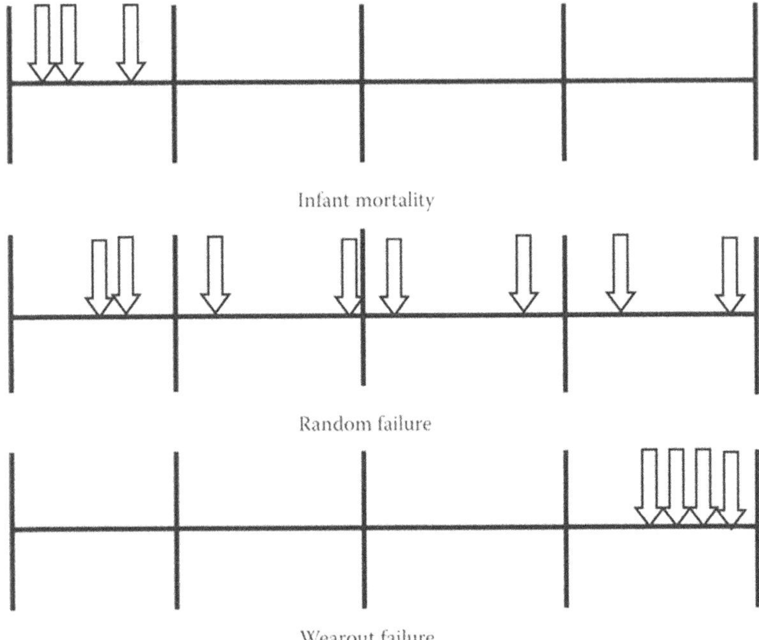

FIGURE 1.2 Three failure patterns.

The Negative Exponential Distribution

This distribution is shown in Figure 1.3. The negative exponential distribution has the form

$$f(t) = \lambda e^{-\lambda t}, \tag{1.3}$$

and its corresponding reliability function has the form

$$R(t) = e^{-\lambda t}, \tag{1.4}$$

where λ is the failure rate. It is therefore seen to be a *single-parameter distribution.* Furthermore, the parameter λ is seen to be a constant. The reciprocal of this failure rate is the MTBF. The symbol normally used for MTBF is θ. θ is also therefore the mean of the distribution.

The negative exponential failure distribution describes the phenomenon of random failure, which implies that the history of previous failures is irrelevant and cannot be of any use in predicting when the next failure might occur. Random failure is unfortunately a reality in many systems, and the consequence is that scheduled or preventive maintenance is not possible in these cases.

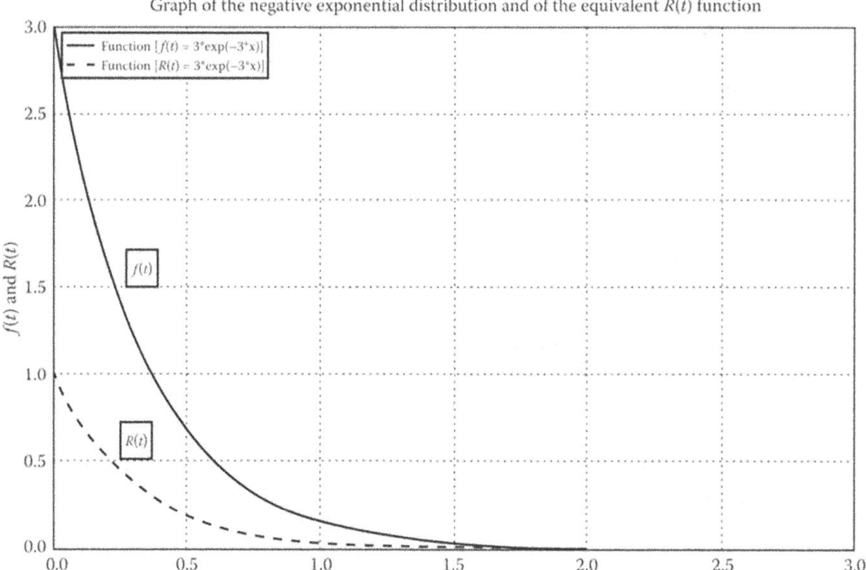

FIGURE 1.3 The negative exponential function $f(t)$ and the corresponding reliability function $R(t)$, with $\lambda = 3$.

THE MATHEMATICS OF RANDOMNESS

A constant failure rate implies that the system has no memory. Therefore, one cannot predict when failures might occur on the basis of history. Consider the following example: The lifetimes of certain engineering devices follow a negative exponential distribution with a mean life of ten years. What is the probability of failure by the end of the succeeding year for devices that have already lasted:

Three years?
Eight years?

For the three-year case, the chance of failure before three years is given by $F(3) = 1 - \exp(-\lambda t)$, where

$$\lambda = (1/10) = 0.1$$
$$F(3) = 1 - \exp(-0.1 \times 3)$$
$$= 0.2592.$$

And the chance of failure after four years is

$$F(4) = 1 - \exp(-0.1 \times 4)$$
$$= 0.3297.$$

Thus, the probability of a failure during the year is

$$P(\text{failure in the fourth year}) = \frac{F(4) - F(3)}{1 - F(3)} = \frac{0.3297 - 0.2592}{1 - 0.2592} = 0.0951.$$

A similar calculation will show that
P(failure in the ninth year) = 0.0951.

Hence, for the negative exponential distribution, the age of an item is immaterial in determining whether the item will last until the end of the next period. This is bad news if it is predictability of failure that we are after, but it is good news as regards cannibalisation. Parts with constant failure rate can be reused as if they were new. In fact, this is the case with cell phones and computers. The chips in them are not new. They have been subject to a "burn-in" process that eliminates those that fail by infant mortality. The rest that survive are sold as brand new even though they might have accumulated many hours of burn-in.

The Normal Distribution

The normal distribution is a two-parameter distribution, defined as

$$f(t) = \frac{1}{\sigma\sqrt{2\pi}} \exp - \left[\frac{(t - \theta)^2}{2\sigma^2} \right], \qquad 1.5$$

where θ is the mean of t, and σ is the standard deviation about the mean.

The standard deviation is a measure of the spread of the distribution along the t axis and is defined as the distance from the mean to the point of inflection of the curve, projected down onto the t axis. (In contrast to this, $\theta = 1/\lambda$ in the negative exponential distribution and is both the mean and the standard deviation of the distribution.)

The normal distribution is presented as an $f(t)$ function in Figure 1.4.

When we think of wearout failures, we automatically think of failures produced by abrasion, such as in brake and clutch linings, which wear away. But there are other forms of wearout failures as well. The most common kinds of wearout failures are as follows:

- Abrasion
- Corrosion
- Fatigue
- Creep

The Weibull Distribution

The Weibull distribution can be defined by the equation

$$f(t) = \frac{\beta t^{\beta - 1}}{\eta^{2}} \exp\left[-\left(\frac{t}{\eta} \right)^{\beta} \right], \qquad 1.6$$

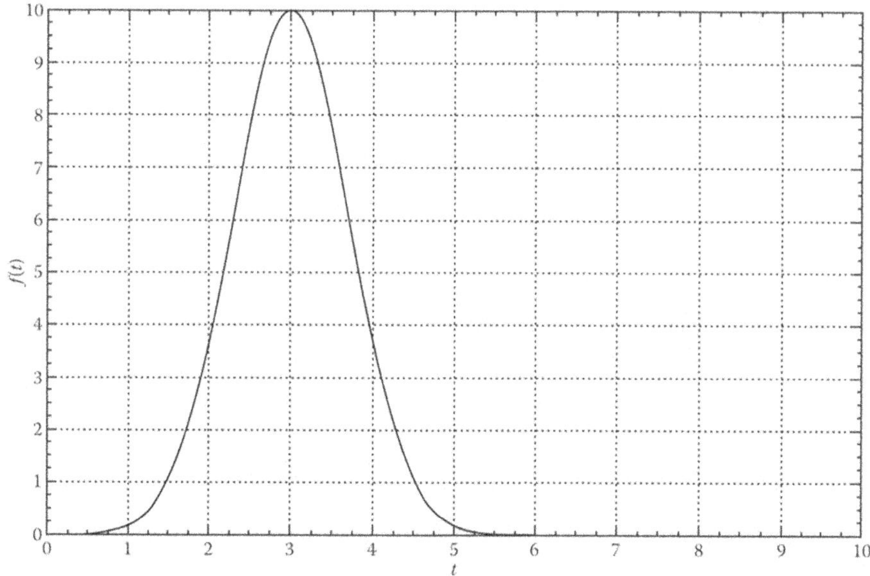

FIGURE 1.4 Normal distribution with θ = 3 and σ = 0.25.

where

$$f(t) = \frac{dF(t)}{dt}. \qquad 1.7$$

It can be seen from Figure 1.5 that the Weibull distribution alters its shape, dependent on the value taken by β.

- If β is less than 1, the curve is very skewed and very steep and corresponds to infant mortality failure.
- If β = 1, the shape of the curve is that of random failure.
- If β is large, say, greater than 2, then the distribution imitates the normal distribution.

Thus, we see that the Weibull distribution can imitate the other three distributions we have discussed and hence can model infant mortality, random failure, and wearout failure.

We also see that the Weibull distribution is a three-parameter distribution:

- Gamma, or γ, is called the location parameter. The value of γ will move the distribution along the t axis. If γ is positive, the values of γ must be subtracted from t before the equation is developed.
- Beta, or β, is called the shape parameter. As can be seen from Figure 1.5, changes in the value of β can dramatically affect the shape of the function.

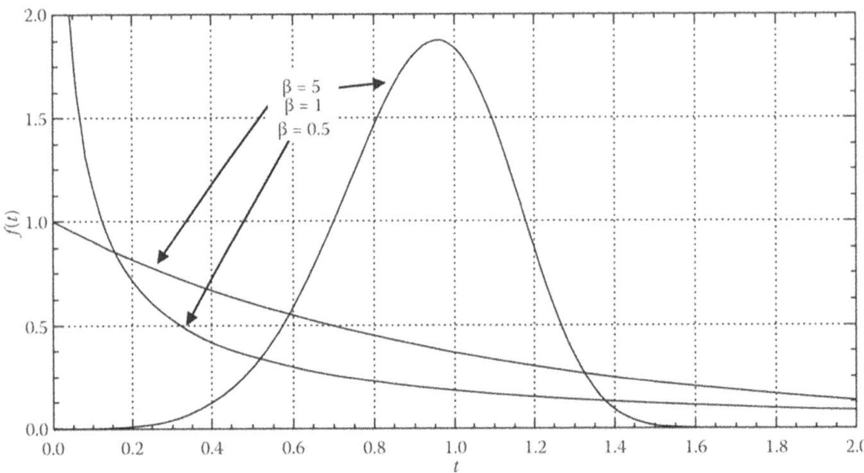

FIGURE 1.5 The Weibull distribution with $\gamma = 0$ and $\eta = 1$.

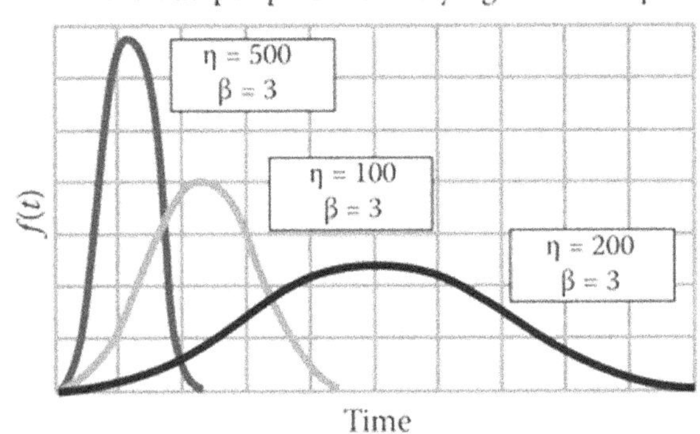

FIGURE 1.6 Graph indicating the effect of η on the Weibull function.

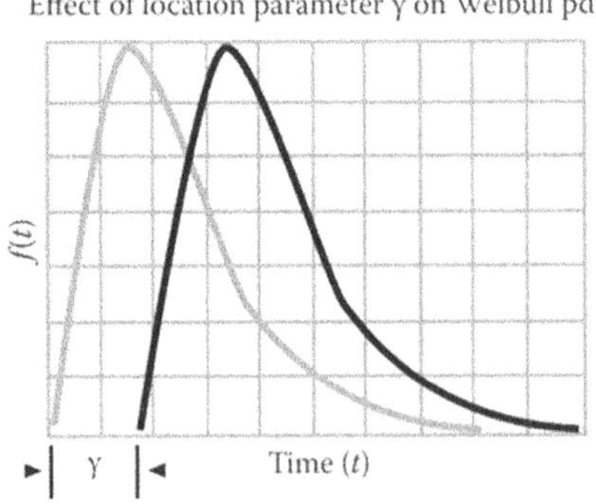

FIGURE 1.7 Graph indicating the effect of γ on the Weibull function.

- Eta, or η, is called the scale parameter. It works rather like the standard deviation of the normal distribution. The effect that η has on the distribution, when other values are kept constant, is given in Figure 1.6, while Figure 1.7 shows the effect of γ on the distribution when η and β are held constant.

The Bathtub Curve

A common model used in reliability engineering is the bathtub curve, which is normally shown as a plot of *failure rate* against *time*, as shown in Figure 1.8. Note that bathtubs are usually plotted with failure rate as the abscissa, not probability of failure. Different terminology is used in Figure 1.8: The early failure period corresponds to what has been called infant mortality previously. The intrinsic failure period has been called the random failure period. This period is also sometimes known as the useful life period. The third period is the wearout failure period, as before. The similarity to the cross section of a bathtub is obvious.

Figure 1.9 goes into more detail, depicting the bathtub as the addition of the failure rate curves for the three periods already described.

The bathtub curve is applicable not only to a population of components but also to systems. It is a useful model, but it does not always occur in practice. In this connection, the following quote can be noted from the *Engineering Statistics Handbook* webpage: https://www.itl.nist.gov/div898/handbook/

> The bathtub curve also applies (based on much empirical evidence) to repairable systems. In this case, the vertical axis is the Repair Rate or the Rate of Occurrence of Failures (ROCOF).

Weibull Analysis

Weibull analysis is used to analyse failure data. It will show us if a set of failures is from the first, second, or third phase of the bathtub. It can only be used for one failure mode at a time. Often, the data contain more than one failure mode, and these must

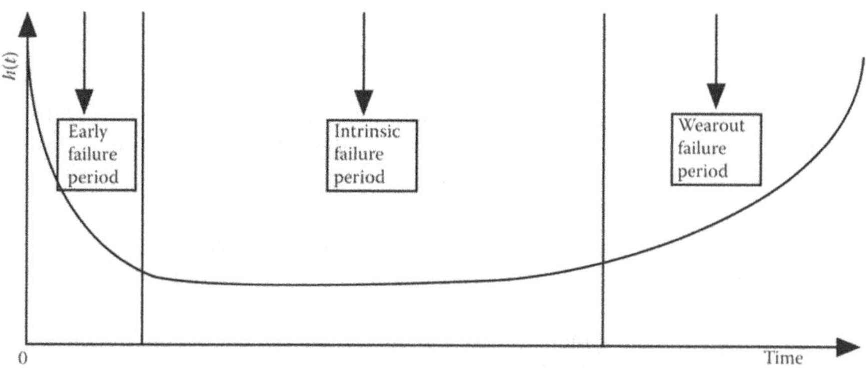

FIGURE 1.8 The bathtub curve.

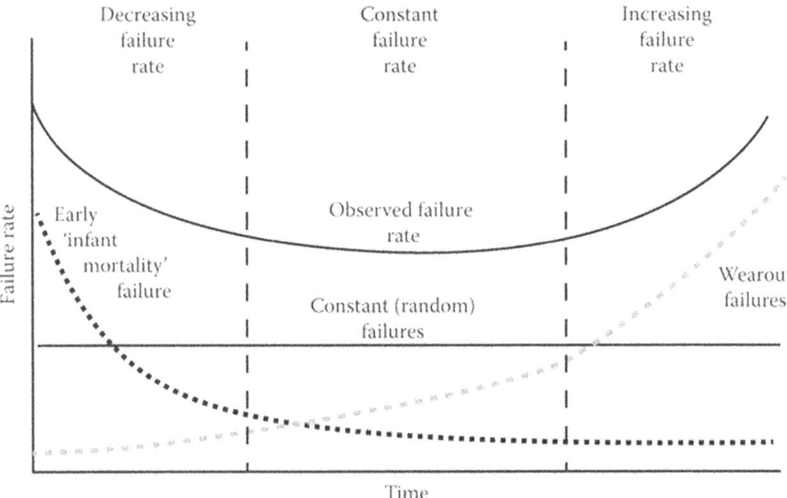

FIGURE 1.9 The bathtub curve, shown as a combination of three failure rate curves.

be separated before any analysis can be done. Although it is a superficially simple technique, there is quite an amount of experience necessary to draw the proper conclusions from the analysis.

The technique was invented by Prof W. Weibull in the middle of the last century, before the advent of the computer. The analysis was then performed using special graph paper. Such techniques are no longer necessary with the advent of several programs downloadable from the internet.

One such program is Weibull-DR21, which can be downloaded in a fully operational version for a 45-day trial from the internet. It can be purchased if required by using a credit card.

EXAMPLE 1.1 DEALING WITH SUSPENDED DATA

When using Weibull analysis, suspended data are usually involved (Table 1.1). By suspended data, we understand data associated with items that have been taken out of service before failing, or items that are still in operation at the time the analysis is done, or items that have failed in another mode. If only the data from the failed items were used in the analysis, we would clearly be losing valuable information on the lifetime properties of the population, and we would obtain a pessimistic view of the reliability of the item.

A plot using the Weibull-DR21 software is given in Figure 1.10.

We see that the results are a β value of 5.8 with a γ of −622 hours. This represents what is known as the shelf life phenomenon. There is a probability that some of the items have failed before they have been put into service, while the lives of the remainder are shortened. This can be the case for rubber components that have not been stored properly, or for paints, adhesives, and batteries. The η value is 1,760 hours.

TABLE 1.1
Life Data with Suspensions

Item	Life (Hours)	Failure (F) or Suspension (S)
A	500	F
B	660	F
C	800	S
D	820	S
E	900	F
F	920	F
G	940	S
H	1,100	F
I	1,200	F
J	1,400	S

Weibull-ease™ Weibull distribution summary report

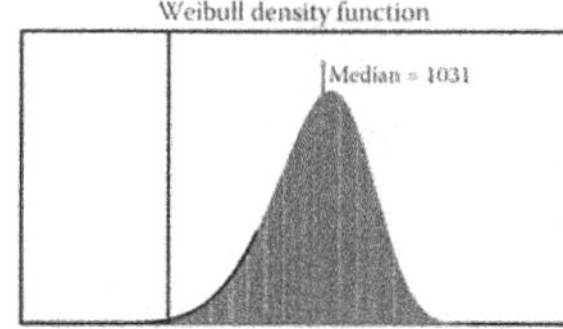

Mode name = Suspended data
Type units = Hours
Data points = 6
Units exposed = 10
Shape, β = 5.882
Scale, η = 1760
γ, offset = −622.92
Mean (MTTF) = 1008.20
Correlation = 0.99156

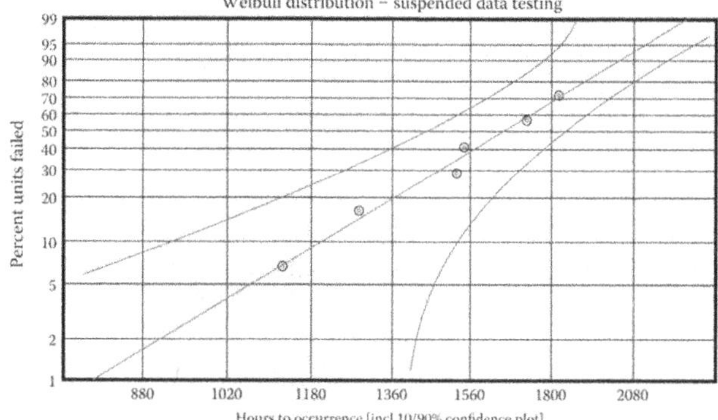

R%	F%	90% Conf	* 50% Conf	10% Conf
99	1	−169.93	182.27	778.50
98	2	−55.52	283.76	795.61
95	5	131.89	439.38	840.55
90	10	302.05	577.67	901.91
80	20	498.62	741.03	998.77
70	30	629.26	854.24	1079
60	40	732.32	947.26	1152
50	50	821.22	1031	1221
40	60	902.78	1111	1292
30	70	981.67	1194	1369
20	80	1062	1286	1459
10	90	1152	1405	1585
5	95	1205	1498	1690
2	98	1242	1597	1807
1	99	1256	1659	1883

Date 01-20-2014
Time 19:10

Signed ______________

FIGURE 1.10 Weibull plot.

EXAMPLE 1.2 GAS TURBINE BLADE COATING

An important point about Weibull analysis is that the results can never be separated from one's engineering knowledge of the problem at hand. The following example will show how this is done.

A protective coating for gas turbine blades was life-tested under normal operating conditions. Ten blades were randomly selected and life-tested. The hours to failure were recorded as shown in Table 1.2. Two of the coatings had not failed when the test was terminated at 5,000 hours. These are end-of-test suspensions and must be included in the data; otherwise, the MTTF of the blades will be pessimistic. The data should also be ranked from the lowest to the highest value, but the software will do this automatically.

TABLE 1.2
Blade Coating Life Data

Blade Number	Hours to Failure of Coating
1	3,200
2	4,700
3	1,800
4	2,400
5	3,900
6	2,800
7	4,400
8	3,600
9 (Not Failed)	5,000
10 (Not Failed)	5,000

When these data points were inserted in the software, the results were as follows. The Weibull-Ease printout is not shown here—the student is to prepare his own and compare the results with the figures for the following Weibull parameters:

$\beta = 3.2$
$\eta = 4{,}450$ hours
$\gamma = 0$
MTTF = 3,985
Correlation = 0.985

What do we learn from this?

- We have a high β value, indicating wearout failure. This is good and to be expected. If we had read a β value of 1 or less, we would have had concerns—a coating like this should fail by wearout, and it has.

- The correlation is high at 0.985, giving us confidence in the derived β value.
- However, γ is zero, meaning, that there is a small probability of failure immediately on start-up.
- The mean life of the coatings is 4,450 hours.

So is this good or bad news? It depends. If this is an airliner engine, the results are not good enough. The coating is there to protect the blade, and while we have no data on blade life after coating damage, the fact that a coating could fail after a very short while is probably not acceptable. However, if these blades are for a cruise missile engine, then the parameters are probably good enough. A cruise missile engine only has to work for a small number of hours—perhaps as low as four hours. On the other hand, for a land-based turbine, we would also need to know more about the duty. So we see that we must combine the Weibull results with our knowledge of the technology before we can say that the results are good or bad.

Problems with Weibull

Study the data in Table 1.3.

The Weibull plot for the data in Table 1.3 is shown in Figure 1.11. It will be seen that the β value is approximately 1.4, and the correlation seems quite good at above 0.93. But as we have emphasised before, one cannot divorce Weibull analysis from engineering principles. In fact, the first three failures are from one distribution, and the remainder is from another. A visual examination of the Weibull-Ease plot should cause the analyst to become suspicious—the data are by no means a straight line. If we separate the data as separate distributions, we obtain the following:

TABLE 1.3
Failure Data (Mixed Modes of Failure)

Serial No.	Time (Hours)	Failure or Suspension	Number Off
1	800	F	1
2	1,100	F	1
3	1,300	F	1
4	1,400	F	1
5	1,410	F	1
6	1,420	F	1
7	1,500	S	13

The students should plot the two distributions in Tables 1.4 and 1.5 for themselves and should find that:

- For failure mode 1, β = 3.933, with a high correlation of 0.99546.
- For failure mode 2, β = 1021, and the correlation is 0.98583

Weibull-ease™ Weibull distribution summary report

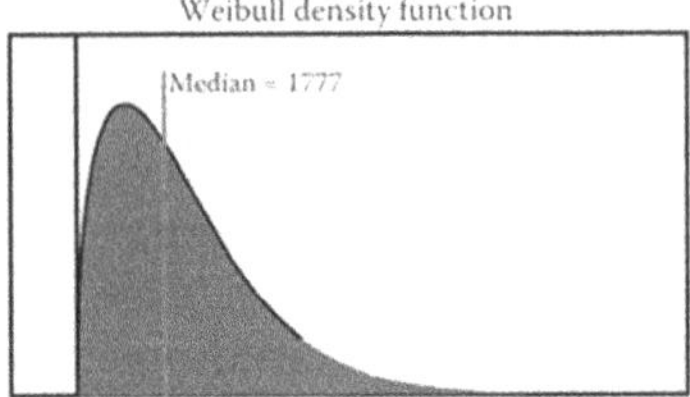

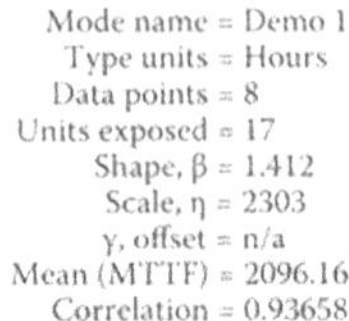

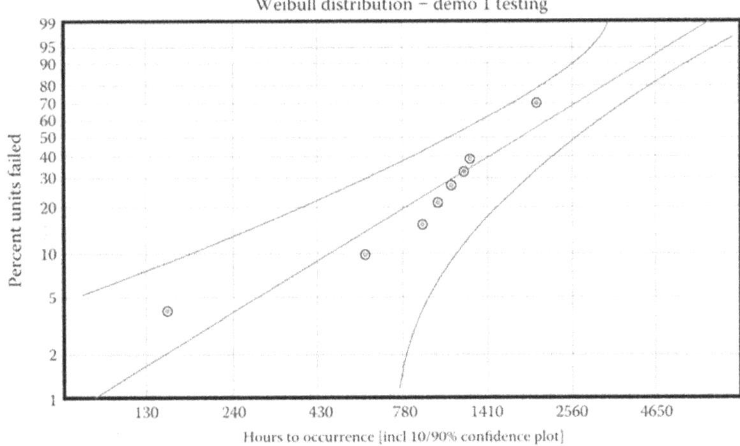

R%	F%	90% Conf	50% Conf	10% Conf
99	1	9.75	88.68	753.55
98	2	24.62	145.39	802.30
95	5	78.81	281.22	937.13
90	10	179.36	468.14	1140
80	20	390.39	796.39	1511
70	30	609.74	1110	1874
60	40	840.92	1432	2250
50	50	1090	1777	2660
40	60	1366	2165	3129
30	70	1683	2627	3699
20	80	2067	3226	4465
10	90	2579	4157	5726
5	95	2933	5009	6963
2	98	3213	6050	8583
1	99	3325	6791	9788

Date 02-12-2014
Time 08:23

Signed ________________

FIGURE 1.11 A plot of two failure modes.

What are we to make of these results? For distribution 1, we have some sort of wearout failure that the other items did not experience. We need to investigate this further. For distribution 2, we have three items failing practically simultaneously. This could be because of some overload condition. The suspended items, 13 in all, did not fail at the time these data were collected, namely, 1,500 hours, but might have

TABLE 1.4
Failure Mode 1

Failure No.	Time (Hours)	Failure or Suspension	Number Off
1	800	F	1
2	1,100	F	1
3	1,300	F	1

TABLE 1.5
Failure Mode 2

Failure No.	Time (Hours)	Failure or Suspension	Number Off
4	1,400	F	1
5	1,410	F	1
6	1,420	F	1
7	1,500	S	13

failed very soon thereafter if this was an overload condition. Once again, both the failed and unfailed items need to be examined to establish the mode of failure and if failure appears imminent in the unfailed items.

Weibull Cautions

So we have seen that there are problems with the implementation of the Weibull method that an analyst must be aware of. These include the following:

- Curved Weibull plots, which may be attributed to the plotting of more than one failure mode.
- Curved Weibull plots attributed to non-zero γ values.
- Weibull may only be used on *components*, that is, on items that are replaced. (There are exceptions to this rule, which are not appropriate for this level, of course.) *Maintained systems* are generally not amenable to Weibull analysis unless they always exhibit the same failure mode after maintenance.

Using Weibull When Very Little Data Are Available

When data are scarce, a Weibull library may be consulted. Several are published on the internet. A partial listing from such a library is given in Table 1.6. It may be noticed from this list that the β values are perhaps unexpectedly low. For example, all the bearings have β values less than 2. Bearings supposedly fail by wearout and hence should have β values greater than 2. Hence, practice does not always follow theory.

TABLE 1.6 PARTIAL LIST FROM A WEIBULL LIBRARY

Item	β Values (Weibull Shape Factor)			η Values (Weibull Characteristic Life, Hours)		
Low	Typical	High	Low	Typical	High	
Components						
Ball bearings	0.7	1.3	3.5	14,000	40,000	250,000
Roller bearings	0.7	1.3	3.5	9,000	50,000	125,000
Sleeve bearings	0.7	1	3	10,000	50,000	143,000
Belts, drive	0.5	1.2	2.8	9,000	30,000	91,000
Bellows, hydraulic	0.5	1.3	3	14,000	50,000	100,000
Bolts	0.5	3	10	125,000	300,000	100,000,000

Knowing β gives us the ability to perform all sorts of predictions. See, for example, the Test Design/Planning example in the Weibull-DR21 software. Here, knowing from previous tests that the β value is 2 and the MTTF is 1,000 hours, we can design a test as specified in the example—that is, for five new items put on test and no failures allowed, what must be the duration of the test? The answer is 766 hours, which means that there is a 90% probability that a sample that passes this test comes from a population with an MTTF of at least 1,000 hours.

ASSIGNMENTS

Assignment 1.1: Weibull Familiarisation

Download the Weibull-DR21 software and familiarise yourself with it. Work through the following examples, making sure that you understand the techniques and are able to discuss them if necessary:

- Weibull Demo
- Test Design Plans
- Bearing Calculations

Assignment 1.2: Weibull Problem

Plot the following data in your Weibull program (all data are in hours): 150, 190, 220, 275, 300, 350, 425, 475. Note the β value. Now, press the Calculate Offset button. Notice how the shape of the distribution changes as there is now a failure-free operating period (FFOP) and a lower β value. Explain the meaning of the FFOP. Describe how you would estimate the FFOP if the software could not perform this calculation, simply by examining the plot.

Assignment 1.3: Another Weibull Problem

Plot the following data in your Weibull program (all figures are in hours): 70, 120, 130, 165, 200, and 210. Note the β value and once again see how the offset function changes the β value but also that the offset is now negative. Explain what this means.

Assignment 1.4: Spring Design

Study the following two sets of data, which are for two different designs for a special spring (Table 1.7).

Using your Weibull analysis program, compute the Weibull parameters. Also find what the reliability is for both sets of data at 400,000 cycles (50% figure). On this basis, which design is better?

If you require a failure percentage not greater than 1% at 90% confidence, and if the average use of the spring is 8,000 cycles per year, what length of guarantee might be appropriate?

Assignment 1.5: Weibull Analysis of Rolling Element Bearings

In theory, rolling element bearings should fail by wearout and hence should exhibit high β values of the order of 2 and above. However, when one studies the databases on the internet and in textbooks, low β values, of the order of 1.3, are usually quoted. Investigate this phenomenon and write a short report on your findings.

Suggested effort: Your report should include, but is not limited to, the following:

- Title and author page
- Statement of originality of the work
- Summary
- Table of contents
- Table of illustrations
- Table of tables
- Introduction

TABLE 1.7
Spring Data

Design A		Design B	
Number	Cycles to Failure	Number	Cycles to Failure
1	726,044	1	529,082
2	615,432	2	729,000
3	807,863	3	650,000
4	755,000	4	445,834
5	508,000	5	343,280
6	848,953	6	959,900
7	384,558	7	730,049
8	666,600	8	973,224
9	555,201	9	258,006
10	483,337	10	730,008

- History of the technology
- Theoretical Weibull values for rolling element bearings
- Actual Weibull values for rolling element bearings
- Discussion on the reasons for the low β values
- Appendices:
  1. References
  2. Others as may be required

Photographs and drawings are to be included if possible, including photographs of failures. Reports are to be typewritten, single spacing in Times New Roman, 12 point. Margins and headings are to be similar to those in this document.

Quality of presentation and depth of thinking are required rather than a set number of pages. But as a guide, a report to fulfil the aforementioned requirements will probably be not less than five pages long, including illustrations, tables, photographs, and so on.

Assignment 1.6: New Era Fertilizer Plant

Complete Case 1.1 that follows.

CASE 1.1 WEIBULL ANALYSIS AT THE NEW ERA FERTILIZER PLANT

Author's Note: Here is a case showing how Weibull analysis can be used to assist in a root cause analysis investigation. However, a careful study of the numbers will lead to the correct decision even without the use of Weibull analysis.

Mark Francis, the maintenance consultant, had been called in to assist with RCM on the New Era Fertilizer Plant's coal mill. This was a vertical spindle mill capable of grinding coal to powder consistency at the rate of 70 tons per hour. A cross section of a similar mill, but by a different manufacturer, is shown in Figure 1.12.

Most of the items on the mill were long-life components, subject to the wear of the coal grinding process over periods of thousands and tens of thousands of hours. But one set of items was proving particularly troublesome. This was the set of rubber bellows that allowed for the movement of the hydraulic rams that applied pressure to the upper loading table so that the coal could be ground. (These bellows are not apparent in Figure 1.12 as a result of the sectioning of that figure.) The internals of the mill resembled a huge ball race, with the coal pouring into the hole in the upper "race" or loading table and being ground by the balls while being moved by centrifugal force to the outer edge of the lower "race" or table as it rotated. The top table did not rotate and was loaded by the hydraulic rams, acting through levers, as shown in Figure 1.13. The inside of the mill was pressurised to approximately 1 atm to enable the ground coal

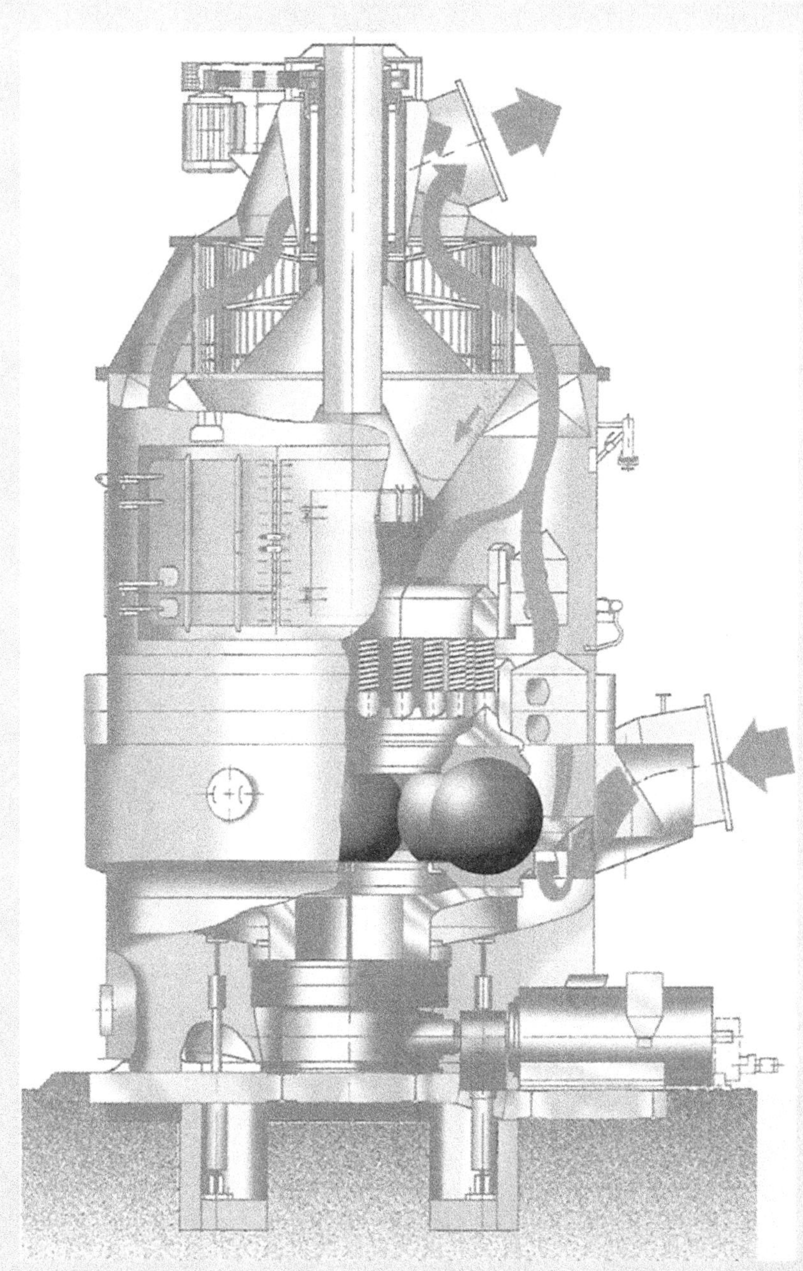

FIGURE 1.12 Vertical spindle coal mill similar to the type described in the case. (Courtesy of Claudius Peters GmbH.)

to be swept up out of the mill. The interior of the mill was at approximately 100°C, to assist with the drying of the coal. The powdered coal was then used in the manufacture of ammonia, which in turn was used in the manufacture of fertiliser.

The rubber bellows were subject to a fairly arduous pressure-and-temperature regime. The bellows had to allow for the movement of the rams and contain the pressure of the air inside the mill. Eight rams were arranged circumferentially around the cylindrical outer casing of the mill. The bellows were of "fir tree" configuration, with eight convolutions and upper and lower diameters of approximately 125 and 250 mm, respectively. The upper diameter attached to a spigot on the mill casing, and the lower diameter attached to the pushrod of the hydraulic ram. The bellows were secured top and bottom by large hose clips, similar to those used on car radiator hoses. A part section through one of the bellows is shown in Figure 1.13.

Failure of a bellows occurred as a circumferential crack in the base of the convolutions. When it failed, the mill had to be shut down immediately, as hot coal dust blew out of the crack in immense quantities.

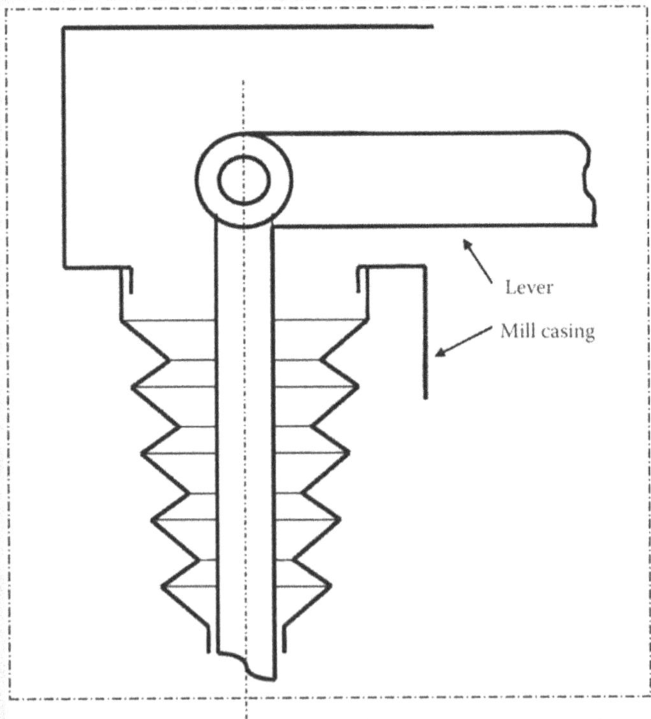

FIGURE 1.13 Part section through one of the bellows.

Peter Nobel was the maintenance superintendent under whom the coal mill fell. "We have up to now replaced the bellows as a set when one has failed," he said in reply to Mark's question.

"Do you think that's the right policy—block replacement?" enquired Mark further.

"Man, I don't want any more trouble with bellows than I have already!" exclaimed Peter.

Mark studied the replacement intervals for the bellows in the operating log. The last three cycles had been accurately recorded: All bellows replaced on the failure of one at the following times: 600, 150, and 2,000 hours. The latest replacement had just occurred, and the mill had only been back in service for a few hours.

ASSIGNMENT

Analyse the failure data using Weibull analysis. What recommendations do you think Mark should make, and what provisos should accompany the recommendations? Second, what has this exercise shown concerning the power of the Weibull technique?

Assignment 1.7: The Life History of a Hillman Vogue Sedan

Complete Case 1.2.

CASE 1.2 THE LIFE HISTORY OF A HILLMAN VOGUE SEDAN

This case reveals something that most textbooks ignore! What is it?

This 1976 car (Figure 1.14) was produced right at the end of the model's production life. It was in some ways a "parts bin special," using parts from other models, for example, the clutch-and-exhaust system. The reliability was not in the same class as the 1972 model owned previously. Data on failures were kept fairly rigorously over the 17-year lifespan of the car, and the results are given following.

All maintenance was done by the owner, except where specified. Two points to be noted are as follows:

- Those were the days in which cars were simple enough for owner maintenance to be an option.
- The inflation in the South African economy since the 1970s has been horrific!

Table 1.8 shows the complete repair record for the car, with normal servicing—for example, spark plugs and ignition points—excluded.

FIGURE 1.14 Vogue sedan.

TABLE 1.8
Repair Record

Kilometres	Repair Action
42,300	New clutch
47,000	New silencer, one new tyre
50,000	Wheel alignment
62,000	Clutch master cylinder kit, one new tyre, straighten bent rim
62,000	Adjust position of LH front wheel—move forward 15 mm
64,350	Wheel alignment, set valve clearance
66,500	Two new front tyres
75,000	Replace original MS exhaust system with SS, clean alternator brush
80,000	Replace alternator brush and slip plate
87,000	New clutch master cylinder kit and new slave cylinder kit
88,000	Replace front wheel bearings, disc brake pads, and rear brake shoes
89,500	Recondition fuel pump, clean carburettor filter, new distributor cap
89,500	Retrofit in-line fuel filter
91,000	Adjust valves, replace accelerator cable, and replace water pump
97,000	Clean carburettor jets
111,500	Rebuild differential
116,000	Replace front brake disc pads
120,000	Upholstery repair

TABLE 1.8 *(Continued)*
Repair Record

Kilometres	Repair Action
124,000	New fuel pump
125,000	New battery, four new tyres
126,000	Replace water pump
135,000	Replace windscreen, replace rubber seal for boot lid
135,500	Replace wheel bearings
136,000	Panel beating, one new tyre
140,500	New head gasket, valve grind
140,900	New clutch plate, engine rebuild
145,000	New silencer
149,000	Two new tyres, propshaft replacement
155,000	New radiator, new fan
160,000	New starter solenoid
165,000	New rear brake shoes, replace starter
170,000	Blown head gasket—car sold as scrap

The car was seen to enjoy a failure-free operating period of 40,000 km, covering approximately 3½ years' life. It then suffered a premature clutch failure. Minor failures or servicing operations then followed in a fairly random pattern throughout much of the rest of the car's life. We do, however, observe clusters of failures, such as those between 60,000 and 70,000 km. In this particular case, the car had been driven through a pothole (rare on South African roads in those days, and drivers then were not as vigilant for those sort of things as they are now). This incident resulted in a bent wheel rim, a burst tyre, and a bending of the subframe and of the swingarm of the MacPherson strut suspension, as shown in Figure 1.15.[1] The rim was straightened with a 4 lb hammer. The suspension was restored by cutting 15 mm of the rubber block anchoring the rod that held the swingarm. The wheels were then realigned by a professional.

LINEAR PLOT OF CORRECTIVE ACTIONS

Figure 1.16 shows the plot of corrective actions on a linear timescale. This is also sometimes called an Asher plot. As we study this figure, we see the failure-free operating period of more than 40,000 km, followed by failures that are effectively random until disinvestment. There is no apparent third phase to the bathtub. So what is happening, and what caused the disinvestment?

A professional rebuild of the engine was undertaken at 140,500 km after the engine suffered a blown head gasket. Repair thereafter included a new propshaft at 149,000 km, and a new radiator and fan were necessitated when the old fan came loose and propelled itself into the radiator. This was done at 155,000 km. The next repair was of the starter motor solenoid at 160,000 km, followed by a complete starter replacement at 165,000 km. These lives are given in Table 1.9.

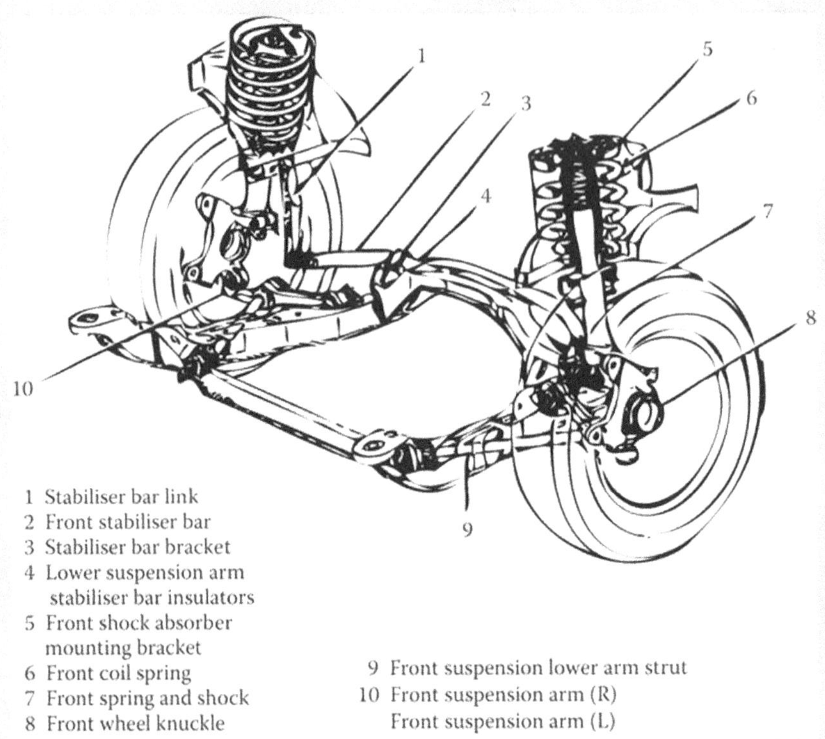

FIGURE 1.15 MacPherson strut front suspension and subframe.

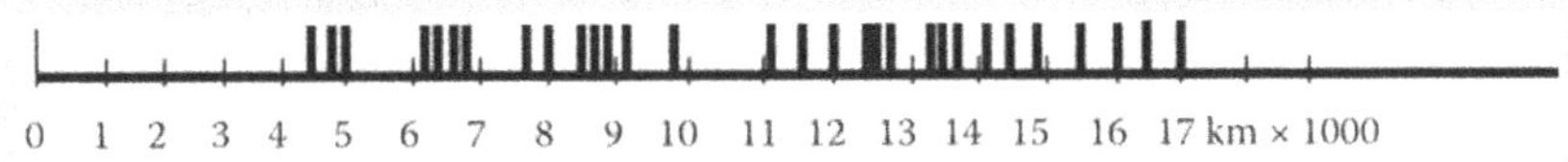

FIGURE 1.16 Linear plot of corrective actions: Scale in kilometres × 1,000.

Finally, the new head gasket, installed at 140,500 km, blew at 170,000 km. This was accompanied by a ruptured radiator hose. Whether the hose first ruptured, leading to water loss, which led to overheating, which led in turn to gasket failure, or whether the cylinder head gasket first failed, is unknown. Because of the high cost of the projected repair, coming on the tail of the one so near in the past, it was decided to dispose of the vehicle.

TABLE 1.9
Component Lives of the 1976 Vehicle

Component	Lives (km)	Lives for the 1972 Version of the Same Vehicle
Clutch plate	42,000; 98,600	120,000
Wheel bearings	88,000; 47,000	110,000
Brake pads	28,000	
Rear brake shoes	88,000; 77,000	107,000
Clutch master cyl. kit	62,000; 25,000	
Clutch slave cyl. kit	87,000	129,000
Silencer, mild steel	31,000	
Silencer, stainless steel	67,000	
Water pump	91,000	
Fuel pump	89,000	34,000
Head gasket	140,000	122,000

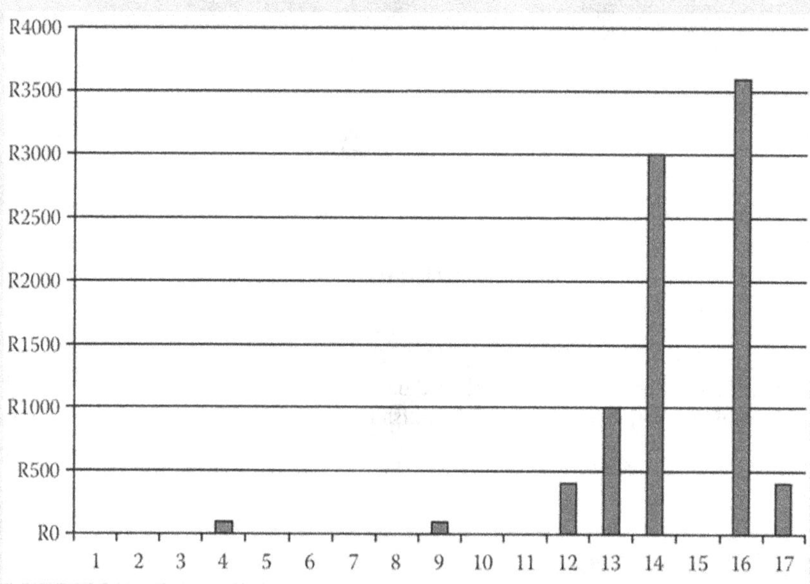

FIGURE 1.17 Major repair cost versus years of service.

ASSIGNMENT

1. Study the repair record and other information given prior. This simple case illustrates several reliability engineering concepts, including the following:
   - Infant mortality failure
   - Incomplete repair
   - Life extension
   - Indication of failure
   - Retrofit
   - Visual inspection
   - Preventive measure or proxy replacement (that is, replacing something so that something else does not fail)
   - Root cause analysis
   - Find these in the case.
2. Furthermore, study the prior case and comment on the results. What about the classic bathtub curve discussed elsewhere? Why does it not appear here? Are the textbooks in error? Particularly, what is the significance of Figure 1.17?
3. Comment on the fact that the 1972 car, which was practically the same as the 1976 version, except for some minor styling changes, was so much more reliable.

NOTE

1. **MacPherson strut:** An independent suspension system with combined spring and damper set-up. Compact, lightweight, and cheap to build, it comprises a vertical telescopic shock absorber that links to the upper control arm to dictate the wheel position. It is cheaper than multilink layouts, but not as robust. Developed in the 1940s by Ford's Earl S. MacPherson, it was introduced on the 1949 French Ford Vedette, and then the Consul, Zephyr, and Zodiac in the English Ford range.

2 Reliability Fundamentals II
System Reliability

As stated in Chapter 1, Robert Lusser has been called the father of reliability engineering following his work on the V1 missile for the German Air Force in World War II. A drawing of the missile is shown in Figure 2.1.

The V1 might be considered as the first cruise missile, but as a weapons system, it was at first notoriously unreliable. (A mission success rate of less than 40% has been reported.) What Lusser did was to construct the first known reliability block diagram, or RBD, showing the various subsystems of the weapon as several blocks joined together in series. The reliability of the weapon system would be the product of the reliability of the blocks. Lusser's model might have looked as shown in Figure 2.2.

What the diagram demonstrates is the fact that if a system consists of a series of subsystems, each with no backup, or in reliability engineering terms, with no redundancy, then the reliability of the system is the product of the reliabilities of the subsystems. This is known as Lusser's law or the product rule for series systems. (This is only true, however, if failures in the blocks are *statistically independent*, that is, the failure of one part does not change the failure probability of any other part.)

The product rule is expressed mathematically as:

$$R_s = (R_1 \times R_2 \times \ldots \times R_n) \tag{2.1}$$

For example, if the required reliability of the weapon was to be 99%, then each of the blocks in the diagram would need to be $\sqrt[5]{0.99}$, or 0.99799. If this level of reliability were not achievable in some of the blocks, it would mean that the reliabilities of the other blocks would have to be even higher. What Lusser had done was enable his fellow engineers to set *targets* for the various subsystems. It was now up to the engineers to achieve these targets by using redesign, development testing, and whatever other techniques were available at that time.

Each of the blocks can be developed further into lower-level subsystems. For example, we can see from the illustration that the guidance system can be further subdivided, as shown here:

If the guidance subsystem were to consist of only the five blocks shown in Figure 2.3, then for a subsystem reliability of 0.99799, the individual blocks would have to have a reliability of $\sqrt[5]{0.99799} = 0.99959$. So the lower we go in the hierarchy, or indenture level, as it is sometimes called, the more reliable the subsystems or components have to be.

DOI: 10.1201/9781003326489-2

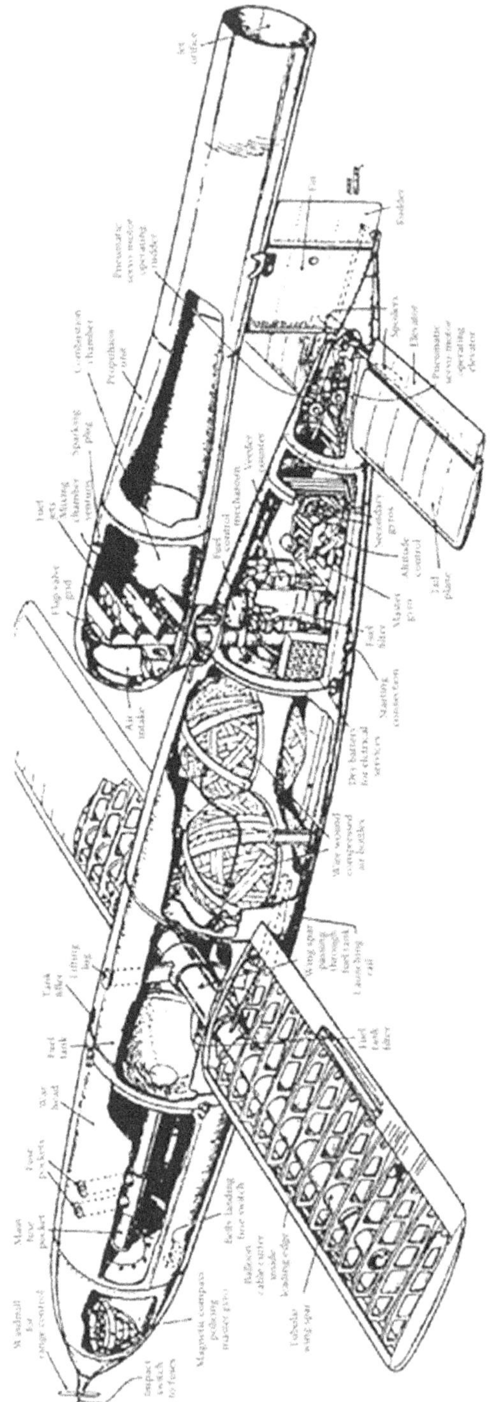

FIGURE 2.1 The Fiesler 103, otherwise known as the V1. (Picture courtesy of Flight International.)

FIGURE 2.2 High-level reliability block diagram.

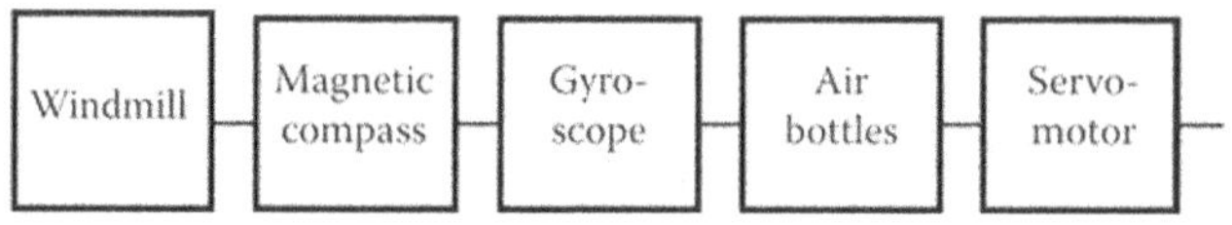

FIGURE 2.3 Typical RBD at the next indenture level after Figure 2.2.

A NOTE OF CAUTION

Reliability prediction of designed systems is not an exact science, as there are too many imponderables to be considered. Reliability prediction should be used for comparative purposes; for example, how will the reliability be changed if we add redundancy to one of the subsystems? Or how does this system configuration compare with an alternative? The figures derived from a reliability prediction should be treated with caution. In the words of Patrick O'Connor (2012):

> An accurate prediction of the reliability of a new product, before it is manufactured or marketed, is obviously highly desirable. Depending upon the product and its market, advance knowledge of reliability would allow accurate forecasts to be made of support costs, spares requirements, warranty costs, marketability etc. . . . In fact, a reliability prediction can rarely be made with high accuracy or confidence. Nevertheless, it can often provide an adequate basis for forecasting of dependent factors such as life cycle costs. Reliability prediction can also be valuable as part of the study and design processes, for comparing options and for highlighting critical reliability features of designs.

What is essential to remember is that reliability engineering techniques can never be separated from the basic engineering of the system in question. Good reliability engineers are good engineers first, with a sound basis in mechanical, electrical, aeronautical, or other engineering disciplines, who can also understand and apply reliability techniques. *There is no substitute for a thorough understanding of, and interest in, the underlying technology of the system under consideration.*

To quote O'Connor (2012) again:

> Since reliability is affected strongly by human-related factors such as training and motivation of design and test engineers, quality of production, and maintenance skills, these factors must also be taken into account. In many cases they can be much more significant than past data. Therefore reliability databases must always be treated with caution as a basis for predicting the reliability of new systems.

In fact, this author goes further to assert that reliability problems are *always* human-related. Someone specified, designed, manufactured, operated, or maintained the plant incorrectly. The human culprit, or culprits, may not always be identifiable, but they are always there. The plant is neutral; it is not emotional, stupid, forgetful, ignorant, or spiteful.

SYSTEM CONFIGURATIONS

REDUNDANCY

Very often, redundancy is encountered. The reliability engineering use of the term coincides with the dictionary definition meaning "surplus to immediate requirements." Redundancy is encountered when it is not possible to secure sufficient reliability by having only one unit. Redundancy has a powerful effect in increasing the reliability of a system, but it comes at a cost—the upfront doubling of the capital cost.

There are two types of redundancy, active and standby (or passive). These are shown in Figures 2.4 and 2.5.

For a redundant system, either active or standby, the reliability is:

$$R_s = 1 - F_1F_2 \tag{2.2}$$

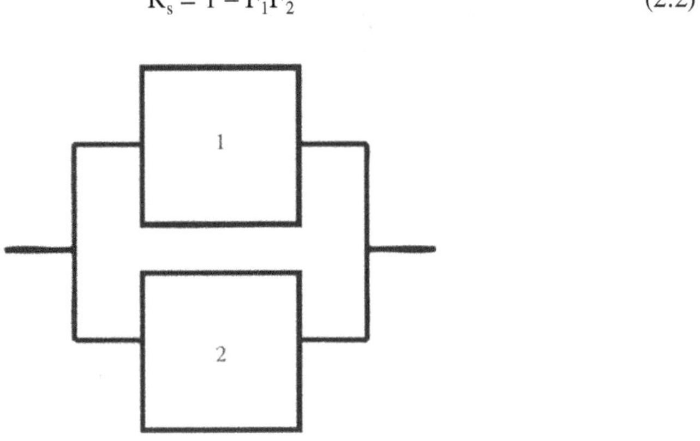

FIGURE 2.4 An example of active redundancy.

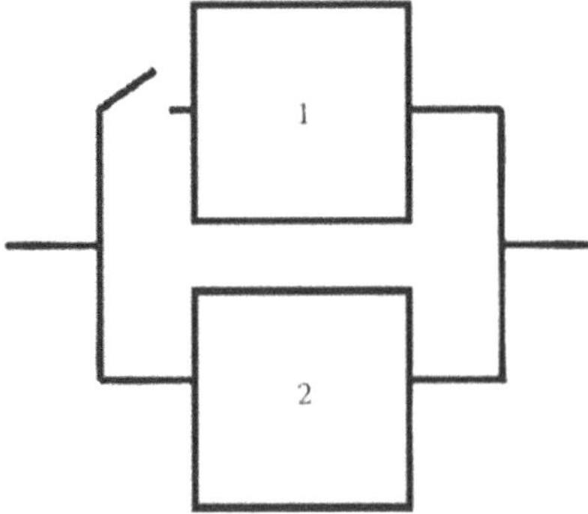

FIGURE 2.5 An example of standby or passive redundancy.

There are advantages and disadvantages to both forms of redundancy. In active redundancy, there is no switching circuit to give trouble. The two units are permanently on load, both running at 50% capacity. If one unit fails, the other unit ramps up to 100% load immediately and the downstream system is unaffected. The system of two high-voltage transformers in a transmission substation is an example of this configuration.

In the case of the standby system, one unit is working and the other is idle. If the working unit fails, there must be a sensing-and-switching system to bring in the standby unit. This sensing-and-switching system could be a human operator or an electronic system. This system itself could be unreliable. Mineral conveyors are often designed with standby redundancy.

The table that follows is a summary of the advantages and disadvantages of both active and standby redundancy.

So it will be seen from Table 2.1 that standby redundancy could be less reliable in practice than active redundancy. In practice, the equation for system reliability applies to both. In any event, redundancy to achieve reliability is often the last choice of the designer, because as mentioned, it doubles the capital cost. Better to ensure that the base unit is reliable enough, if at all possible.

M-OUT-OF-N REDUNDANCY

Another situation in which redundancy is resorted to is when one cannot purchase large-enough units to allow 100% capacity in a process. A classic example of this is in the provision of three electrically driven feed pumps each of 50% capacity on many power stations, two of which supply the water to the boiler, and one is a standby. In this case, the redundancy is less than 100%—at 50%, in fact. The reason for this was originally that although one could purchase pumps of 100% capacity, motors of the required size were not available. (This is no longer a restriction.)

Such a system as referred to prior is also called a 2-out-of-3 system. One also encounters 3-out-of-4 systems and others. The question is how to evaluate the reliability of such systems. There are formulas for this, based on the binomial distribution,

TABLE 2.1
Advantages and Disadvantages of Active and Standby Systems of Redundancy

System	Active	Standby
Advantages	Automatic pick-up of load if one fails	Idle unit does not wear out.
Disadvantages	Both components active, means both are wearing out	Some modes of failure occur when plant is idle, e.g.: • corrosion • brinelling of rolling element bearings when idle Sensing and detection unit may itself be unreliable. Parts of the unit may be cannibalised when idle—the unit then refuses to start when required, or leaks oil, or whatever, depending what parts were removed.

1
1 1
1 2 1
1 3 3 1
1 4 6 4 1
1 5 10 10 5 1

FIGURE 2.6 Pascal's triangle.

but the author has always found it more useful to simply apply Pascal's triangle, as shown in Figure 2.2.

Pascal's triangle is easily constructed. In every row, any number is the sum of the two numbers immediately above it, except for the ones that make up the border. So if, for example, we have a 2-out-of-3 system and the reliabilities of the components are equal at R_c each, the equation for the reliability of the system is:

$$R_s = 1\left(R_c^3\right) + 3(R_c^2\ F_c^1) \tag{2.3}$$

Which equation is formed from the first two terms of Pascal's triangle, for the row beginning 1, 3. . . . The decision rule is, therefore, for a 2-out-of-**3** system, enter Pascal's triangle at the row with a **3** in it. Then, as it is a **2**-out-of-3 system, take the first **two** numbers, 1 and 3, as the coefficients of the **two** required terms in your expression. Furthermore, as it is a 2-out-of-3 system, the first *R* term is *cubed*, whilst the second *R* term is squared and is multiplied by an F^1 term. (The indices of the *R* and *F* terms must always add up to **3**.)

Another way of looking at this is that equation 2.3 represents all the ways in which at least two units are working.

Similarly, for a 2-out-of-4 system, the equation is:

$$R_s = 1(R_C^4) + 4(R_c^3 F_c^1) + 6(R_c^2 F_c^2) \tag{2.4}$$

And so on. In some metallurgical processing plants, the writer has encountered 7-out-of-8 systems.

The alternative to Pascal's triangle is to use the binomial formula, as follows:

$$(R + F)^n = R^n + nR^{(n-1)}F + \frac{n(n-1)}{1!2!}R^{(n-2)}F^2 + \frac{n!}{x!(n-x)!}R^{(n-x)}F^x + \tag{2.5}$$

Notice how, taking for a 2-out-of-3 system, the first two terms of the previous equation yields

$$R_s = R^3 + 3R^2F$$

Where R^3 is the probability that all components work.
And $3R^2F$ is the probability that two components work.

As we need two of the three components to work in a 2-out-of-3 system, the previous equation describes the reality of such a system, which is the same as equation 2.2, when the component reliabilities are equal.

When the component reliabilities are not equal, then the following applies, since the sum of R and F is always unity:

$$(R_1 + F_1)(R_2 + F_2)(R_3 + F_3) = 1 \text{ (i.e. } 1 \times 1 \times 1 = 1)$$

Expanding:

$$R_1R_2R_3 + (R_1R_2F_3 + R_1F_2R_3 + F_1R_2R_3) + (R_1F_2F_3 + F_1R_2F_3 + F_1F_2R_3) + F_1F_2F_3 = 1 \quad (2.6)$$

Which describes the condition:

(Probability of all three working) + (probability of two working) + (probability of one working) + (probability of none working), which covers all possible cases and hence is equal to unity.

Once again, for a 2-out-of-3 case, we need the first two terms:

$$Rs = R_1R_2R_3 + (R_1R_2F_3 + R_1F_2R_3 + F_1R_2R_3)$$

Now, let us see if the equation we previously stated for a simple redundant system also follows Pascal's triangle. That is, for a 1-out-of-2 system, Pascal's triangle is given as:

1

1 1

1 2 1

Entering the triangle as before at the line with a 2 in it, all states for a redundant system are described by:

$$R^2 + 2RF + F^2$$

The first two terms represent (probability of both working) + (probability of one working):

$$\begin{aligned} R_s &= (1 - F)^2 + 2(1 - F)F \\ &= 1 - 2F + F^2 + 2F - 2F \\ &= 1 - F^2 \end{aligned}$$

Which is the same as equation (2.2), given previously.

SYSTEM RELIABILITY PREDICTION

There are various methods of system reliability prediction. The simplest and most common is the decomposition method, whereby a system of series and parallel (i.e.

redundant) units are systematically combined until only one unit remains. Consider, for example, the following (Figure 2.7):

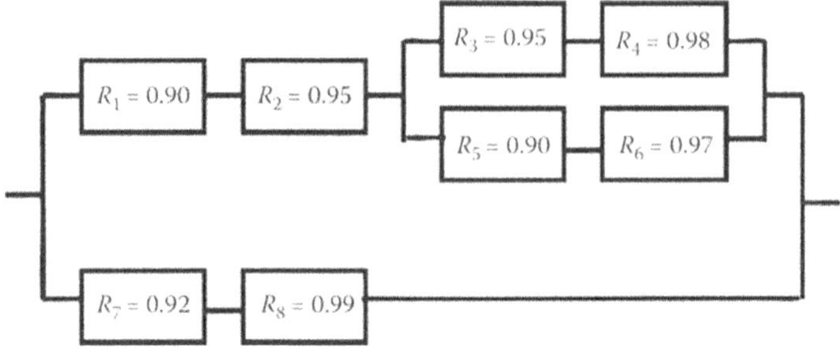

FIGURE 2.7 Series-parallel circuit.

Taking all the series networks first:

$$R_1 \times R_2 = 0.90 \times 0.95 = 0.855$$
$$R_3 \times R_4 = 0.95 \times 0.98 = 0.931$$
$$R_5 \times R_6 = 0.90 \times 0.97 = 0.873$$
$$R_7 \times R_8 = 0.92 \times 0.99 = 0.910$$

The resulting model looks like this (Figure 2.8):

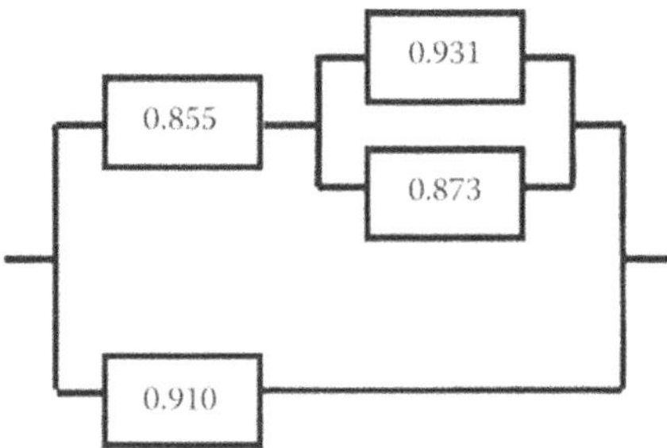

FIGURE 2.8 Decomposition.

Decomposing the redundant section, we get:

$$\begin{aligned} R_R &= 1 - (1 - 0.931)\,(1 - 0.873) \\ &= 1 - (0.069)\,(0.127) \\ &= 1 - 0.008763 \\ &= 0.9912 \end{aligned}$$

Our model now looks like this (Figure 2.9):

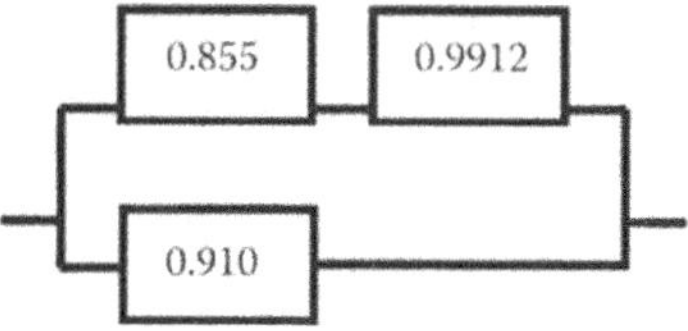

TAKEOUT: FIGURE 2.9 Second decomposition.

Continuing in this manner, we eventually get the answer as:

$$Rs = 0.986$$

AVAILABILITY AND MAINTAINABILITY

Up to now, we have discussed the reliability of equipment with no thought given to repairing it after it has failed. Consideration will now be given to the maintenance of failed components and systems.

The governing parameter for maintained systems is availability, which, as we defined it in Chapter 1, is the result one gets from the combination of reliability and maintainability of the system.

In practice, maintainability is probably the most neglected of the reliability engineering parameters. Many of us could cite cases of poor maintainability in consumer and commercial products. Some include:

- The nuclear power station where a wall must be broken down for the re-tubing of a turbine condenser.
- The luxury car that when the in-line six-cylinder engine was replaced by a V8, it became necessary to remove the front wheel to get at one of the spark plugs.
- Certain domestic appliances that, because it is a very price-sensitive market, represent the ultimate in cheap-to-make, but made hard to repair in the process.

In fact, increases in maintainability become a fruitful source of improvement in availability upgrade studies. For example, good industrial engineering support helps with the making of special jigs to reduce changeover time in manufacturing processes, items often neglected by the suppliers of the systems. Here follow some simple ideas for maintainability improvement. They may sound obvious, but they are often neglected.

- Avoid complexity.
- Emphasise accessibility in specifications and designs. For example:
  - The parts with the lowest reliability should be the most accessible.
  - Consideration should always be given to the use of quick-release fasteners and captive screws. Loose screws can be dropped inside equipment and cause faults later.

  - Instruments should be able to be withdrawn from the front of the panel. Compare the ease of replacement of instruments in a light aircraft compared to a car dashboard!
- It may be necessary to provide for adjustment of equipment in service. Access must be provided for this adjustment.
- Built-in test equipment can be provided for the automatic diagnosis of faults.
- Plug-in components assist maintainability. The plug-in module can be replaced and sent for repair at a specially equipped workshop, with no time constraint.
- Redundancy can be used as an aid to maintainability just as it is an aid to reliability.
- Adequate maintenance instructions should always be provided. It is often possible to incorporate essential circuit diagrams, etc., on the equipment itself. Manuals should make liberal use of illustrations.
- Standard components should be used wherever possible. This not only simplifies the spares situation, but it is also probable that the maintenance staff will be more familiar with standard parts.
- The use of special tools should be properly considered. They can be either a hindrance to maintainability, if not available at all locations, or an aid to maintainability, if properly controlled and available.

THE MAINTAINABILITY EQUATION

The maintainability equation, which is derived from the Poisson distribution, is:

$$\mathrm{M}\,(\mathrm{t}) = 1 - e^{-\mu t_m} \tag{2.7}$$

Where *M (t)* is the probability that a failed system will be repaired in a given time.

M is the maintenance action rate.
t_m is the maintenance time constraint.

It can therefore be seen that maintainability, which is the probability that a system will be repaired within the maintenance time constraint, is analogous to reliability, which is the probability that an item will survive for a certain time. Mathematically, however, it is analogous to unreliability, the probability that an item will not survive for a stated time. In this comparison, *maintenance action rate* μ is analogous to *failure rate* λ. This is summarised in Table 2.2

THE EQUATIONS FOR SYSTEM AVAILABILITY

The good news is that these equations are identical in form to the equations for reliability. By simply substituting *R* by *A* and *F* by *U*, the equations previously given for reliability can be used for availability predictions, where *A* = availability and *U* = unavailability. Hence, Pascal's triangle, for example, can be used for availability calculations as well. Hence, we have, for a series system:

$$A_s = A_1 \times A_2 \times \ldots A_n \tag{2.8}$$

TABLE 2.2
Reliability and Maintainability Compared

Term	Definition	Formula
Reliability	The probability that a component or system will survive for a given time	$R(t) = e^{-\lambda t}$
Unreliability	The probability that a component or system will fail before a given time	$F(t) = 1 - e^{-\lambda t}$
Maintainability	The probability that a failed system will be repaired within a given time	$M(t) = 1 - e^{-\mu t_m}$
"Unmaintainability" (a term not used)	The probability that a failed system will not be repaired within a given time	$e^{-\mu t_m}$

For a redundant 1-out-of-2 system:

$$A_s = 1 - (U_1 \times U_2) \tag{2.9}$$

Where A = availability.
And U = unavailability.

And as before:

$$A + U = 1 \tag{2.10}$$

STORAGE CAPACITY (FIGURE 2.10)

There are, however, extra mathematical insights that are necessary in the analysis and prediction of availability which are not necessary in the case of reliability. The first of these is *storage capacity*. Applying upstream, downstream, or intermediate storage, as required, can have a powerful effect on the availability of a system. The equation for storage is given here:

$$\theta' = \theta e^{\left(\frac{ST}{\varnothing_A}\right)} \tag{2.11}$$

Where

ST = storage capacity in time units
Θ = MTBF of upstream plant
θ' = modified MTBF of upstream plant
Φ = MTTR of upstream plant

Equation 2.11 can be derived from the maintainability equation (2.7), with a suitable change of nomenclature.

Since availability is $\theta/(\theta + \Phi)$, inserting the new θ' into this equation gives us a new virtual availability for the upstream plant. The effect can be dramatic. Assume the availability of the MTBF and the MTTR for both the upstream and downstream

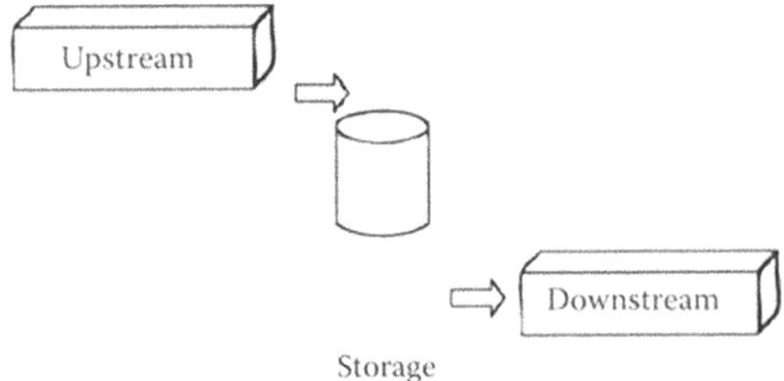

FIGURE 2.10 Effect of storage capacity on availability.

systems are 100 days and 5 days each, giving availabilities of ~ 95.25% and therefore a combined availability of 0.9072. If the storage capacity is 10 days, then by equation 2.11,

$$\begin{aligned}\theta' &= \theta\, e^{ST/\varnothing} \\ &= 100e^{10/5} \\ &= 738 \text{ days}\end{aligned}$$

This yields a virtual upstream availability of 738/(738 + 5) = 99.33%. The overall availability of the system is now 0.9933 × 0.95.25 = 0.9461, an increase of 3.89%. With 8,760 hours in a year, this translates to an extra 340 hours, or about 14 extra days of available plant. Storage is in fact one of the cheapest ways of effecting large availability increases.

One must be careful as to what is "upstream" and what is "downstream." For a plant supplying product to a warehouse, then to the consumer, the model is as in the previous figure. But consider the case of a coal-fired power station supplying electricity and the ashing plant breaks down and is unable to convey the ash away to the ash dump. To continue to supply electricity, the boiler is now made to dump ash on the floor of the boiler house, which can accommodate about 24 hours of ash. In the model as shown in Figure 2.6, the modified MTBF would now be applied to the "downstream" section, the ashing plant in this case. And the storage section would be the boiler house floor. But as a general rule, there are problems with storage. Storage of intermediate or final product means capital tied up on the plant floor and hence goes against the modern lean manufacturing paradigm of minimising work in progress wherever possible. Secondly, if the product stored is hazardous, it will not be allowed in terms of any HAZOP study (hazard and operability study). HAZOP as a technique is discussed later in the course.

REDUCED-CAPACITY STATES

Up to now, we have only considered systems that are either up or down, that is, they run at 100% capacity or they switch off. In practice, some systems can run at intermediate loads. Three 50% capacity units are shown in the following figure. For the 100% system capacity case, this was discussed before as a 2-out-of-3 system.

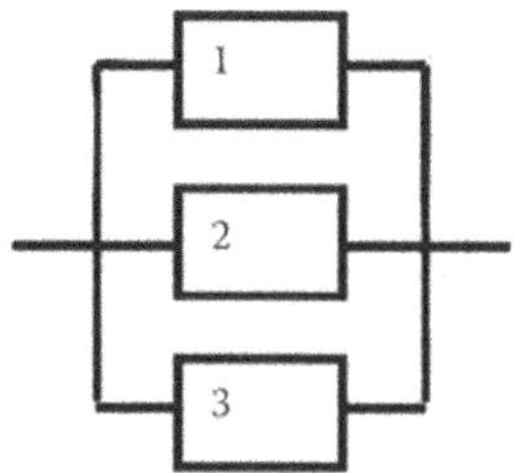

FIGURE 2.11 Three 50% units with reduced-capacity states allowed.

TABLE 2.3
Reduced-Capacity States for Three 50% Units in Parallel

Component 1	Component 2	Component 3	Capacity Loss
Available	Available	Available	0
Unavailable	Available	Available	0
Available	Unavailable	Available	0
Available	Available	Unavailable	0
Available	Unavailable	Unavailable	0.5
Unavailable	Available	Unavailable	0.5
Unavailable	Unavailable	Available	0.5
Unavailable	Unavailable	Unavailable	1.0

However, if the system is capable of running at reduced capacity, we have the following (refer to Figure 2.11 and Table 2.3).

We can now represent the previous table by the following equation:

$$U_s = (0.5)A_1U_2U_3 + (0.5)U_1A_2U_3 + (0.5)U_1U_2A_3 + U_1U_2U_3 \quad (2.12)$$

If $U_1 = U_2 = U_3 = U_c$, then:

$$U_s = 1.5A_cU_c^2 + U_c^3$$

If $A_c = 0.9$, $U_c = 0.1$, then:

$$U_s = 0.0145 \text{ and } A_s = 0.9855$$

OTHER FORMS OF SYSTEM RELIABILITY ANALYSIS

The decomposition method, discussed earlier, is the most common form of system reliability analysis. However, there are many cases where the method cannot be used. One is if the failure and repair rates of the individual subsystems are not *statistically independent*. In other words, if the failure of one item affects the failure or repair rate of another item, the method cannot be used.

As a second example, the method cannot handle configurations which are not simple series or parallel systems. Even a system as simple as the one shown later

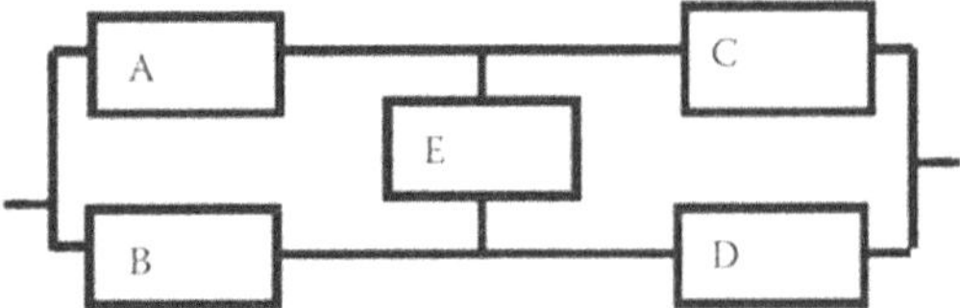

FIGURE 2.12 Bridge circuit.

needs special treatment. It is known as a bridge circuit configuration and is commonly encountered in many systems (Figure 2.12).

This configuration can be analysed in two ways.

METHOD 1: DECONSTRUCTION METHOD

The circuit is replaced by two circuits, as shown in Figure 2.13:

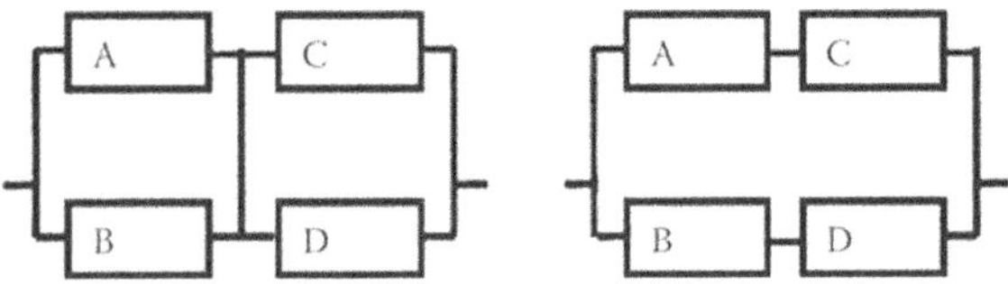

FIGURE 2.13 The bridge circuit deconstructed.

For example, if we have $R_c = 0.9$ for each component, then we have:

$$0.9801 \times 0.9 + 0.9639 \times 0.1$$
$$= 0.97848 \sim 97.8\%$$

METHOD 2: CUT SETS

For this method, we redraw the system as shown in Figure 2.14. The definition of a *cut set* is that if any one item of the cut set is activated, then the set is activated. It is very important to make sure the cut sets are minimal, that is, that they do not have more items in the set than necessary to achieve the previous definition.

If we compare the cut set diagram in Figure 2.10 with the original bridge circuit in Figure 2.8, we will see that they are logically equivalent. Looking at the first cut

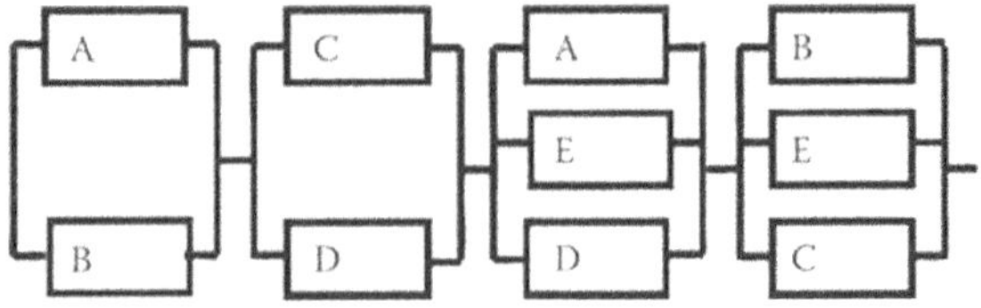

FIGURE 2.14 The bridge circuit reconstructed as cut sets.

set, we see that if A and B are both down and the other cut sets are up, the system will be down. If either A or B are restored, the system is up. Similarly with the other cut sets.

If we once again assume the *R* for all the components is 0.9, then the answer for R_s for the cut set model is

$$R_s = 0.99 \times 0.99 \times 0.999 \times 0.999$$
$$\text{i.e. } R_s = 0.97814$$

It will be noticed that this answer is slightly less than the 0.97848 achieved by the previous method in Figure 2.9. The answer in Figure 2.9 is correct, while the answer using cut sets is slightly pessimistic. The reason for this is that there are more blocks in the cut set model than in reality, and as each of the blocks contains a value less than unity, it is inevitable that the resulting answer will be slightly pessimistic. For systems with realistic reliabilities, the difference is minimal. Only when one encounters very poor reliability in the components or subsystems is there the possibility of a significant error using the cut set method. The cut set method is equally applicable to availability calculations as well.

MONTE CARLO SIMULATION

Even with the versatility provided by the cut set method, there will be situations where none of the simple methods, using simple equations, will do for system reliability/availability prediction. It is then necessary to resort to simulation, usually Monte Carlo simulation, but some reliability engineering packages use Markov simulation, which is more mathematically elegant but is a voracious consumer of computer memory and time. Monte Carlo simulation usually suffices. In most packages, an ABD is drawn in the usual way, and then a failure distribution and a repair distribution of one's choice is inserted into every block in the ABD. A clock then starts to run, and the distributions are sampled every ten minutes in real time, or every fraction of a second in computer time. The status is reported at every interval. Sampling the distributions causes some components to fail. The relevant repair distributions are then sampled to see when these components will be repaired. The availability of the system is continually assessed. The scenario can run for, say, a year in real time, which might mean a few hours in the computer. The scenario can then be rerun as often as thought necessary. A report is generated on the availability of the system over, say, a year.

MARKOV SIMULATION

For Markov simulation, a suite of states is determined in a space-state diagram. For example, for a two-component system, with two possible modes for the components, that is, working or failed, up or down, we have the following, as shown in Figure 2.15:

Figure 2.13 shows the Markov diagram with two components and two modes. To get from one state to the other, the transition vectors are used, which are the failure rates and the repair rates of the components.

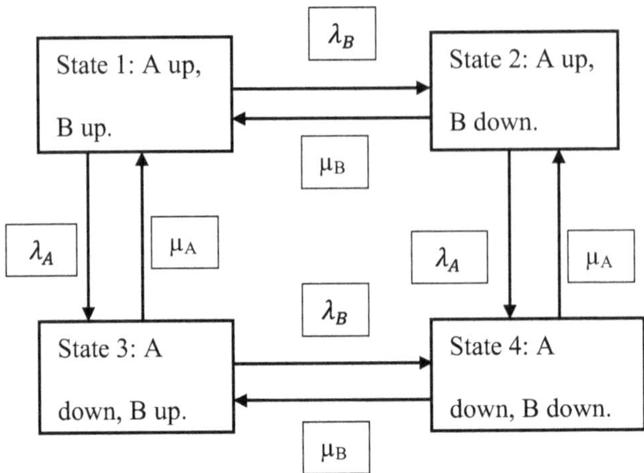

FIGURE 2.15 Markov diagram for a two-component system. (Diagonal vectors omitted.)

For three components and two modes, we would have eight states, as follows:

A up, B up, C up
A up, B up, C down
A up, B down, C up
A up, B down, C down
A down, B up, C up
A down, B up, C down
A down, B down, C up
A down, B down, C down

For two components and three modes, for example, up, down, idle, we have nine states:

A up, B up
A up, B down
A up, B idle
A down, B up
A down, B idle
A down, B down
A idle, B idle
A idle, B up
A idle, B down

Therefore, the complexity of a Markov model $\mathbf{C}_\mathbf{m}$ is given by:

$$\mathbf{C}_\mathbf{m} \propto \mathbf{m}^\mathbf{n} \tag{2.13}$$

Where $\boldsymbol{m}$ is the number of modes and $\boldsymbol{n}$ is the number of components.

By comparison, the Monte Carlo model will have a complexity as shown here:

$$\mathbf{C_{mc} \propto m \times n} \tag{2.14}$$

So for a Markov model with 100 components and two modes (which is not unreasonable in a real-life situation), the degree of complexity is proportional to 2^{100}, which is a very big number indeed—1.26×10^{30} in fact. In comparison, the degree of complexity for the equivalent Monte Carlo model would be proportional to 2×100. Another problem with Markov is that it assumes random failure and repair. This can be overcome, but at the expense of further complication. So in general, Markov modelling remains an elegant mathematical curiosity, while Monte Carlo is reliability engineering's workhorse.

For Monte Carlo modelling, then, we can deduce the following:

Modifications can then be made—adding redundancy, adding work crews to speed up times to repair, etc., until an optimum solution is obtained. The system has no limitations, such as statistical independence of failure. Information can be added that if item A fails, the failure rate of item B is doubled, for example. The cost of unavailability can be entered to assess whether the upfront capital conditions will be worthwhile in terms of the savings or extra profits over the remaining life cycle.

It is possible to simulate almost any situation, provided the necessary information is available. Most reliability engineering software packages include a simulation option.

INTERCONNECTIONS

A simple way to increase availability or reliability at low cost is by means of interconnections. This has already been demonstrated in Figure 2.12 and 2.13. With the interconnection, the reliability of the system is 0.9801, while without the interconnection, the reliability is 0.9639, a difference of about 1.7% over the 0.9639 in this case. Electronic circuits make use of many interconnections as an almost zero-cost way of improving reliability. Mechanical systems can also benefit. For example, a string of conveyors may have a redundant string installed alongside, as shown in Figure 2.16. For this configuration, assuming all the blocks have equal reliability and that one is allowed only one interconnection, where should one put it? (See Assignment 2.2.)

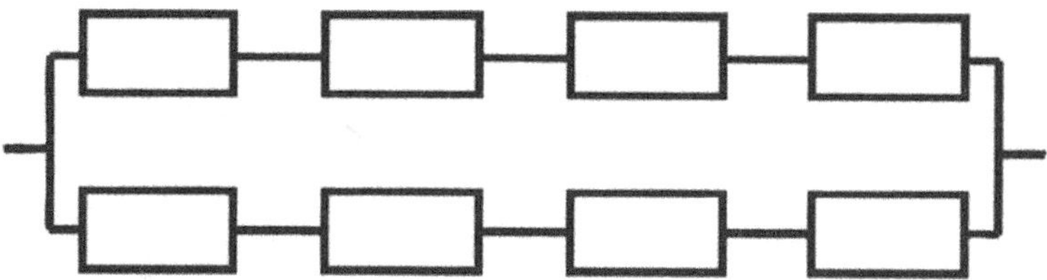

FIGURE 2.16 Parallel system of conveyors.

LAPLACE ANALYSIS

We saw in Chapter 1 how Weibull analysis should generally be applied to components only, that is, things that are discarded rather than repaired. This was only confirmed as late as in the 1980s. Even the great quality specialist Juran made the mistake in one of his textbooks of applying Weibull analysis to maintained air-conditioning units. The whistle was blown by Harold Asher and Harry Feingold when their 1984 book was published, *Repairable Systems Reliability*. When asked what form of analysis should be applied to maintained systems, they suggested Laplace analysis.

The Laplace statistic is defined as follows:

$$U = \frac{\dfrac{\sum_{i=1}^{m-1} T_i}{m-1} - \dfrac{T_m}{2}}{T_m \sqrt{\dfrac{1}{12(m-1)}}} \tag{2.15}$$

Asher and Feingold describe the following answer in their book. The times to failure are in chronological sequence, not ranked as in Weibull. (Table 2.4)

Now, this is clearly a system on its last legs—each time it is restored, it runs for a shorter time than before. The Laplace statistic confirms this:

$$U = \frac{\dfrac{1811}{6} - \dfrac{410}{2}}{410\sqrt{\dfrac{1}{12(7-1)}}}$$

$$= 2.0$$

Now 2.0 is greater than 1.96, which is the cut-off value for statistically significant evidence of deterioration at the 5% two-sided confidence level. For a system that was improving, our U statistic would have had to be less than –1.96. So negative is good; positive is bad.

TABLE 2.4
Laplace Analysis

Time to Failure	Cumulative Time = T_i	Notes
177	177	
65	177 + 65 = 242	
51	242 + 51 = 293	$\Sigma = 1811$
43	336	
32	368	
27	395	
15	410	$410 = T_m$

TABLE 2.5
U Values for Various Degrees of Confidence

Degree of Confidence	U value
90%	+/–1.65
95%	+/–1.96
99%	+/–2.58

The problem with Laplace analysis is that there is this wide deadband from –1.96 to +1.96. If our *U* value is within that band, we can say nothing—we cannot say that the system is improving or deteriorating. We can, however, reduce the width of the deadband by deciding on lower limits of confidence. The 95% limit means we are sure of the result we get 95 times out of a hundred. If we were to drop our confidence to 90%, then by reference to the table of the normal distribution in Appendix I, we arrive at a figure of 1.65. If we wanted greater certainty, say, 99%, we would have to have limits of +/–2.58!

We can summarise this as shown in Table 2.5.

All we are doing is rearranging the information to suit the degree of confidence required. This confidence level might be specified in a specification, or we might decide on it ourselves. Are you satisfied with being right 9 times out of 10 (90% confidence)? Or are you more cautious and want to be right 99 times out of a 100? This depends on one's own *risk profile.*

The traditional confidence limits of 99, 95, and 90% are from the pre-computer days of printed tables in statistics textbooks. These days, any limit is possible—refer to the Weibull-Ease program, for example. The Laplace test is a valuable index of plant and of maintenance management performance and may be implemented as a key performance indicator (KPI) for a maintenance manager. Because of the deadband, however, sometimes the index does not tell us anything. It should be used together with other KPIs, such as the availability of plant (and sections of plant) plotted on a monthly basis.

FAILURE MODES AND EFFECTS ANALYSIS (FMEA) AND FAILURE MODES AND EFFECTS CRITICALITY ANALYSIS (FMECA)

History

FMEA and FMECA are reliability assurance techniques. The first recorded official use of FMEAs dates from the late forties.[1] The US Armed Forces Military Procedures Document MIL-P-1627 was issued in 1949, detailing procedures for conducting an FMECA. This was in response to problems experienced with contractors, who were supplying products with poor reliability and quality. This is a generic problem—designers are trained to optimise the design of a product and to thereby ensure its efficiency, power consumption, and other engineering parameters. They are not generally trained to look at their designs in a negative way—that is, seeing how they might fail. Indeed, this has been the writer's personal experience. In an attempt by his company to introduce such techniques into specifications for the buying department, the writer was met with incredulity by some suppliers. One, the chief designer of very large motors for a large international company protested by saying, "But our products are reliable," despite evidence to the contrary by the users.

In 1967, the Society of Automotive Engineers (SAE) released the first civil publication to address FMECA for civil aircraft. The Ford Motor Company began using FMEA in the 1970s after safety problems and legal issues experienced with one of its models. In Europe, the International Electrotechnical Commission published IEC 812 (now IEC 60812) in 1985, addressing both FMEA and FMECA for general use. The British Standards Institute published BS 5760–5 in 1991 for the same purpose. The American Military standard MIL-STD-1629A was cancelled without replacement in 1998 but nonetheless remains in wide use for military and space applications today.

FLAVOURS OF FMEA/FMECA

There are various forms of FMEA developed by various organisations. The original RCM technique, discussed later in this book, uses a very simple four-column format of function, functional failure, failure mode, and failure effect. A more complex format is given later, which is the format suggested by the ASQ, or the American Society for Quality. Note that functional failure and failure mode in the RCM format mean something different in the ASQ format, corresponding rather to failure mode and failure cause.

In the example given in Table 2.6 we see that the process function is to drill a hole of a certain depth. There are two failure modes: hole too deep and hole not deep enough. For the first case, there is one potential cause of failure, and on the second case, there are two. The severity of the failure and the probability of occurrence are also given, on numerical scales. The current process controls are then given, which are attempts to prevent the failures, followed by the probability of detecting the failure.

The severity, probability of occurrence, and probability of detection are then multiplied together to determine the risk priority number (RPN). The higher the number, the more serious the failure. There is also usually a specified lower limit to the RPN, below which no further action is required. In this example, the RPNs 45 and 63 would appear to be below the limit, as no further action is taken for those failure modes. For the case of the broken drill with an RPN of 225, further action is taken, namely, to install tool detectors, which then reduces the RPN to 25.

PROCESS VS DESIGN OR PRODUCT FMEA/FMECA

The previous example is of a *process* FMECA, highlighting things that could go wrong during a manufacturing process. The other main type of FMECA is a *design*

TABLE 2.6
Typical FMECA Structure

FAILURE MODE AND EFFECTS ANALYSIS			
Item: Drill Hole	Responsible: A N Other	No:	123456
Current	Prepared by: A N Other	Page :	1 of 1
Core Team:	FMEA Date (Original): 2014/01/01	Rev:	1

Process Function	Failure Mode	Effect(s) of Failure	Severity	Cause(s) of Failure	Probability of Occurrence	Current Process Controls	Detection Prob.	RPN	Recommended Action(s)	Responsibility and Target Completion	Action Results: Actions Taken	Action Results: Severity	Action Results: Occurrences	Action Results: Detection	Action Results: RPN
Drill Blind Hole	Hole too deep	Break through bottom of plate	7	Improper machine setup	3	Operator training and instructions	3	63							0
	Hole not deep enough	Incomplete thread form	5	Improper machine setup	3	Operator training and instructions	3	45							0
			5	Broken Drill	5	None	9	225	Install Tool Detectors	J. Doe	2014/03/01	5	5	1	25

FMECA, done early during the design phase, to highlight how a design might fail in service even if correctly made.

What is also required with such FMEAs and FMECAs is a set of evaluation criteria to define and select the appropriate numbers in the columns, such as severity, probability of failure, and probability of detection. These criteria tend to be industry-specific. Here in Table 2.7 is an example from the motor industry that might be applicable for the severity index in the FMECA mentioned prior.

This is adapted from McDermott et al. (2009).

Table 2.8 gives generic values for the "occurrence" column in a process FMEA, once again adapted from the motor industry.

TABLE 2.7
Severity Evaluation Criteria for Process FMECAs for the US Automobile Industry

Effect	Severity of Effect on Process	Rank
Failure to meet safety or regulatory requirements	May endanger operator without warning.	10
Failure to meet safety or regulatory requirements	May endanger operator with warning.	9
Major disruption	100% of product is scrap/line shutdown.	8
Significant disruption	Some product scrapped/line slowdown, additional manpower required.	7
Moderate disruption	100% of product requires rework offline.	6
Moderate disruption	Some product requires rework offline.	5
Moderate disruption	100% of product requires rework online.	4
Moderate disruption	Some product requires rework online.	3
Minor disruption	Slight inconvenience to operator or process.	2
No effect	No discernible effect.	1

TABLE 2.8
Evaluation Criteria for the Occurrence Column of a Process FMECA

Probability of Failure	Occurrences	Rank
Very high	100 per thousand or greater	10
High	50 per thousand	9
High	20 per thousand	8
High	10 per thousand	7
Moderate	2 per thousand	6
Moderate	0.5 per thousand	5
Moderate	0.1 per thousand	4
Low	0.01 per thousand	3
Low	0.001 per thousand	2
Very low	Failure eliminated through preventive control	1

TABLE 2.9
Detection Criteria for a Process FMECA

Probability of Detection	Criteria	Rank
Almost impossible	No current process control.	10
Very remote	Not easily detected. Random audits are done.	9
Remote	Detection is by post-processing by the operator by visual/tactile/auditory means.	8
Very low	Operator use of attribute gauging post-process, e.g. go-no-go gauge.	7
Low	Operator use of attribute gauging in-process.	6
Moderate	Automated controls for gauging detect defective part and notify operator.	5
Moderately high	Automated controls for gauging detect defective part post-process and prevent further operation.	4
High	Automated controls for gauging detect defective part in-process and prevent further operation.	3
Very high	Automated controls that will detect error and prevent part being made.	2
Almost certain	Item has been error-proofed by process or product design.	1

The next table (Table 2.9) lists detection criteria for a Process FMECA

Using the prior three tables, 2.7, 2.8, and 2.9, consult the FMECA in Table 2.6 again and see if you can concur with the ratings.

DESIGN FMECA

A design FMEA may have the same layout as a process FMEA, but the "rating" columns will be different. Typical examples are given in Table 2.10 for a design FMECA. The third column, "exposure," might not be necessary, depending on the circumstances.

WHICH TEMPLATES TO USE?

This depends on circumstances—different industries have developed different templates, as described prior. One might want to use a template specific to one's industry, or one might want to develop one's own.

BASIC ANALYSIS PROCEDURE FOR AN FMEA OR FMECA

The basic steps for performing an FMEA/FMECA analysis include:

- Assemble the team.
- Establish the ground rules.

TABLE 2.10
Risk Evaluation Criteria for a Design FMEA

Probability		Consequence or Severity		Exposure	
Can it happen?		What injury or damage could we expect?		How often are people currently exposed to the hazard?	
Well expected/happens often	7	Catastrophic disaster Multiple fatalities +R5 million damage	7	Continuous/full shift	7
Quite possible	6	Fatality Severe or permanent disability R1–5 million damage	6	Frequently/daily	6
Unusual, but has happened before	5	Very serious injury (ICU) R500,000–R1 million damage	5	Often/weekly	5
Remotely possible; has happened elsewhere	4	Serious injury (ICU) R100,000—R500,000 damage	4	Occasionally/monthly	4
Very unlikely but conceivable; has not happened yet	3	Disabling injury R1,000–R100,000 damage	3	Unusual/yearly	3
Practically impossible; one in a million chance	2	Treat and return injury Below R1,000 damage	2	Rare/within five years	2
Virtually impossible; full certainty	1	Near miss Some light damage	1	Remote/beyond five years	1

- Brainstorm to establish all possible failure modes.[2]
- Gather and review relevant information.
- Identify the item or process to be analysed.
- Identify the function(s), failure(s), effect(s), cause(s), and control(s) for each item or process to be analysed.
- Evaluate the risk associated with the issues identified by the analysis.
- Prioritise and assign corrective actions.
- Perform corrective actions and re-evaluate risk.
- Distribute, review, and update the analysis, as appropriate.

ADVANTAGES OF THE FMEA/FMECA PROCESS

FMECA is a very systematic method to identify all failures and their criticality. Based on this analysis, it is easy to prioritise design improvements or plan for maintenance and repair. It is a method to effectively systematise the information from expert sources and so helps improve a design.

LIMITATIONS OF THE FMEA/FMECA PROCESS

The multiplication of severity, occurrence, and detection rankings to determine RPNs is a flawed process. In each column for severity, occurrence, or detection, the rank numbers are *ordinal,* that is, they only say that one rank is better or worse than another, but not by how much. For example, a severity ranking of 2 might not be twice as bad as a ranking of 1. Yet the process of multiplying three ordinal numbers together implies this. So we may have a certain severity rating, 5, say, which when multiplied by numbers 1 and 1 for the other two columns does not make it onto a criticality list when actually it should be on the list.

There is no real answer to this problem other than to review each RPN critically to ensure no critical failure is neglected.

FMEA OF A SCRAPER WINCH

There are several manufacturers of scraper winches worldwide. Such machines are used in quarries and underground mines to scrap blasted rock along passageways and the like. They are skid-mounted and fixed in place. It is often required to move them as development of the mine proceeds. They are extremely ruggedly built but still sometimes fail because of abuse, particularly in underground applications. For example, they have been known to be rolled down incline shafts instead of being winched or self-winched into a new position. One make of scraper winch is Galison, drawings of which are shown in Figures 2.17 and 2.18. Table 2.11 is the key to the numbered parts. These illustrations are reproduced with permission of the Galison Company.

OPERATION

The winch operates as follows:

1. The motor pinion transmits torque to the lay shaft gear.
2. The lay shaft pinion transmits torque to the main shaft gear.
3. The main shaft pinions transmits torque to the idler gears.

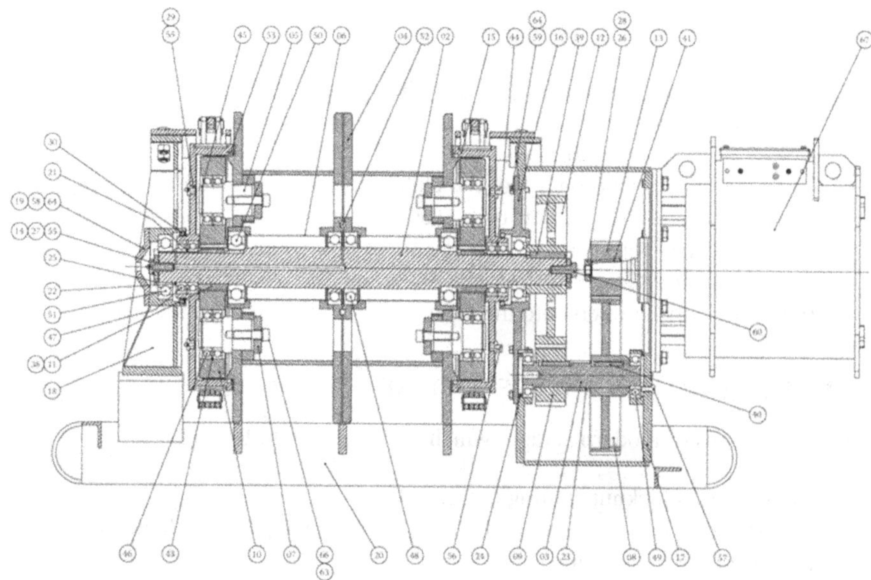

FIGURE 2.17 Galison scraper winch.

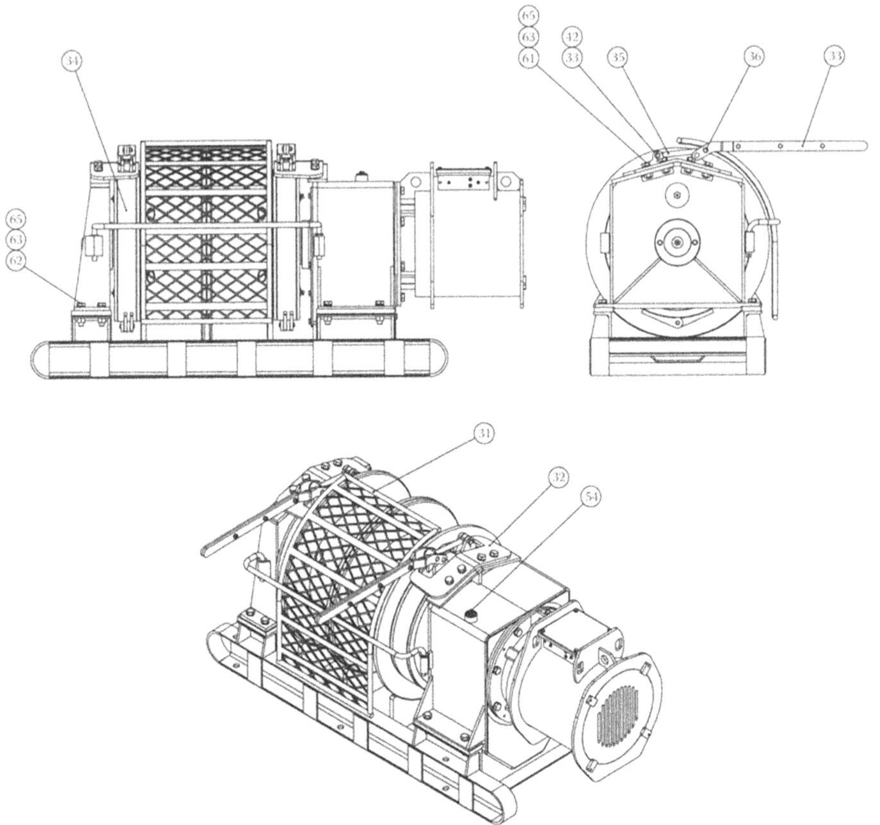

FIGURE 2.18 Galison scraper winch.

4. With the clutch band disengaged, the idler gears rotate about their own axes only and transmit torque to the clutch gear, which rotates about its own axis (which is the main shaft axis).
5. With the clutch band engaged, the clutch gear remains stationary and the idler gears rotate about their own axes and about the main shaft axis. The idler pins therefore rotate about the main shaft axis, which acts on the drums, causing the drums to rotate about the main shaft axis. The rope on the drum is then coiled onto the drum, and the scraper is pulled.

COMMON FAILURES AND RECENT IMPROVEMENTS

Common failures of modern scraper winches include the following:

1. Motor pinion locknut coming loose.
2. Motor pinion key dislodges and falls out, leading to gearbox damage.
3. Motor pinion coming loose.
4. Pedestal bearing failure.
5. Motor failures, such as phase-to-phase or phase-to-earth faults.

TABLE 2.11
Key to Numbered Parts

Item No.	Description	Qty	Item No.	Description	Qty
01	Drum assembly	02	35	Adjusting screw	02
02	Main shaft	01	36	Adjusting nut	02
03	Lay shaft	01	37	Clutch swivel pin	02
04	Rope drum	02	38	Key, main shaft pinion	02
052	Idler stud	04	39	Key, main shaft	01
06	Filling piece	02	40	Key, layshaft	02
07	Washer	01	41	Key, motor pinion	02
08	Gear first motion	01	42	Split pin	04
09	Pinion second motion	01	43	Internal circlip	04
10	Idler gear	02	44	Internal circlip	02
11	Main shaft pinion	02	45	External circlip	04
12	Gear second motion	01	46	Bearing	08
13	Pinion motor	01	47	Bearing	04
14	Screw	01	48	Bearing	02
15	Clutch gear	02	49	Bearing	02
16	Gearbox cover	01	50	Bearing	02
17	Gearbox	01	51	Bearing	02
18	Pedestal	01	52	Grease packing, drum	01
19	Pedestal cover	01	53	Grease packing, clutch	02
20	Base frame	01	54	Plug	01
21	Spacer	02	55	Grease nipple	03
22	Spacer	02	56	Grease nipple	02
23	Spacer	01	57	Hex head screw	01
24	Spacer	01	58	Hex head screw	02
25	Washer	02	59	Hex head screw	10
26	Washer	01	60	Hex head screw	09
27	Washer	02	61	Hex head screw	08
28	Locknut, motor	02	62	Hex head screw	10
29	Plug screw	04	63	Spring washer	26
30	Oil seal	02	64	Spring washer	13
31	Rope guard	01	65	Hex nut	18
32	Stop plate	02	66	Socket head cap screw	04
33	Clutch handle	02	67	Motor	01
34	Clutch band	02			

Recent design improvements include the following:

1. The gearbox used to have an oil plug, which led to gearbox lubrication contamination and objects dropped into the gearbox, leading to gearbox damage. The gearbox is now a sealed unit.
2. The motor shaft is tapered, and so are the pinion bore and key. There is also a lock washer and lock nut to secure the motor pinion.

3. Duplex bearings, which can withstand greater loads than single bearings, and oil seals are fitted for the clutch gear bearings and the main shaft bearings.
4. The pedestal bearing is easily accessible for the replacement thereof.
5. In some cases, the ropes end up wedged between the drums. Between the drums is a curved flat bar section to prevent the rope from coiling between the drums.
6. Frequent motor failures. Modern winch motors are purposely designed and built to operate at larger slip angles, due to the variable loading that they experience during the cleaning/ore scraping process.
7. Pressed sleeves are fitted to the shafts to locate the gears and bearings.
8. All interference fit components are factory-pressed with a 100-ton press.
9. The modern scraper winch is of a very robust design in order to survive underground transport and operations.

ASSIGNMENT

Produce a design FMECA for a *scraper winch*. Mining people should all be familiar with this piece of equipment. There are several manufacturers who may be contacted for information via their websites. A visit to one or more such manufacturers would probably be beneficial. Also, consult data bases for information on the failures of such equipment, if you can locate any. Include background material on the product—drawings, photographs, description of its operation, and details of failures, if possible. The scraper winch is a very mature piece of technology, with probably not much opportunity for improvement in its design or operation. Use the FMECA, therefore, to demonstrate how the design has catered for and prevented various failure modes.

FAULT TREE ANALYSIS

Fault tree analysis (FTA) was originally developed in 1962 at Bell Laboratories by H. A. Watson, under a US Air Force contract to evaluate the Minuteman I Intercontinental Ballistic Missile (ICBM) Launch Control System. The use of fault trees has since gained widespread acceptance. The Boeing aircraft company began using FTA for civil aircraft design around 1966. Today, FTA is widely used in system safety and reliability engineering and in all major fields of engineering.

Within the nuclear power industry, the US Nuclear Regulatory Commission began using probabilistic risk assessment (PRA) methods, including FTA, in 1975 and significantly expanded PRA research following the 1979 incident at Three Mile Island.

Following process industry disasters such as the 1974 Flixborough explosion, the 1984 Bhopal disaster, and the 1988 Piper Alpha explosion, FTA has been used extensively in the process industries as well, along with other techniques, such as HAZOP (hazard and operability analysis).

Fault tree analysis really has two main uses:

1. As a predictive tool to determine the probability of something happening, like the failure of a large engineering system. In such cases, it is constructed with AND gates and OR gates.
2. As a post hoc tool in incident investigation, where it describes the events leading up to an incident in a logical fashion. In this case, the fault tree only has AND gates.

In Figure 2.19 is an example of a very simple fault tree, with just a few events described in a few blocks. In practice, a real fault tree may have hundreds of blocks. It will be seen that there is a top event to the tree; in this case it is "Plant Destroyed." When performing fault tree analysis, it is important that all members who contribute to the tree agree as to what the top event is. It will also be seen that there are two types of gate in the tree—AND gates and OR gates. All the events under an AND gate must occur for the event above the AND gate to occur. With an OR gate, only one of the events under the gate need occur.

When used in incident investigation, all the gates are AND gates, because if the investigation has been done correctly, there are no "loose ends"—no either/or scenarios. The truth has been established.

Some texts on fault trees provide for complex diagrams with rectangles, circles, diamonds, etc. to provide for basic failures, dependent failures, and the like. This approach is not followed here. All that is necessary for a clear presentation are rectangles of the events and AND and OR gate symbols for the logic.

Actually, the fault tree and the RBD are mathematical transforms of each other, as is demonstrated next. It is seen that the OR gate corresponds to a series RBD. The mathematical expression describing each is given after, and the expression for the

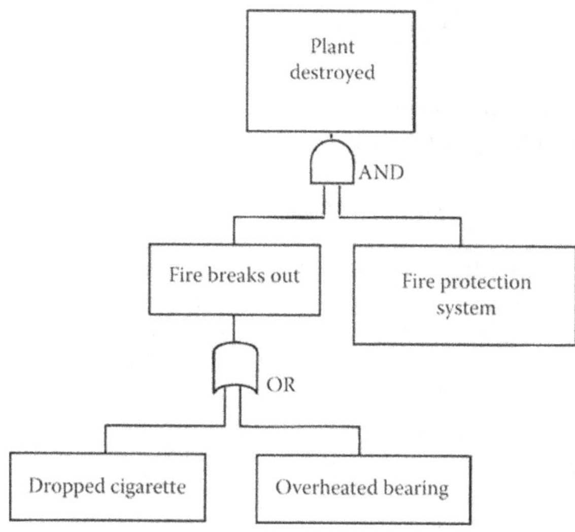

FIGURE 2.19 Simple fault tree.

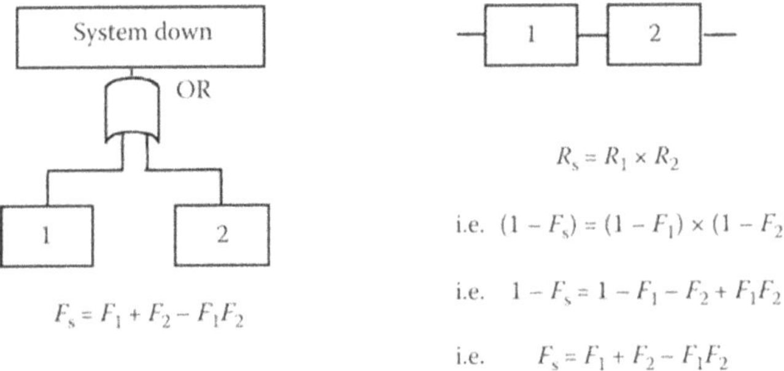

FIGURE 2.20 Correspondence between an OR gate and a series RBD.

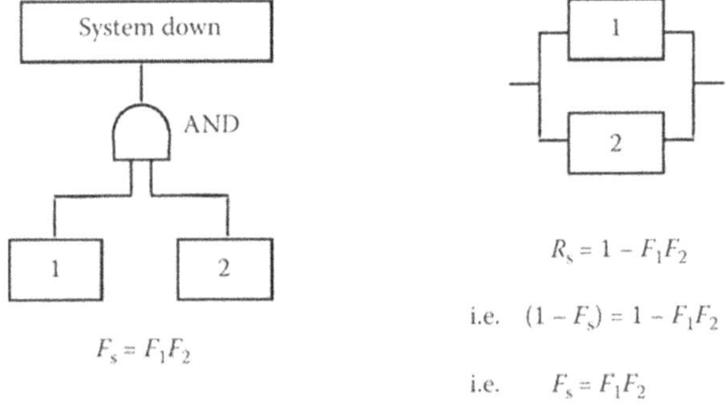

FIGURE 2.21 Correspondence between an AND gate and a parallel RBD.

RBD is also shown converted from R symbols to F symbols, where the expression is shown to be identical to the expression for the fault tree (Figure 2.20).

Likewise, the AND gate in a fault tree corresponds to a parallel RBD, as shown in Figure 2.21.

So if the two methods are mathematically equivalent, which one should one use? This can sometimes be simply a matter of personal preference. The fault tree is failure-based, whilst the RBD is success-based. This may lead people, depending on what they want to achieve, to choose one or the other method. However, the advantage of the fault tree over the RBD is that it is easy to put events into the blocks, like human interventions—for example, "Operator fails to close valve." A second advantage is that the fault tree provides an easily understood picture of an event, either one being predicted or one that has happened. This is often a boon for busy managers, who do not have time to read the full report of a reliability prediction or of an incident investigation. On the other hand, for availability studies, the ABD is preferable to a fault tree presentation. Sometimes, both methods are used in prediction studies of complex systems, and software programs sometimes provide the user with a choice of which to use.

ASSIGNMENT

Study the following text, compile a drawing of the system, and then complete the fault tree and calculate the answer required.

A building basement is equipped with a sump pump to prevent flooding. The following scenario has been decided on: During heavy rainstorms, flooding may still occur as the rate of water inflow exceeds the system's capacity. Flooding may also occur for smaller inflows if the system itself fails. The system consists of a primary pump operated by mains power and a battery-powered auxiliary pump. This auxiliary pump will cut in automatically if the mains power fails. Two failure modes exist for each pump. For the main pump, the modes are pump failure itself or mains power failure. For the auxiliary pump, the modes are pump failure or battery is drained. There are three contingencies for battery drainage: because water inflow is for a period longer than the battery capacity, and if the power outage exceeds the battery capacity, and if the building superintendent does not take remedial action (e.g. replacing the battery with a freshly charged one).

The annual probabilities of the events previously described are given in Table 2.12. Assume that all the aforementioned probabilities are statistically independent.

- Draw the fault tree.
- Determine the annual probability of failure.
- Be prepared to discuss your solution in class and to explain exactly what your answer means.
- For budget reasons only, a single item in the system may be changed or modified. Therefore, what single action should be taken in a redesign of the system to improve reliability?

TABLE 2.12
Probabilities of Failure for Pump System

Event	Description	Probability
A	Water inflow exceeds the pump system's capacity.	0.05
B	Water inflow occurs within the system's capacity.	0.95
C	Power outage occurs.	0.1
D	Primary pump failure.	0.1
E	Superintendent fails to take remedial action.	0.2
F	Length of power outage exceeds battery capacity.	0.05
G	Period of inflow exceeds battery capacity.	0.50
H	Backup pump failure.	0.05

ADDITIONAL ASSIGNMENTS

Assignment 2.1

The V1

Using the models shown in Figures 2.2 and 2.3 as a guide, develop your own RBD for the V1 as shown in Figure 2.1. Include all the features annotated in the figure. Define the number of systems and their subsystems and components yourself. In other words, you may have more blocks than are shown in Figures 2.2 and 2.3. You may also require a third level of model. Assume the reliability required of the weapon is 99%. *Reliability* will here be defined as the same as mission success rate. Assume no counter-attacks by the other side. Assume all subsystems and components are to have equal reliability. What will the required reliability be at the lowest indenture level?

Assignment 2.2

The Parallel System of Conveyors

If we are only allowed one interconnection in Figure 2.11, and assuming all the blocks, each representing a conveyor, have the same value of reliability or availability, where is the best place to put the interconnection, or does it not matter?

Assignment 2.3

Availability Upgrade

Assume that you are the maintenance manager of a large production facility. Your superior has tasked you with upgrading the availability of the facility. He has given you a target and a time in which to achieve it. Typically, this could be a 5% improvement over the next two years for a system which is already well managed. There are many possible initiatives you could take, but not all would be applicable to your plant. But for the exercise, simply list all the possible initiatives that could be incorporated into an availability upgrade programme. Your list should include 20 or more items.

Assignment 2.4

Laplace Calculation

Consider the following times to failure of a maintained system:

100, 250, 3,000, 450, 900, 1,030, 1,000, 2,500

Analyse using the Laplace test and comment on the results.

Assignment 2.5

A System Availability Prediction

Construct the ABD from Tables 2.13 and 2.14. Sections A to G are in series with each other. Both tables give information that will assist you in computing the system availability. To assist you, it is suggested that the cells marked with question marks should be filled in. The answer will then be the product of the terms in the A_{ss} column in Table 2. The Weibull paper is included at the end of this chapter for the calculation for items A and G, for those whose Weibull-Ease program has expired and those who want to familiarise themselves with old-fashioned graphical methods. In any event, for anyone who does not have a Weibull program, the paper can prove useful. Its use is demonstrated in Appendix 1. Anyone know how to use a slide rule? J

ABBREVIATIONS

A_{ss} = subsystem availability
FFT = fault finding time
LT = logistic time
MDT = mean down time
MTBF = mean time between failures
MTTR = mean time to repair
* For component A, 10% have failed by 2,000 hours, and the beta value is 1.5.
** The storage allows component A to fail and the system to keep operating for the length of storage time available.
*** For component G, the failure data is in the following second table.

TABLE 2.13
Data for Components in the ABD

Component Type	Configuration	MTBF, hrs	FFT, hrs	LT, hrs	MTTR, hrs	MDT, hrs	Availability	Ass	MTBF′	A′
A	Single component	*?				750 hours		?	?	?
B	Storage = 750 hours **									
C	Two components in active redundancy; each unit 100% capacity						0.9	?		
D	Single component	5,000	12	10	88	?	?	?		
E	2 out of 3 for full load; reduced capacity not acceptable	10,000				1,000	?	?		
F	Reduced capacity acceptable; 2 × 50% units						0.9			
G	Single component	***?				48	?	?		

TABLE 2.14
Failure Data for Item G

Hours	Failure or Suspension	No of Items Following Suspended Set	Increment	Failure No.	Mean Rank
700	F			?	?
750	S				
750	S				
750	S				
1,000	F	?	?	?	?
1,250	F			?	?
1,500	F			?	?
1,700	F			?	?
1,700	S				
1,700	S				

NOTES

1. Many of the techniques we use today came out of the WWII period and the Cold War that immediately followed. For example, PERT, or program evaluation and review technique, was developed for the *project management* of the US nuclear submarine programme in the 1950s. As a second example, value engineering was developed during WWII by the General Electric Company to optimise the value/cost function of products. And we have already seen how reliability engineering itself developed in WWII.
2. Brainstorming is a simple process which can be used in a variety of contexts. During the brainstorm, members of the team must refrain from criticism and allow all members to propose any idea they can think of, even if it seems far-fetched to others. This lack of critical censure encourages all team members to creatively come up with new ideas. At the end of the brainstorming session, critical thinking is restored, and those ideas that are deemed too improbable are weeded out.

3 Maintenance Optimisation

"During World War II, a thousand B-29 bombers were deployed to islands in the Pacific. Assigned to each machine was a crew of 75 maintenance specialists—engine people, sheet metal men, radar techs, etc. Maintenance proved to be a serious bottleneck. When General Curtis LeMay took command in 1945, he changed the system. Among 1000 engine mechanics, maybe 50 really knew what they were doing, 800 were only procedures trained and another 150 were useless. LeMay created separate departments, organised assembly-line style, with service jobs run by the people who knew what they were doing. The top 50 engine men analysed problems and decided what needed doing. The 800 carried out the planned work, while the luckless 150 fetched parts and ran the engine hoists and tractors. Everybody learned something and service became faster and more effective. The percentage of aircraft available for missions rose significantly."

Kevin Cameron, *Top Dead Center*, 2007, p 318

MAINTENANCE—RAISON D'ÊTRE

The simple reason maintenance is necessary is that we cannot make systems reliable enough to work without failure for their entire life cycle. Those very same natural laws and forces which enable engineers to construct things work against them as well. For example, a phenomenon such as friction, necessary to allow torque transmission in clutches, is also abrading the clutch linings.

And as the quotation on the title page of this chapter shows, maintenance needs to be organised. The better the organisation, the higher the reliability and the lower the life cycle cost of the system.

This chapter will describe two common techniques to optimise the maintenance process, reliability-centred maintenance and total productive maintenance. It will also show how the results from such processes must be used in the planning and scheduling of maintenance.

KNOW YOUR PLANT, KEEP IT GOOD AS NEW

What is also important to remember is that no number of management and engineering techniques can replace the foundation, which in this author's opinion is *plant knowledge*. A tragedy in South Africa today is the number of large, sophisticated, and even dangerous plants that are managed by people who do not understand the

DOI: 10.1201/9781003326489-3

technology of the plant. Poets and politicians may run supermarkets and stadiums, but it is dangerous for a nuclear power plant, for example, to be run by anyone other than an experienced nuclear engineer. Hence this author's foundational rule for plant operation and maintenance: *Know your plant and keep it good as new.*

By this we mean to understand the physical laws by which the plant works—the dynamics, statics, fluid mechanics, thermodynamics, etc. that engineers learned at university and must at all times try to put to practical use. It also means knowing how the plant might fail—what the failure mechanisms are.

The second part of the rule, to keep the plant good as new, is perhaps more controversial. We are not saying simply to keep production rates and efficiencies up, but also the aesthetics of the plant. Various chief financial officers will baulk at that, but it has been the writer's experience in the dozens of maintenance audits that he has performed locally and internationally that a plant's aesthetic condition is an indicator of the financial state of the company. More often than not, it is a sign that the plant is being properly managed, not that the plant is managing the managers. A good profit is being made so that there is money available for clean windows and painted roofs.

However, it also has to be borne in mind that the way a system is maintained can affect the way it can fail. We will discuss this topic next.

The first policy is to replace all lamps before they fail, as shown in Figure 3.1. This is a workable policy for items like streetlamps, as they are known to fail according to a predictable normal distribution. But it might not be economical to do so—consider the lights in the right-hand side of the distribution. Under this policy, these lights would still have a lot of life left in them, which would be lost on replacement. So an economic evaluation would have to be done as well.

With regards to Figure 3.2, the alternative policy is to only replace lights when they fail. Here there would be extra costs to consider compared to Figure 3.1—a vehicle would have to patrol the entire municipal area every night to discover and

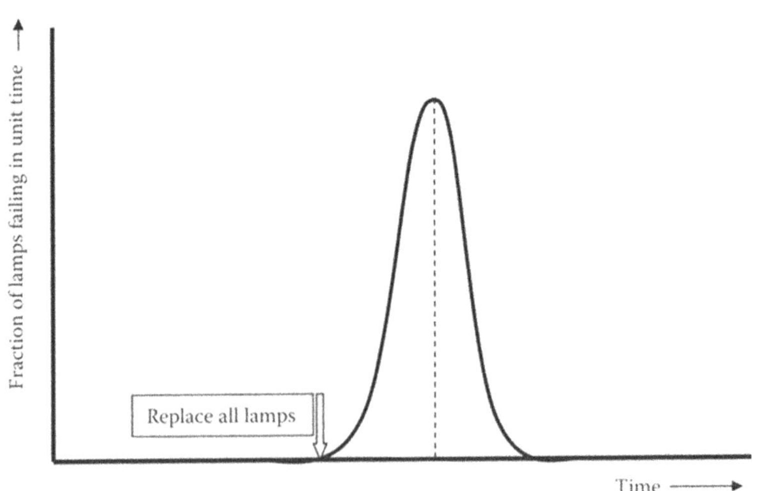

FIGURE 3.1 Maintenance policy no. 1: replace all lamps before any fail.

replace failed units. But there is another consideration as well—notice what happens to the failure distributions of the lights over time. We can see that the first *installation* distribution is a spike at time zero, which coincides with the vertical axis of the graph. This is followed by the first failure distribution, which becomes the second installation distribution. What follows it is the second failure distribution, etc. Notice how the standard deviations of the failure distributions increase and that the peaks simultaneously reduce in height. Furthermore, notice the dotted line joining the peaks of the distributions. It is starting to look like a negative exponential distribution. The system has degenerated into one in which the lamps fail randomly. Our method of maintaining the system has led to the failure behaviour of the system changing. Whatever predictability was in the system to start with has been lost.

Despite this fact, replacement at failure might still be the best policy when all the economics are considered. But this Philips scenario illustrates the point for a very simple, understandable case that complex systems sometimes degenerate in a random failure pattern.

This same tendency to randomness was noted years ago in a study done by the Rand Corporation in the USA on engines in the bus fleet of the Greyhound Bus Company. Such engines were reliable after overhaul for two overhauls. After that, failures on the road were unavoidable. In this case, there had been planned maintenance for two cycles, after which failures in service would still occur. The mechanics of this situation are not the same as in the Philips case but were probably due to long-term failure distributions manifesting after very high mileages which were not catered for in the planned maintenance schedules.

A third example which we must consider was work done by the US military after the Second World War. In an attempt to confirm the maintenance schedules for various items of equipment (jeeps and radars come to mind), one set of equipment was maintained according to the schedules, and another set, the control group, was not maintained at all. After a certain period, the number of working units in each group was counted. The persons who conducted the experiment confidently expected that the units in the maintained group would outnumber those in the control group. To their dismay, they found that the reverse was the case.

What had probably happened here was a lot of disturbance maintenance—unskilled, conscripted mechanics damaging parts as they replaced them, combined with overmaintaining of the units. This was still under the now-discredited paradigm that the more parts that are replaced, the more often, the more reliable the units should be.

Does this study prove that we should not apply scheduled maintenance to anything? Not at all! Scheduled maintenance has its place, but it must be applied only where it is of benefit. If not, other forms of maintenance must be applied. This is the basis of reliability-centred maintenance, which we shall discuss next.

Figure 3.2 is the result of some early research into maintenance policy and how it affects failure. The Philips Company in the Netherlands was involved in this work in the middle of the last century.

Consider the replacement of lamps in a large system such as street lighting. Such lamps do fail according to a normal distribution. Let us assume the mean of the

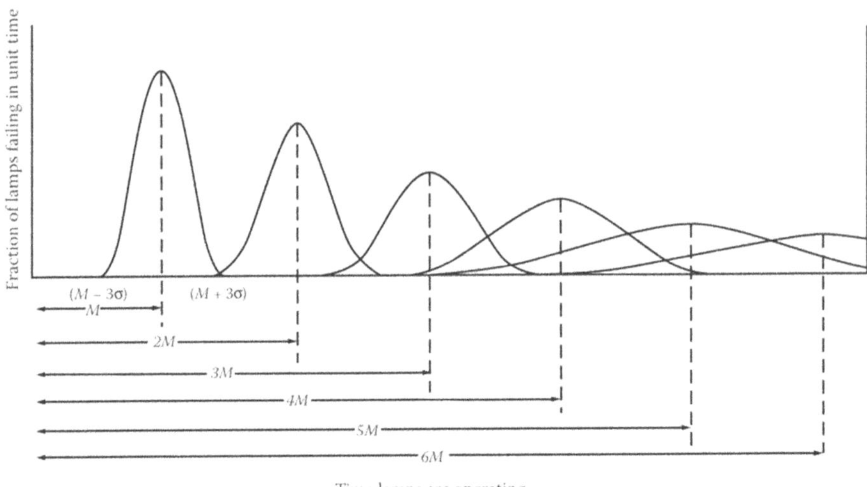

FIGURE 3.2 Change in the normal failure distribution when parts are replaced at failure.

distribution is M and the standard deviation is σ. If all lamps are installed at time zero, then the installation distribution is simply the vertical axis at time zero.

The lamps can be expected to start to fail at a time (M-3σ), and the failure distribution will then look like the first normal distribution in the figure. We now have two choices. There are two maintenance policies that we can follow:

1. Replace all lamps before any fail. Then we will not have any failures in service.
2. Replace lamps as they fail. With this policy, the first failure distribution becomes the second installation distribution. This leads, in turn, to the second failure distribution, and so on. The result is demonstrated in Figure 3.2. It can be shown that the standard deviation of the second failure distribution is $\sqrt{2}\sigma$. If we repeat the replacement process several times, the frequency distributions become progressively lower and wider and also overlap more and more. This means that before one generation of lamps have all failed, some of those that have already been replaced need replacing again. Eventually, the failure rate becomes constant and all predictability has been lost. This occurs at about the time $M^2/3\sigma$. Notice how a curve drawn through the peaks of the normal distributions starts to look like the negative exponential distribution.

Which policy is correct, 1 or 2? It depends on many factors. For example, if the standard deviation is large and the cost of individual lamps is high, then much life and money will be wasted if Policy 1 is followed, with all lamps being replaced before there are any failures. However, the alternative has costs of its own, viz the need to maintain a vehicle and crew who must continually, every night, travel the city, looking for and replacing failed lamps.

Accurate costing would have to be done and accurate failure knowledge would be required before a final decision on policy could be made.

What this simple but accurate model demonstrates is that the way one maintains a system affects the way it fails. In the previous case, if Policy 2 is chosen, the initial predictability that one has of failures disappears completely.

Sometimes the degeneration of wearout failures into a constant failure rate pattern is unavoidable. For example, Davis of the Rand Corporation showed years ago that bus motors exhibited this degeneration into constant failure rate after their second overhaul. Internal combustion engines have many parts, each with their own normal distribution, with widely varying means and standard deviations. This can lead to the degeneration of the engine into a constant failure rate system, as not all parts are replaced at an overhaul. Some parts with large means and large standard deviations are left unrepaired at overhauls. In fact, it is practically impossible to restore a complex system to good as new, because this would mean replacing everything.

This degeneration of complex systems into constant failure rate machines has been proven for systems other than bus motors, viz linotype machines, hand power tools, and radars, among others.

We thus see that the opportunities for the application of preventive maintenance are less than we might have supposed. For scheduled maintenance to work, the following five conditions must apply:

1. The failure rate must be increasing.
2. The total cost to replace before failure must be less than the total cost to replace after failure. For total cost we must include labour, materials, and outage cost.
3. The failure rate after replacement must be lower than the failure rate before replacement.
4. One must be able to predict, at least approximately, when failure is likely to occur. This requires accurate data recording of past failures and analysis thereof using Weibull analysis.
5. There must be no infant mortality spike—that is, the maintenance or replacement operation must be perfect. Infant mortality introduced by a maintenance action is not to be tolerated.

DATA COLLECTION

The type of data collected and the amount collected is of major importance in any maintenance function. If proper thought is not given to data collection beforehand, the wrong data might be collected.

Another thing to remember is that data collection systems, more often than not, degenerate with time. The irony is that a maintenance data system itself needs maintenance! The reasons for this include:

- A lack of motivation on the part of those required to do the collecting.
- Allied to this is the feeling that the data is of no or little use.

- Even well-maintained data collection systems may collapse when there is a change in hardware or software. The IT personnel usually take the easiest option, which is to dump the data stored over past years, with the excuse that it could not be transferred to the new system or that it was too difficult/time-consuming/expensive to do so. It is imperative that these excuses be rejected. The author has seen good data, accumulated over years, dumped more than once during software or hardware changes. The IT function is to be reminded that there is one system where this never happens—salaries and wages! The reason being that there would be an absolute uproar if data on salaries, etc. was ever dumped.

What type of data, then, should be collected? The data collected must be able to be used to answer the following questions:

- What should the maintenance policy be for each piece of important equipment? How much scheduled maintenance? How much predictive maintenance using condition monitoring?
- What should the frequency be for each maintenance action?
- When is the equipment not required for operational use?
- What are the chief weaknesses of each item?
- Are some manufacturers better than others regarding reliability, maintainability, operating cost, and maintenance cost?
- Which artisans are good at their job, and which require further training or alternative deployment?
- How large should the maintenance complement be?
- Is the skill mix correct?
- What factors contribute to downtime, including spares availability?

With the aforementioned factors in mind, the following data should be collected.

THE BIG FIVE

What the author calls the "big five" of maintenance data collection are, for every piece of relevant equipment:

1. Cost of repair (material)
2. Cost of repair (labour)
3. Cost of repair (downtime apportionment)
4. Uptime (a better term than *time to failure*, as shutdown might not be due to failure)
5. Downtime; might be profitable to split further into:
   - Actual repair time
   - Logistic time (travelling, waiting at stores, etc.)
   - Fault finding time

The next point to consider is who collects this data and how. If the artisans have access to a computer screen and keyboard, then perhaps they should enter what data they can themselves. The data entry template should:

- Be as brief as possible.
- Be quite clear as to what is required. The enterer must never be caused to ask, "What does this mean exactly?"
- Include a menu of failure descriptions, with tick boxes.
- Have space for a description to be inserted.

Having said this, it is worthwhile considering the wisdom of Caplen (1972), given in the pre-computer age in the last century:

> Form filling on a large scale is expensive and the information collected is of no value unless it is analysed and action taken on it. Maintenance technicians tend naturally to be practical men, who are more interested in getting on with the next breakdown than in filling in a form about the one they have just done, and this leads to forms being filled in from memory, perhaps once a week. Inaccurate information results and this can be worse than no information at all.

Before adding to the paperwork, it is therefore worthwhile to consider how much information is already available from other sources. For example:

- An analysis of stores requisitions gives an indication of:
  - Which parts are failing.
  - In what quantity.
  - If there are any patterns to the data, for example, seasonal or geographical.

In practice, whether the site is new or existing, there will be some strategy in place for maintenance, very often based on manufacturers' recommendations. It is probable that there will either be overmaintenance, based on the prior information, where the plant runs smoothly but at high maintenance cost, or if there may be much run-to-failure, where both maintenance and downtime costs are high. We should start to collect data from whatever baseline. When enough data is in place, we can begin to optimise the maintenance system.

MAINTENANCE OPTIMISATION

There are two methods of maintenance optimisation in general use in industry at the time of writing: reliability-centred maintenance, or RCM, and total productive maintenance, or TPM. RCM is an American technique, and TPM is Japanese. This author will always recommend RCM to any organisation attempting to optimise its maintenance. TPM, however, is another question, being the Japanese total quality approach to maintenance, involving as it does the entire organisation. It is very much a human-centred approach, whereas RCM concentrates on the hardware alone. RCM assumes that the organisation had the correct skills in sufficient number, the correct spares policy, the correct infrastructure, etc. On this basis, it optimises the plant.

However, having done that, the maintenance schedules and plans generated can be used to backfit whatever is necessary with regard to infrastructure, skills, stores, etc.

TPM, on the other hand, relies on constant interventions, meetings, flag waving, slogan setting, and human motivation, along with a set of hard measures, to measure the effectiveness of the maintenance function. It is, to some extent, culturally biased but can be adapted to cultures other than the Japanese. RCM needs no such adaptation and can be applied anywhere at any time.

There are also definite philosophical differences between the two techniques. For example, RCM allows run-to-failure if that is the best policy economically. In contrast, TPM never tolerates run-to-failure. This makes using RCM as an analysis model within a TPM structure problematic.

An ideal solution might be to first apply RCM and then follow later with TPM. The two techniques are now described in more detail.

RCM

RCM is probably the most popular method for the implementation of maintenance strategy in many companies. RCM was developed by Nowlan and Heap of United Airlines in the 1970s as a method of optimising the maintenance of the Boeing 747 airliner. The claim is made that if RCM was not resorted to, the 747 would have spent so much time in the hangar, with so many parts being replaced unnecessarily, that it could not have made a profit. This is because previous maintenance regimes for aircraft were based on the paradigm that short-term scheduled replacement was the best policy. RCM leads instead to the design of a maintenance programme that considers the reliability characteristics of the equipment. If the item of equipment will respond well to short-term scheduled replacement, then that is what will be done. If it does not, then some other appropriate maintenance action must be undertaken.

RCM is a logical process that separates out items of plant which will respond to some sort of scheduled maintenance from those that will not. The scheduled maintenance is divided into three categories, scheduled inspection, scheduled rework, and scheduled replacement. If an item does not respond to one of these three, there are two options left: run-to-failure or redesign. Redesign often means replacing the item with a part that will respond to scheduled maintenance.

The application of RCM in the aircraft industry was based on Nowlan and Heap's research into equipment item failures. This research led them to the conclusion that only 11% of items on an aircraft responded to any form of scheduled maintenance. This implied that there were only 11% of maintained items that exhibited an increasing failure rate over time.

It is now known, through the work of Asher and Feingold (1984), that this figure of 11% was certainly too low. Although their book does not go into the details of the analysis, Nowlan and Heap had probably committed the classic error of applying Weibull analysis to the failures of maintained items, which, unless the failure rate increases after each repair, will result in the wrong answer. Nevertheless, the application of RCM still proved of tremendous benefit in the airline industry and is now standard practice, under the title of MSG (for maintenance steering group) rather than RCM.

THE RCM PROCESS

RCM is a team effort and has all the good and bad aspects of similar team efforts, such as value engineering, root cause analysis, etc. The good aspect is that all relevant knowledge is brought to bear on the problem. The possible bad aspect is that if the process is not properly managed, the saying becomes true that a camel is a horse designed by a committee.

Once a section of plant has been selected to analyse, a committee is formed. The committee should consist of a chairperson (who must be an expert in RCM), a secretary, and company representatives from maintenance, engineering, and operations. A representative from the manufacturers of the equipment might also be invited, as might other users of the equipment from outside the company. If the RCM process is not understood, the first stage in the process will be a training workshop. This generally takes two days. Often, the chairman will give the training. It is vitally important that the chairman is a good manager, with not only RCM skills but project management skills as well. The secretary is needed to record all decisions and to arrange the delegation of work to committee members for those questions arising out of the process that cannot be immediately answered and have to be left over to the next session. Even with modern software programs, these tasks should not be left to the chairman.

The committee should be kept as small as possible, consistent with the need to bring all relevant knowledge and experience to bear on the problem. Specialists might be co-opted to the meeting. The central core committee should not be much more than about five persons. The committee will normally sit one day a week until the job is complete. The amount of work is considerable. As an example, a large outdoor electrical substation consisting of high-voltage transformers, breakers, current transformers, voltage transformers, and other equipment could take a team of six eight days to analyse. This represents 48 man-days of work for skilled, expensive personnel. It is therefore essential that management support is secured at the outset and is ongoing over the life of the analysis. Two case studies at the end of this chapter describe the management of the RCM process, and common mistakes that can be made.

The RCM process has the following steps:

- Decide what equipment to analyse.
- Determine what functions each item of equipment performs.
- Prepare a failure modes and effects analysis (FMEA) for each item.
- For every line in the FMEA, enter the RCM decision diagram and perform the RCM analysis for that particular failure mode.
- Enter the results of the analysis on the RCM decision worksheet.

Repeat the process for the next line in the FMEA.

The process is time-consuming and mentally arduous if done correctly. Furthermore, Nowlan and Heap did not give much guidance on how to select the items to be analysed. This led to a disenchantment with the RCM process by many who attempted it. It is vitally necessary to apply some criticality sort to the plant so that time is not wasted on analysing trivial items, items that never fail, or items where failure has insignificant consequences. Two criticality sorting processes that can be applied:

- Rank equipment according to how much of the maintenance budget is consumed by each item.
- Determine a simple risk ranking, with *risk* being defined as probability of failure × consequence of failure. Such a list need not be statistically complex—high, medium, and low values of probability and consequences can be determined and assigned numerical values of 3, 2, and 1. A set of numbers result as 9, 6, 4, 3, 2, and 1. Apply RCM to the 9s and the 6s first and to the lower numbers if time and budget allows.

The fact that RCM is a time-consuming and expensive process has led to the development of other forms of RCM, some of which are acceptable and some of which are not really theoretically or practically adequate. Users in the field should take note of the maxim "Let the buyer beware" before selecting the final product to be used.

Once the equipment to be analysed is selected, functional analysis must be done on the item to ascertain what functions it performs. The FMEA is then developed, which has columns for function, functional failure, failure mode, and failure effect. Each function can be lost in possibly more than one way, hence the need for the "failure mode" column. An example of an FMEA is shown in Table 1.2.4.1, which follows. Notice the nomenclature. Various lines in the diagram are designated as 1A1, 1A2, 2B1, etc. This shorthand will be transferred to the RCM decision worksheet after the RCM decision diagram has been used. For every line in the FMEA, the RCM decision diagram must be consulted and the process pursued until an answer is arrived at. The FMEA is then entered again at the next line.

In the simple example chosen, it has been decided that the fuel pump of a land-based vehicle has three possible failure modes. We then move from the FMEA to the decision diagram and then record the results in the decision worksheet.

THE RCM DECISION DIAGRAM

The original form of the RCM decision diagram is given in Figure 3.4. The diagram is very well thought out, and the language concise. Some definition of the airline terminology is required as it is encountered on the diagram.

ON-CONDITION INSPECTION TASK

The item is inspected, and depending on the condition in which it is found, either no action is taken or some specific action, such as replacement or rework, is done. In these times, the whole field of condition monitoring has come to be included in this term, "on-condition inspection."

SCHEDULED REWORK

The item is repaired and reused.

SCHEDULED DISCARD

The item is replaced.

ACRONYMS

Certain acronyms in the diagram include the following:

- LL: life limit—used with the scheduled discard category
- OC: on-condition inspection
- RW: rework

A simple example of an RCM analysis will now be given (Table 3.3). It will be seen that the FMEA has only three lines. In a real-life exercise, there might be many more.

THE RCM FAILURE MODES AND EFFECTS ANALYSIS

Item to be analysed: electric fuel pump.

General description of item: In a full-scale analysis, such details as model number, number in service, make, service conditions, reference drawings, etc. would be provided here.

The reader is encouraged to review the FMEA, proceed to the decision diagram, and see if he or she agrees with the results in the decision worksheet.

TABLE 3.1
FMEA for a Fuel Pump for a Land-Based Vehicle

Function (F)	Functional Failure (FF)	Failure Mode (FM)	Failure Effect (FE)
1. Supply fuel to engine.	A. No fuel supply.	1. Blocked filter.	Engine will not run.
		2. Seized shaft.	Engine will not run.
2. Supply pressure signal to on-board computer.	A. Low pressure.	1. Partly blocked filter	Fuel-air mixture will be wrong; engine will not run properly.

TABLE 3.2
Example of an RCM Decision Worksheet for the Fuel Pump Example

F	FF	FM	1	2	3	4	5	6	7	8	9	10	11	12	13	14	15	16	Proposed Task	Period
1	A	1	Y	N	Y					Y									Check filter for dirt.	2 wks
1	A	2	Y	N	Y					N	N	N							None, run-to-failure.	
2	A	1	Y	N	Y					Y									Check filter for dirt.	2 wks

TABLE 3.3
RCM Analysis of an Electrical Breaker

Failure Mode	Proposed Task	Frequency
1. Burnt trip coils	Run-to-failure (RTF)	-
2. Contacts welded together	Insignificant failure mode	-
3. Mechanism jammed due to dirt ingress	Inspection during battery inspection	Once per month
4. Loss of DC supply	Refer to battery analysis	
5. Loose wiring to trip coil	RTF	
6. Tripping mechanism liable to stick	Regrease bearings and rollers	Once every three years, in conjunction with breaker lubrication
7. Loose trip spring	Insignificant failure mode	
8. Not opening on all phases	Insignificant failure mode	
9. Mechanism jammed due to humidity (failed heater)	Test operation of the heater	Once per month
10. Loss of DC supply	Refer to battery	
11. Mechanism jammed by dirt	Inspect when inspecting battery	Once per month
12. Burnt closing coils	RTF	
13. Loose wiring to closing coil	Refer to battery	
14. Closing mechanism faulty	RTF	
15. Loose closing spring	Insignificant failure mode	
16. Internal flashover	Perform pole-contacts service	After 48 operations
17. Loss of oil	Check level in gauge glass	Once per month

ALTERNATIVE AND MODIFIED FORMS OF RCM

- **RCM II:** This is a form of RCM developed by the late John Moubray, a leading RCM consultant and founder of the Aladon organisation. It addresses environmental concerns by including a question on possible environmental damage. The wording of the questions is also simplified (some would say dumbed down and made less clear). Apart from these two aspects and the fact that Moubray's decision diagram reads from left to right instead of top to bottom, there is not much difference between RCM II and classical RCM.
- **MSG 3:** This is the development of RCM currently used in the aircraft industry. As already stated, MSG stands for Maintenance Steering Group. MSG 3 is essentially the same as classic RCM, except that some questions are asked about lubrication.
- **SCRM:** The Electric Power Research Institute, or EPRI, has developed its own form of RCM for the analysis of electrical transmission and distribution systems, which includes templates of typical failure modes and suggested corrective actions and application frequencies. It is known as streamlined RCM, or SCRM.
- **Eighty-Twenty RCM:** This is a process developed by the US Navy, which, as its name suggests, applies Pareto's rule to the choice of which systems are to be analysed. Templates of typical failures, failure frequencies, etc. are also included in this system.

- **The SAE Standard:** In 2000, a technical committee sponsored by the Society of Automotive Engineers (SAE) produced a standard on reliability-centred maintenance. The previously referred-to John Moubray was involved, and the committee represented American commercial and military interests. This standard requires that "full effort," or classical RCM, must be used for any programme used to increase machinery reliability through a logical, disciplined approach to maintenance. All contractors to US government institutions are obliged to use the standard and therefore to use classical RCM. This development reflects the concern which had existed with regard to hybrid, simplified RCM systems.
- **PM Optimisation:** Another modified form of RCM is known as PMO, or PM optimisation. This is the form that BHP Billiton has in fact standardised on. It is the creation of Steve Turner, an Australian professional engineer previously trained in RCM II.

SUMMARY OF THE RCM OUTPUT

All that RCM provides at first is an optimised list of tasks and frequencies for the equipment that has been analysed. However, this list, when implemented, will indicate deficiencies in many areas of the maintenance system, for example, a lack of skills, inadequate facilities, etc. In fact, RCM is an excellent vehicle for maintenance budget justification. Having optimised the maintenance using RCM, clear needs are demonstrated to the company's management.

AN RCM EXAMPLE

The following example indicates the extent of savings which can be achieved with RCM. A large 44kV electrical substation was analysed. The following list of equipment was included in the analysis:

- Isolators
- Circuit breakers
- Surge arrestors
- Batteries
- Chargers
- 44kV feeder
- Voltage transformers
- Current transformers
- HV transformers
  - Oil
  - Bushings
  - Main tank
  - Core and windings
  - Bucholtz unit
  - Tap changer
  - Conservator
  - Cooling system

The results of the RCM analysis on the circuit breakers are given in Figure 3.3 as an example. The circuit breaker is chosen as it is the most maintenance-intensive component.

ECONOMIC EVALUATION

The group of ten persons took 11 days to perform the RCM analysis. This cost the company R61,000. Performing the maintenance in the future according to RCM principles would cost the company R220,000 over the next 12 years, the remaining life of the station. By contrast, if the previous maintenance strategy was to be continued, the cost would be R927,000. This indicates the savings that can be made with a dedicated RCM programme.

TOTAL PRODUCTIVE MAINTENANCE, TPM

First, a note on English grammar: The technique should be entitled *totally* productive maintenance. Although it is bad English, the title total productive maintenance is now so ingrained in the literature that we will use it from now on.

The author will always advise the use of RCM to optimise a facility's maintenance but will hesitate to recommend TPM. This is because TPM is much more complicated, involves the entire organisation, and costs a large amount of money to implement and sustain. Nevertheless, some companies can and will benefit by the introduction of TPM.

There are different flavours of TPM. Originally developed for motorcar assembly plants, it has now been available in a version for process plants as well.

TPM can be said to be the Japanese total quality approach applied to maintenance. As such, it is very much a people-driven approach, in contrast to RCM, which considers only hardware, at least initially.

TPM has been promoted by the Japan Institute of Plant Maintenance since 1971. It seems to have developed from work done on Toyota assembly lines, although Toyota claims no credit for it. (Toyota, a very innovative firm, does claim credit for several engineering and management innovations, e.g. lean manufacturing and the five whys method of root cause analysis, to name two.)

To quote from a Japanese TPM expert, Tokutaro Suzuki (1994):

> *Preventive or Scheduled Maintenance* was introduced in Japan from America in the 1950's, when Japanese industries were getting back on their feet after WW2. *Productive Maintenance*, developed in the 1960's, incorporates such disciplines as maintenance prevention design, reliability and maintainability engineering and engineering economics to enhance economic efficiency of equipment investment for the entire life of the equipment.

Despite his rather-dense writing style, Mr Suzuki describes the essence of TPM quite well.

Another Japanese TPM expert, Kunio Shirose, puts it this way (1992):

> TPM is a set of activities for restoring equipment to its optimal conditions and changing the work environment to maintain those conditions.

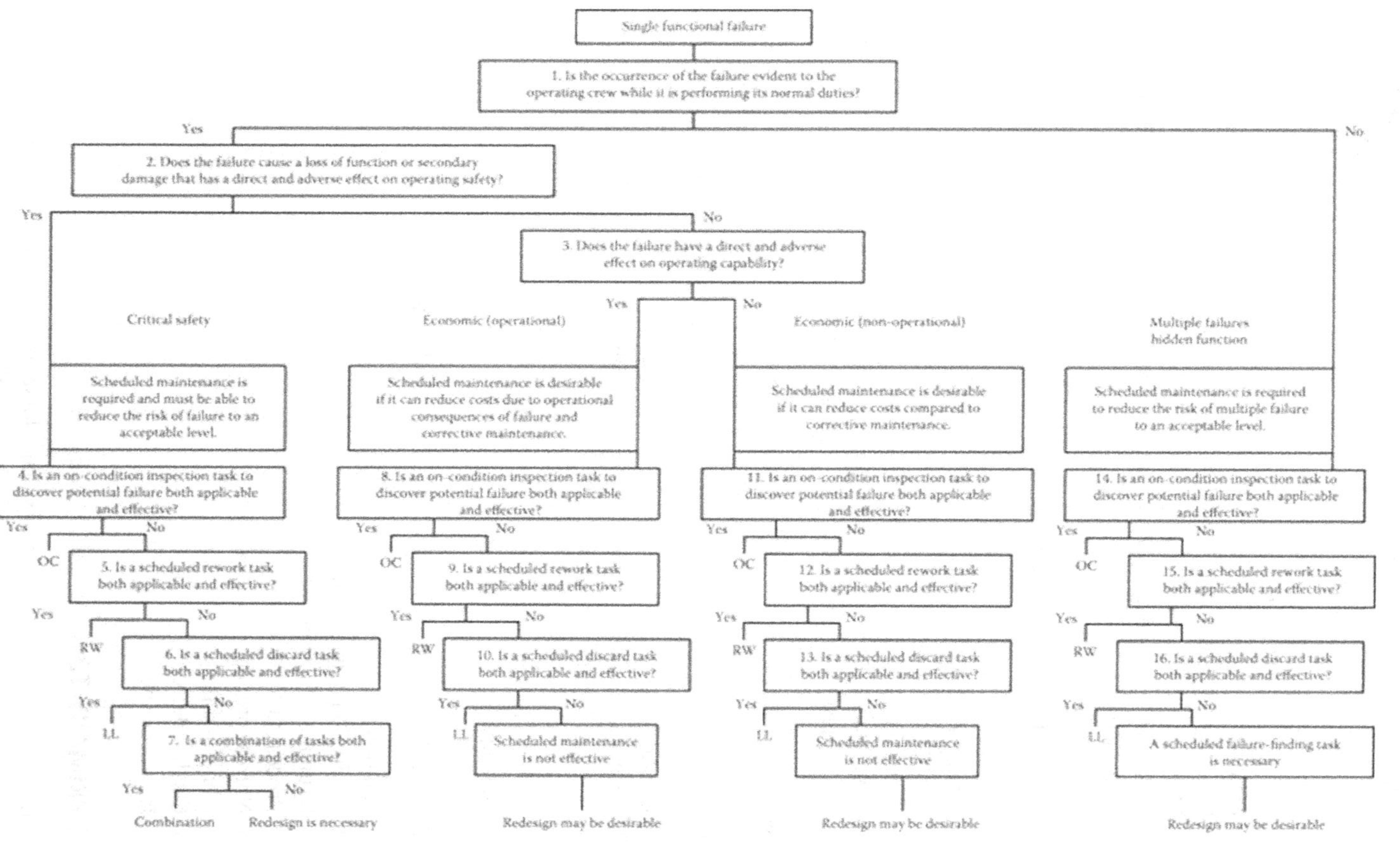

FIGURE 3.3 Nowlan and Heap's original RCM decision diagram.

He continues to say that if one restores the equipment and modifies the work environment to make it more congenial, the behaviour of the personnel will change for the better as well.

The aims of TPM are said to be:

- Maximise equipment effectiveness.
- Optimise the maintenance system to achieve the aforementioned.
- Make sure the entire company is involved.
- Make sure that all employees are involved, by developing autonomous small-group activities.

THE FIVE PILLARS DEFINITION OF TPM

1. Increase equipment effectiveness by eliminating the six major losses.
2. Train operators to do low-level maintenance (known as autonomous maintenance).
3. Planned maintenance must be in place.
4. A maintenance prevention plan must be in place.
5. There must be training of operators in plant fundamentals.

We will now elaborate on some of the points mentioned earlier.

THE SIX MAJOR LOSSES (MANUFACTURING INDUSTRY)

1. Breakdowns
2. Set-ups and adjustments to production machinery
3. Idling and minor stoppages
4. Rework due to production defects
5. Reduced production output
6. Start-ups

THE EIGHT MAJOR LOSSES (PROCESS INDUSTRY)

1. Shutdowns
2. Production adjustments
3. Equipment failures
4. Process failures
5. Normal production losses
6. Abnormal production losses
7. Quality defects
8. Reprocessing

THE PRIMARY TPM METRIC: OEE

Whether we are talking about the six major losses or the eight major losses, the index that is used to assess improvements in them is called overall equipment effectiveness in the case of manufacturing industry and overall plant effectiveness in the case of the process industries. Usually shortened to the acronyms OEE or OPE, the formula in both cases is the same:

$$\text{OEE} = \text{Availability} \times \text{Production Rate} \times \text{Quality Rate} \tag{3.1}$$

THE GOALS OF TPM

These can be summarised as follows:

- Zero breakdowns
- Zero adjustments
- Zero idling
- Zero defects

For process industries, another could be added: zero problems during recommissioning. It is the established practice in some industries to have the plant recommission itself—leaks at start-up, etc. are tolerated and corrected as they manifest. TPM would not allow such practices.

To return to the aforementioned four "zeroes" mentioned. How does TPM aim to achieve them? With regard to *zero breakdowns*, the following programme is adopted.

ZERO BREAKDOWNS

- **Step 1: Cleaning of the Plant.** This is to be done every day, by the operators concerned. The act of cleaning leads to the discovery of defects. For example, a crack might indicate the presence of vibration, which might then be corrected by the tightening of mounting bolts. Leaks might indicate the need for other corrective measures. Another important result of these actions is the fact that pride in the workplace starts to develop.

  To emphasise the importance of this point, the first time the plant is cleaned, the board members do it, while the operators stand and watch! In a typical TPM promotional video, the board, all dressed in clean overalls with the company logo displayed, and with all the necessary safety gear, clean the plant, then return to their offices. This is one aspect of TPM that cultures other than the Japanese might find difficult to implement!
- **Step 2: Adherence to Operating Conditions.** One of the best ways to reduce a plant's life is to thrash it for short-term production gains. When senior management enjoys bonuses based on short-term production targets, the tendency to overtrade the plant might be hard to break. But this TPM rule, like the previous one, is imminently sensible.
- **Step 3: Restoration of the Plant to Its Original Condition.** This is a non-negotiable in TPM. But to restore the plant to good as new, although once again an admirable motive, might simply be too expensive. Hence the need for proper budgeting and project management to assess the feasibility of TPM before it is instituted.

- **Step 4: Correct Design Weaknesses.** Here, TPM prescribes a structured programme to improve the plant from a maintenance point of view, with the aim to eliminate recurring maintenance problems.
- **Step 5: Increase the Operators' Maintenance Skills.** An extensive education programme is required to allow operators to understand the plant in their care and to take the necessary corrective actions if required. It is essential in a unionised environment to get the unions on one's side before instituting such a programme. This programme will include training of operators to understand the equipment that they are responsible for, training them to identify abnormalities and be able to make minor repairs. To this end, they are trained in the following, as required:

    - Basic maintenance philosophy
    - Bolted joints
    - Keys and keyways
    - Bearings
    - Power transmission by chains, belts, and gears
    - Basic hydraulics
    - Basic pneumatics

    It is therefore seen that the amount of training given is very extensive. The point is to make the operator like a mother to its child, the machine. A mother can care for her child and treat minor problems. However, she also realises when a problem is beyond her, and then she calls the doctor. The doctor in this analogy is the artisan.
- **Step 6: Artisan Training.** And apart from increasing operator skill levels in maintenance, the maintenance personnel are also trained in specialist maintenance tasks so that their skill level is also improved. In fact, the aim is to bring all the artisans up to the same high level of skill and performance. In the writer's experience, this is seldom possible.

    If the above steps are taken and are successful, then the back of the TPM programme is broken. But remember, these steps are only for the first of the big zeroes—zero breakdowns. So the programme must proceed to attack the other three big zeroes. The next one is zero adjustments.

ZERO ADJUSTMENTS

This is, to a large extent, a matter of operations rather than maintenance, but it is included as part of TPM. The first step is to distinguish between what TPM calls external setup and internal setup. With regard to external setup, have equipment ready at the scene of the change. With regard to internal setup, the following is required:

- Standardised work procedures
- Work allocation procedures
- Standardised jigs and tools
- Reduced weight of dies (to facilitate handling)
- Robust designs to eliminate damage

ZERO IDLING AND ZERO MINOR STOPPAGES

Idling occurs when the flow of work in progress is interrupted but the equipment keeps working. To minimise this condition, it is necessary to do the following:

- Install the relevant alarms.
- Ensure operator vigilance.
- Measure the time lost in idling.

Stoppages occur when a problem is detected by instrumentation and the process is shut down. As for the preious one:

- Ensure operator vigilance.
- Measure time lost.
- Avoid overproduction.
- Avoid quality defects.

The last two aforementioned points should of course have already been covered in previous initiatives.

SPEED LOSS

This results in lost production caused when the design speed of the line is not achieved.

- Check the engineering specifications for the correct speed.
- Ascertain if poor maintenance is the cause.

QUALITY PROBLEMS

Chronic quality defects may be caused by:

- Poor maintenance
- Lack of daily cleaning
- Insufficient operator skill
- Neglect of minor defects
- Poor-quality production machinery

So we see that a TPM programme is quite extensive, covering the aims of zero defects, zero adjustments, zero idling, zero speed loss, and zero defects. All of which is admirable. The achievement thereof is by means of an extensive programme.

A budget and schedule must be prepared as for any programme, a TPM director must be appointed, and a TPM office established. The start-up costs can be very high, and the management must understand this and approve it if TPM is to be successful. The start-up costs include:

- Establishment of the TPM office, including TPM director costs and programme launch costs

- Equipment restoration costs, both direct and to cater for extensive industrial engineering input
- Training costs for both operators and artisans

Then the programme can proceed as follows:

1. A formal announcement is made that TPM is going to be introduced into the company.
2. A TPM training and publicity campaign is launched.
3. The TPM promotion organisation is established.
4. The specific baseline and goals are established for the company.
5. The master plan for TPM implementation is drafted.
6. On opening day, there are presentations by senior company members and much fanfare, as well as plant cleaning by senior management.

The core of the programme can now commence:

- Implement maintenance training for operators.
- Implement operator-based maintenance (also called autonomous maintenance).
- Implement a full maintenance strategy, including all four types of maintenance, corrective, planned, predictive, and design-out.
- Implement advanced maintenance skills training for the artisans.
- Focus on plant areas where the greatest improvement can be gained from the aforementioned initiatives.
- Establish quality assurance in maintenance.
- Develop an effective support system in buying and stores.
- Continuously develop the SHEQ system.

The management must maintain continual vigilance during the implementation phase and thereafter. The JIPM states that a successful programme may take three years to implement if the organisation is a big one. The TPM consultants also maintain that provided there is financial and management support throughout, TPM is guaranteed to succeed.

It can be seen from the previous passages that a TPM programme is very extensive and not to be taken lightly. Various aspects of TPM are generic, however, and can be done without a full-blown TPM effort. For example, the maintenance optimisation section of the process, developing the best combination of the four kinds of maintenance, could stand alone. As a second example, there are many companies that use the OEE formula as one of their key performance indicators outside of any TPM initiative.

So the engineering side of TPM is not unique. What is uniquely TPM is the human emphasis—getting everyone involved, informed, and enthusiastic. This it does through top-down goal setting and direction and by bottom-up small-group activities, where workers solve their own problems using techniques such as quality circles, by which the workforce is encouraged to take control of their own destinies as part of the greater company effort. Such programmes are for organisations that are already good but want to move on to excellent. What management must realise, however, is that such programmes are not sustainable without continuous management support. Otherwise, just like the balloons that might be released on the opening day of the programme, the TPM balloon will go soft unless continually topped up. There might also be union problems and cultural problems which have to be overcome.

SEMANTICS IN MAINTENANCE MANAGEMENT: EXPLANATION AND QUALIFICATION

An essential part of any management system is documentation. A vexing problem that often occurs during discussions on documentation is confusion over the meaning of the words used. For example, what is a standard? What is a specification? What is a philosophy?

The reasons for the confusion are several, including the fact that language is not precise. Secondly, various technologies and disciplines use the same word to mean different things. We will be seeking to use the words in an industrial context. It will be necessary, however, in developing an understanding of these words, to recognise that they have been borrowed from other, non-industrial contexts. Thirdly, some confusion is caused simply because of sloppy, incorrect use of a word.

To the extent to which we are able, we should eliminate as much confusion as we can. There are some tools we can use to help us here.

Firstly, we can look at what the word in question originally meant. Many scientific words are made up of Greek and Latin words in the first place. If we study the Greek and Latin meanings, we might get new insight into the English word.

Secondly, we can seek to understand the word in its original context—who used it and why? What was its sphere of use? We will discover that the words we are interested in come, usually, from a few specific sources. These sources are organisations that have been around for a long time, like the church, the military, or the government. We will see that in the history of the development of the word, industry has adopted words from these sources, sometimes keeping the meaning the same, but sometimes modifying it.

Thirdly, we can consult a dictionary that gives us the most usual usage or usages of the word at present.

As a fourth option, we can use what might be termed Kipling's serving-men (Kipling, 1902). These are the questions what, why, when, how, where, and who. Some words are "what" words, some are "how" words, etc.

Finally, we can see whether there is any hierarchy in the words we are interested in. Do some express higher-level concepts (however such concepts might be defined) than others? This will help us in our classification and understanding of terms.

Our first step is to see what dictionaries say about these. We will use the *Oxford Dictionary* (OED), the *Concise Oxford Dictionary* (COED), and the *WordWeb Dictionary*, developed by Princeton University for use with Microsoft programs such as Word. We will refer to this dictionary as *Princeton*. Use is also made of the *Chambers Dictionary of Science and Technology*. The first two dictionaries are useful for studying the definition and original meaning of the words. The OED in particular lists the original and current meanings in a fairly exhaustive fashion. The *WordWeb* reference is used for American alternative meanings and to determine the most modern meaning. The Oxford dictionaries were used as they are generally regarded as setting the standard for British English usage, but not perhaps in the areas of science and technology. Hence the inclusion of the *Chambers Dictionary.*

Where these dictionaries did not supply the meaning, others are referenced. As we study the definitions, we will also look for commonality of meanings between words.

Finally, for the purpose of descriptive clarity, we will call all these items defined later as *instruments*, in the sense that they have a purpose and are used to complete tasks. In a configuration management/documentation context, some of the words and phrases we are interested in are *code of practice*, *creed*, *doctrine*, *guideline*, *instruction(s)*, *logistics*, *manual*, *mission*, *philosophy*, *plan*, *policy*, *programme*, *schedule*, *specification*, *standard*, *strategy*, and *tactics*.

- **Charter:** "A document incorporating an institution and specifying its rights" (*WordWeb*).

  "A document granting privileges to, or recognising the rights of, the people or of certain classes or individuals" (*Oxford*). Universities, for example, are established by charter. The term "performance charter" is sometimes heard of the contract between an employer and an employee, whereby bonuses are awarded subject to the attainment of certain key performance indicators. *Performance contract* is probably a better and less-pretentious term.
- **Code of Practice:** "*(Building, Civil Engineering)* Recommendations (not mandatory) drawn up . . . for design and methods of application" (*Chambers*).
- **Contract:** "A binding agreement between two or more persons that is enforceable by law" (*Princeton*).
- **Creed:** "A brief *summary* of Christian doctrine" (*OED*). Clearly, this is an ecclesiastical word.
- **Doctrine:** The root meaning is from Latin and means "teaching." "A system of beliefs accepted as authoritative" (*Princeton*). Also originally an ecclesiastical word, now also used by the military. The military use of the word is at the highest high level—for example, a nation will have military doctrine indicating that it will defend itself and that an enumerated list of defined threats will be catered for. There is also a political use of the word *doctrine* at the highest level, which is associated with the military doctrine but is not completely synonymous. As an example, one might consider the Monroe Doctrine, as stated by the president of the USA of that name. This doctrine stated that all of North and South America was in the United States' sphere of influence and that the European powers should not interfere. In a more general sense, *doctrine* is synonymous, according to Princeton, with *philosophy*.
- **Guideline:** "A detailed *plan* to guide one in setting standards or determining a course of action" (*Princeton*).
- **Instruction:** "A message describing how something is to be done" (*Princeton*). In the plural, "directions, orders" (*COED*).
- **Logistics:** "The art of moving and quartering troops" (*COED*). "The handling of an operation that involves supplying labour and materials as needed" (Princeton). The original meaning is to quarter, or *lodge*, troops. Here we see that the original meaning is military and that the adopted meaning applies to industry. Furthermore, we see that *lodge* carries the implication of being temporary, as opposed to being *housed*, which has a more permanent connotation. A hunting lodge, for instance, is a place one stays at while hunting but does not live in all the time.

- **Manual:** Here the original meaning is from the Latin for *hand*. In other words, a *manual* is something one carries in one's hand. The implication is that it is "at hand" to help the owner and user thereof. An alternative used in some industries is the Latin phrase *vade mecum*, meaning, "go with me." An alternative given in Princeton is *handbook*. The implication is something that has a lot of useful information in it. It is more likely to be book-sized rather than pamphlet-sized. It might be a set of procedures or instructions or other lower-level documents but might also contain higher-level mission statements and philosophies, for example.
- **Mission:** "The act of sending" (*OED*). "A person's vocation or divinely appointed work in life" (*COED*). "An operation assigned by a higher authority" (*Princeton*). Here we notice a connection. In the case of the *COED* definition, *vocation*, or what we might term the "calling," is by the highest possible authority. The *Princeton* definition allows for lower levels of authority, for example, a bombing mission for an Air Force squadron, where the higher authority might be a general and the receiver of the order a squadron leader (to use the British rank terminology). This is an example where the same word, with the same general meaning, is used at different "levels" by different organisations, in this example, by the church and the military, respectively.

  The word has connotations of idealism and of striving to achieve the highest possible standards. In terms of hierarchy, therefore, it is a "high-level" word. A state might also have a mission—the European powers at one time claimed to have a civilising mission, through empire building, to large sections of the non-European world. The present-day United States might be said to claim to have a mission to democratise the world.

  As regards hierarchy, a mission might be said to be superordinate, that is, even higher than a high-level goal, task, or achievement.
- **Philosophy:** As mentioned prior, this word is defined by *Princeton* as synonymous with *doctrine*, that is, "a system of beliefs accepted as authoritative." It literally means in the Greek "a love of wisdom." The *OED* gives the most common current use of the word as "that department of knowledge which deals with ultimate reality, or with the most general causes and principles of things." We can see here that from a Christian viewpoint, philosophy can be synonymous with doctrine, where Christians would claim that their doctrine represents ultimate truth or reality. In Kipling's terms, this is a "what" word, and also in terms of hierarchy, a high-level word.
- **Plan:** "Way of proceeding" (*COED*). "A series of steps to be carried out" (*Princeton*). "A scheme of action" (*OED*). In terms of our "Kipling classification," this is definitely a "how" word and comes lower in the hierarchy than, say, philosophy. It is a practical rather than an idealistic word.
- **Policy:** Here is an example of a word that has been used incorrectly for so long that the meaning has subtly shifted. The *OED* definition is, "a course of action adopted and pursued by a *government*." Here, "policy" is related to "polity," meaning, "a state of civil order." And hence to *politics*. We see therefore that the origin of *policy* is *political*. Note that according to the OED, it is a course of action. That is, it is synonymous with *plan*. We see further that as the policy is a course of action, we should say *policy document*

when referring to the piece of paper, rather than just *policy*. (Except in the case of insurance, where the dictionaries include the paper as directly being called a policy.) We have established, however, a hierarchical fact: *policies* should be on the same level as *plans*.

- **Programme:** "A series of steps to be carried out or goals to be accomplished" (*Princeton*). Note that in this definition, time is not involved. A time-based document is rather a schedule.
- **Schedule:** "An ordered list of times at which things are planned to occur" (*Princeton*). Hence, schedules have a time element to them. "When" is the operative word.
- **Specification:** "A detailed description of the construction of an engineer's or architect's work" (*COED*). "A detailed description of design criteria" (*Princeton*). Notice here the connection with *design*. The *OED* gives a more all-encompassing definition: "A detailed description of the particulars of some projected work in building, engineering or the like, giving the dimensions, materials, quantities, etc., of the work, together with directions to be followed by the builder or constructor; the document containing this." Hence, unlike *policy*, *specification* can be the description of the document.
- **Standard: "**Degree of excellence required for a particular purpose" (*COED*). "A basis for comparison, a reference point" (*Princeton*). Hence, standard should always be read with specification, rather than standing on its own, rather like policy should be read with document. As with *mission*, there is a connotation of idealism with the word *standard* which is perhaps being lost.
- **Strategy:** This is clearly originally a military term. The *COED* definition is interesting: "The art of so moving or disposing of troops as to impose on the enemy the place, time and conditions for fighting preferred by oneself." Hence, *strategy* has the implication of ensuring an *advantage*. This should be remembered when the word is used in industrial situations. *Princeton* defines the word as: "A branch of military strategy dealing with the planning and conduct of a *war*." Hence, in military terms, *strategy* is lower in the hierarchy than *doctrine*, but higher than *tactics*, which is described next.
- **Tactics:** "The branch of military science dealing with detailed manoeuvres to achieve objectives set by strategy" (*Princeton*). Hence, the hierarchy we observe is that in military terms, *doctrine* refers to the nation, *strategy* to a war, and *tactics* to a battle.

ISO 55000

Note: A more complete discussion of ISO 55000 Is given in Chapter 11.

An attempt has been made to standardise on-maintenance terminology by the recently established ISO 55000 standard. ISO 55000 has definitions for *policy*, *strategy*, and *plan* as follows:

- **Asset management policy:** Overall intentions and direction of an organisation, related to the assets and the framework for the control of asset-related

processes and activities that are derived from and consistent with the organisational strategic plan.
- **Asset management strategy:** Overall long-term action plan for the assets, asset types, and/or asset system(s) that is derived from and consistent with the asset management policy and organisational strategic plan.
- **Organisational strategic plan:** Overall long-term action plan for the organisation that is derived from and embodies its vision, mission, values, business policies, objectives, and the management of its risks.

Here it will be seen that *policy* is more elevated in the hierarchy than *strategy*. Strategy is also inferior to the *business plan* in this schema but is also a plan in itself. Plan, in turn, is said to be derived from and embody both vision, mission, values, objectives, and *policies*, without vision, mission, values, and objectives being defined.

In ISO 55000, the statement is made that "[t]he asset management strategy should demonstrate how the asset management policy is to be implemented." This clearly makes the strategy a "how" document, and the policy a "what" document, and at a higher level.

A figure in ISO 55000 shows the following hierarchy:

- Organisational strategic plan (including vision and mission and other policies, objectives, and the organisational strategy)
- Asset management policy
- Asset management strategy
- Asset management objectives
- Asset management performance/condition targets
- Asset management plans

ISO 55000 states that:

> The asset management policy should drive the asset management system and the asset management strategy should provide the high level description of steps and activities for implementing this policy. The asset management strategy should set out the most appropriate long term course of action for implementing the asset management policy and thereby supporting the delivery of the organisational strategic plan.

It is difficult to decide what this might mean or imply, but it does agree with the other authors quoted in that "steps and activities" are part of strategy.

We may now attempt to make sense of all these definitions by means of a table. The table that follows proposes a hierarchy of words, which set of words are used in each environment, and what type of words they are according to the "Kipling classification," described earlier.

The headings of the table include the level, description of the documents at that level, then four spheres of use viz church, state, military, and industrial. These four spheres of use have been used because one may find dictionary definitions of the words we are interested applied in these spheres. They represent areas of human endeavour where the concepts that this paper deals with have had to be defined. Hence, we can draw on the experience of persons working in those areas. Then follows a column describing the operative words from Kipling's system, and finally there is a column for a typical statement describing documentation at that level.

The classification describes six levels. The title of Table 1 refers to "instruments" rather than "documents," as the item might not be a document, although documentation is generally what is of interest to us here.

Note how the same word might appear at different levels, depending on the sphere of operation; for example, *mission* in the military sense of, say, an Air Force bombing mission is at a lower level than in any other case.

There are also gaps in the table where no dictionary definition has been found to apply in that area of endeavour. For example, the church certainly needs a strategy and a plan, but this is not language usually used in that sphere.

For our purposes in industry, we now have a clear model on how to proceed with a suite of documentation.

At the highest level, we have our vision, which is *what* we are striving to become. It is therefore a "what" word.

At the second level, we have our mission, a "how" instrument described by, "This is how we will achieve our vision."

At the next level, we have our philosophy, the high-level statement of belief, also a "what" instrument, described by, "This is what we believe to be true."

At the intermediate level, we have our strategy, which according to the dictionaries is almost synonymous with *policy*. As we have seen, however, some maintenance authorities split these into two instruments. Both are "how" instruments, describing how the "what" of the previous two levels is to be achieved, with specific emphasis on securing an *advantage* over any opposition or an advantage over our own previous performance. The strategy or policy is described by the two statements, "This is how we will secure an advantage" and "This is how we will win the war." The war in our case, of course, is a metaphorical rather than literal one.

At the lower level, we have all our other documentation, including detailed plans, schedules, instructions, procedures, standard specifications, etc.

A strategy, it seems, in industrial terms will always lead to at least three lower-level documents: an equipment maintenance plan, a resource plan, and a schedule. In other words, two "what" documents and a "when" document.

WRENCH TIME

The term *wrench time* came into prominence with the publication early in the twenty-first century of "Doc" Palmer's book *Maintenance Planning and Scheduling Handbook* (2006). This section is included with the express permission of Doc Palmer himself.

Wrench time is an American expression; the British equivalent might be called spanner time. (This is an engineering fact of life—for so many engineering terms, there exists an American and an English version. This text will use the American term.)

Wrench time is the time craft people actually spend doing the work that the company pays them to do—actually working on equipment. The rest of their day is spent on travelling to the work site, waiting for spares at the store, etc. Doc Palmer's book was based on extensive research at one of the major US power utilities in the state of Florida. What shocked many persons was the very low percentage for wrench

TABLE 3.4
Hierarchy of Instruments in Various Spheres of Use (Industrial Uses in Bold)

Level	Description	Church	State	Military	Industrial	Operative Words	Typical Statements
1		Vision or revelation			**Vision**	What or where	This is what or where we want to be.
2	Idealistic superordinate stated goal	Mission	Mission		**Mission**	How	This is how we will achieve our vision.
3	High-level belief system	Doctrine Creed (as a summary of doctrine)	Doctrine	Doctrine	**Philosophy** (also known as ***values***)	What	This is what we believe to be true and important.
4	Middle-level instrument to attain and maintain the prior belief system		Policy	Strategy	**Policy**	How	This is how we do things around here.
5	As prior, but more specific; will lead to a maintenance plan, a resource plan, and a schedule				**Strategy**	How	This is how we will secure an advantage. This is how we will win the war. This is how we will achieve our mission.
6	Lower-level instruments which support and execute the objectives of level 4			Tactics Missions Logistics	**Specifications** **Instructions** **Procedures** **Maintenance plans** **Resource plans** **Schedules**	What How How What What When	This is how we will win the battle.

time—about 35%! And that was in a public utility—other industries would fare even worse. For example, in some South African underground mines, the logistics are such that workers spend only two hours at the workface.

TAKEOUT: PANEL—WORK SAMPLING

Work sampling is the statistical technique for determining the proportion of time spent by workers in various defined categories of activity (e.g. setting up a machine, assembling two parts, idle . . . etc.). It is as important as all other statistical techniques because it permits quick analysis, recognition, and enhancement of responsibilities, tasks, competencies, and organisational workflows.

Other names used for it are "activity sampling," "occurrence sampling," and "ratio delay study."

In a work sampling study, a large number of observations are made of the workers over an extended period of time. For statistical accuracy, the observations must be taken at random times during the period of study, and the period must be representative of the types of activities performed by the subjects.

TAKEOUT: END OF PANEL

Doc Palmer's (2006) investigation was a properly constituted piece of industrial engineering research using work sampling on a group of 48 persons over seven days. The work classifications used in the study were as follows:

1. Working
2. Waiting
3. Other
4. Unaccountable

These were further subdivided so that there were 23 categories in all (some not used), as follows:

Working

1. Actual working
2. Travelling to and from workplace
3. Setting up
4. Meetings for assignment of work
5. Wrap-up; includes paperwork, cleaning up, end-of-shift meetings, etc.

Waiting

6. Waiting for materials
7. Waiting for tools
8. Waiting for instructions
9. Clearance delay
10. Interference by another trade
11. Other

12. Waiting for operator to explain the fault
13. Weather delay

Other

14. Meetings
15. Training
16. Idle
17. Toilet
18. On break
19. Personal (e.g. visit company doctor)

Unaccountable

23. Unaccountable

The results of the study were shocking. Wrench time for mechanics and welders was as low as 35%! Machinists fared better at 51%. And this is in a well-organised industry in a First World country.

The previous figures are for category 1, actual working. If categories 2 to 5 are added, then the figure is about 60%. Category 2, waiting, accounted for 14% of total time, while category 3, other, accounted for 18%. Category 4, unaccountable, was 8%.

What the very low productivity figure of 35% shows us is that excellent planning and scheduling are necessary to raise or even just to maintain these figures. In fact, Doc Palmer (2006) quotes in his book that a reliability manager of a large international company has stated that:

> "If I had three maintenance persons, I would make one of them a planner."

A colleague of his has stated that:

> "Planned work typically requires only one-third as much labour as unplanned work."

Hence, we arrive at the six principles of planning:

1. Have a separate planning department.
2. The planning department must focus on future work.
3. Configuration management must be in place, with all equipment marked and registered.
4. Planners must have broad trade experience.
5. Planners are responsible for the "what," but the artisan is still responsible for the "how" (so the artisan is not disempowered).
6. Planning performance is assessed by measuring wrench time.

This last point is an extremely significant statement. Wrench time is a key performance indicator, not only for the artisan, but also for the planner. It works both ways! Furthermore, wrench time first demonstrates the need for planning and then measures the effectiveness of the planning.

Finally, good planning will increase wrench time, but only if the planners assign more jobs to the artisans.

ESTIMATING JOB TIMES

Having a good planning department is one thing, but if the times allowed for jobs are incorrect to a large degree, the planning and scheduling process is severely compromised. One authority has stated:

> Nowhere is the application of work measurement likely to generate more productivity improvement and cost reduction than in maintenance. Maintenance represents the largest single variable operating cost in most enterprises when you include physical plant value, maintenance labour, materials and overhead. And yet, maintenance does not receive a proportionate amount of higher management attention.
>
> **(Westerkamp, 2002)**

The methods for setting times for jobs include:

- **Using past history:** This is the most common method, but its only virtue is the fact that the history exists. The times jobs take and the times they should take can be very different
- **Using manufacturer's data:** Perhaps manufacturers of plant can help here, but they usually cannot.
- **Using times from maintainability demonstrations:** This method results in excellent data. Specifications for any new or retrofitted plant should include a requirement for a maintainability demonstration, for example, for a high-pressure multistage pump overhaul. The supplier will do his best to give an optimised time during the demonstration, so as to win the order, especially if the work is on open tender. The company's own workforce will then not be able to develop a historical time which is too long, allowing them to "coast" through the job. However, this approach has probably not been adopted in most plants, and therefore other methods have to be relied on.
- **Using industrial engineering methods:** These are several, varying in complexity and accuracy, but all will take considerable effort to establish realistic maintenance times. One such method is that which was used in the Doc Palmer study, viz work sampling. There are others, all known by the unmemorable names that industrial engineers seem to give their processes. For example, there is methods time measurement, or MTM, which has been used extensively for synthesising times from the basic motions of repetitive, assembly line tasks. It has an advantage over work sampling, in that no stopwatches are involved. In a unionised environment, stopwatch observations are never welcome. Another method is the Maynard Operation Sequence Technique, or MOST, which is simpler to apply than MTM or its derivatives, MTM-2 and MTM-3. For maintenance work, with its complex tasks, all these techniques need to be combined with others to develop universal maintenance standards (UMS). Five levels of data are used to make up the UMS library:

- **Basic Motions:** MTM or MOST, as so-called "predetermined time systems," establishes how long it takes to perform certain basic motions.
- **Basic Operations:** These are *basic motions* that are grouped together to form basic operations common to all trades. For example, part handling. For this operation, we would need to know the weight of the part, the distance moved, etc.
- **Trade or Craft Operations:** Some times are unique to each trade. For example, welding a certain joint.
- **Benchmarks:** Typical benchmarks could include "replace a control switch," replace a set of v-belts, or "troubleshoot a control panel." It is claimed that about 100 benchmarks per trade can be used to synthesise all the possible jobs that need to be done.
- **Spreadsheet Analysis:** With perhaps 14 trades in a large organisation, we might have to deal with 1,400 benchmarks. These are arranged in "slots" as typical jobs. There might be 20 slots, ranging in mean time from ½ an hour to, say, 40 hours. Other jobs that come up are estimated by the planner to fall into one of the slots.

So we have RCM or some similar process to determine the tasks and their frequencies. Some industrial engineering is then required to determine task times. This is a non-trivial exercise but is necessary if an organisation is to progress from good to excellent. A planner can then prepare job times for any job as follows.

All work order times consist of four components:

1. Job preparation time
2. Area travel time
3. Job site time
4. Allowances for rest, personal delays, etc.

The first and second bullets listed prior are developed for each facility during benchmark development. The job site time comes from the nearest similar slot in the spreadsheets. The three components are added together, and the allowances are then applied as a percentage that sum. A simplified example will now illustrate the process.

Times for jobs and their frequency are plotted for a year in Table 3.5. If the jobs have been accurately preplanned, this table will then show the number of artisan hours for each month, for each trade, as shown in the following incomplete table.

TABLE 3.5
Yearly Planned Maintenance Schedule

Month	Jan	Feb	Mar	Apr	May	Jun	Jul	Aug	Sep	Oct	Nov	Dec
Trade												
Mechanic	100	240	360									
Electrician	40	240										
Instrument	40	120										
Welder	40	40										

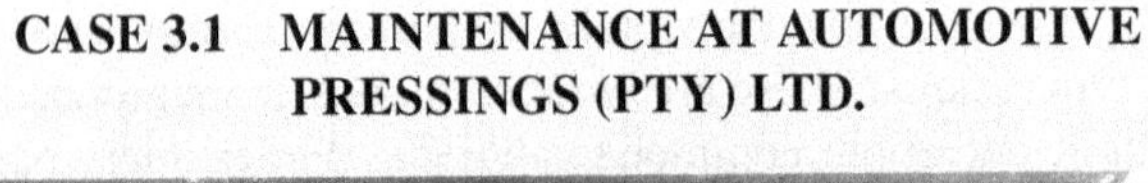

CASE 3.1 MAINTENANCE AT AUTOMOTIVE PRESSINGS (PTY) LTD.

FIGURE 3.0 Untitled picture (machine ai).

Author's Note: I think readers should find this a fascinating case. As regards maintenance management in this case . . . well, there is none. Readers may wonder how this company could make any money at all. Let me assure you, it did. But before First World readers click their tongues and mumble uncomplimentary things about Africa in general and South Africa in particular, I am sure, if they wish to, they could recall some factories in the UK, the USA, and the Commonwealth and elsewhere that exhibited similar characteristics to Automotive Pressings. The case is set in the late twentieth century in an industrial town east of Johannesburg in South Africa.

The case is 100% true, except for the necessary changes of name, time, locality, and product mix, necessary to disguise the identity of the firm and of the people who worked there.[1]

Added to the figures in the table for planned work will be a percentage for breakdown and other work, based on past history.

This table can not only be used to plan jobs for the year but can also be used to justify a maintenance manpower budget.

This table is then broken down into a weekly schedule, which serves as the most important tool for the planner in the weekly planning meeting.

Allan Riley was on the first day of his new job at Automotive Pressings in Germiston, an industrial town east of Johannesburg in South Africa. He had gotten the job after being interviewed by Stanley Horch, one of the families who owned Automotive Pressings. Allan was having his first twinge of doubt about the new job he had chosen over several others because it was the most lucrative.

Allan was standing in the galvanising bay of the factory, where large tanks used in the galvanising process stood in long rows in the gloom. A man approached him, introducing himself as the galvanising bay foreman. When he asked for a requisition to be placed for whitewash for the walls of the building, Allan did not hesitate to oblige. This was, after all, the annual shutdown, and Allan was in charge of building maintenance, among other things. Furthermore, he had an open job number to do all maintenance during the shutdown. He and his team had three weeks to do as much work as possible to rectify problems in the buildings and machinery before production began again in the New Year.

Allan had taken a year's leave from his previous employer to obtain a postgraduate qualification from a local university. When he completed the course, he was dismayed to find that his previous firm had been sold to a competitor and the gentleman's agreement between him and his old boss to re-employ him would not be honoured by the new management. However, there were lots of other jobs around. This one, thought Allan, would help him get more management experience, and he needed factory-floor experience in order to obtain his "government ticket," a unique South African qualification which was required by law if one was to be a factory manager. It was a good thing to have on one's CV.

Allan was a 30-something mechanical engineer, married, with two young children, when he joined Automotive. His previous jobs had been in design engineering and project management.

He had joined Automotive Pressings with the title of works engineer, during their annual shutdown in December. He was nominally in charge of all maintenance at the factory, but the works manager, Eric Nash, had stipulated that the "cut-to-length" machine would remain under Nash himself, due to its critical nature, being the machine on which the entire factory depended. It was also electronically controlled and beyond the ability of the factory's electricians to maintain. The control system was maintained by the supplier, while the mechanical side of the machine was maintained by one of the firm's best mechanics, a man known only by his surname, Santos.

Santos, although nominally under the direct control of the mechanical foreman, Fred Horstmann, seemed to take all his instructions from Nash himself.

All the rest of the fitters and welders reported to Horstmann. The electricians reported to Bernard Dodge, while the machine shop was under the control of Nigel Turner. There was also a small building maintenance team. Horstmann also had a specialist forklift garage under his control, manned by a new mechanic, who had also joined the firm at the same time as Allan. In addition to this, the boiler house was run by an old man, Shorty Ford, who had retired years before from his job as a truck driver for the mines in the area. He had no formal artisan training but in his youth had been a traction engine driver. This is what presumably qualified him in the eyes of the family as a boiler house person. Shorty was also in charge of the compressor installation. Finally, there was Guido Maserati, the plumber.

Automotive Pressings was owned by the Horch family. "Old Man" Horch had immigrated to South Africa from Eastern Europe as a young man and had worked as an itinerant tinker in the Cape before coming to Johannesburg, where he established himself as an ironmonger. With the rapid industrialisation of South Africa during and after the Second World War, Horch moved into manufacturing. Soon he was making body panels and other products for South Africa's motor industry. It was a good market to be in. South Africa in those times was called by some a businessman's paradise—next to no union activity, a workforce called by the editor of a local business newspaper "the most willing in the world," virtually no enforcement of environmental legislation, and sheltered markets with high tariff barriers. Cartel activity sometimes kept prices up as well, a fact once admitted to Allan by a member of Automotive's management.

Old Man Horch had four children—Stanley, Ruth Skoda (also known as Babe by the family), Murray, and Roger. They were all in the business. Stanley handled production and maintenance, Ruth looked after the books, Murray handled the marketing, and Roger seemed to travel to the USA a lot, perhaps to open up new markets, perhaps to transfer money out of the country. (South Africa had strict exchange control at that time.) In addition, Ruth had a son, Aaron, who had the title production manager under his uncle Stanley.

The first impressions that Allan had of Automotive was its size and rather-chaotic nature. The factory covered many hectares and was divided up into about ten different buildings. Allan was struck by the amount of work in progress—stacked around the grounds were huge piles of half-finished products, some of them starting to rust. He remarked to a co-worker, the plant chemist, at how much this must be costing the company. "But the Horches still have plenty of money," came the reply. This certainly seemed to be true—the row of new Mercedes cars in the car park and the Horch houses in the upmarket Johannesburg suburbs of Houghton and Sandton testified to that fact. What amazed Allan therefore was the cramped and dirty condition of the brothers' offices at the factory. After all, they were hard-working people who put in long hours at the plant. "Why not brighten up their working environment a bit?"

thought Allan. The turnover of the firm was a family secret, but it would classify as a medium-sized manufacturing firm.

The three weeks of the shutdown passed quickly. Most of the machinery consisted of huge presses to stamp out car body components, although many other products were made as well. There were some hydraulic presses, but most of the presses were of the flywheel type. There were also many spot welders and metal-forming machinery, such as spinning machines, on which metal products were spun to their final shape. There were polishing bays, enameling ovens, drying ovens, spray booths, and many different types of conveyors. Allan spent much of the three weeks being conducted around the plant by Fred Horstmann on the company-provided bicycles. There was much to absorb, but after three weeks, Allan thought he understood the plant and its complexities.

There was a manual works order system in place. The managers in charge of the various plant sections had to submit a work request to the works engineer. This was given to the relevant foreman, who gave it to the relevant artisan, who used it to draw spares and to book his time. The system could hardly be simplified further and worked well.

The only other pieces of paper, or rather, thin card, were the machine repair records and the artisans' clock cards. The machine repair record cards were not kept up to date. Eric Nash told Allan during their first meeting on the job, "You are supposed to keep the machine record cards up to date—Stanley insists on it. But before you were hired, we had done without a works engineer for months. And I have got better things to do." When Allan studied some of the cards, he found that the latest entries were about three years old, even for machinery which failed monthly. Allan began filling in the cards again, but there was no way to go back in time and fill in the three-year gap. The one time when Stanley Horch asked to see one of the machine record cards, Allan filled it in for the missing three years with bogus information before handing it to Horch, who seemed none the wiser. This action saved a lot of embarrassing questions for Eric Nash and Fred Horstmann.

More worrying to Allan personally was the fact that after working at Automotive for a few months, he found out that his position had been occupied by three different persons in the past eighteen months. Years later, long after he had left the company, he once gave this advice to a younger engineer thinking of changing his job: "When you apply for a job, don't let the company ask all the questions—you ask why this position is vacant that you are applying for and how many people have occupied it in the last few years. You owe that to yourself. It's not only prospective employees who have things to hide." In retrospect, what also struck Allan as odd was that all his other job interviews had been conducted through a personnel agency or he had responded to a company advertisement. For this job, he had replied to an advertisement which just had a box number as an address. "Why did the company not want to advertise its name?" he had thought at that time.

In contrast to the rapid turnover in the works engineer position, other jobs at Automotive were very stable. Eric Nash, Allan's direct boss, had been there about eight years. Fred Horstmann, the mechanical foreman, had been in the firm for at least twenty years, despite having a drinking problem. This was presumably because he had come to know the plant so well, thought Allan. The production foremen who ran the various sections of the plant all had long service, as did many of the production workers.

At the end of the shutdown, when Stanley Horch returned from his annual overseas holiday, he phoned Allan. "Riley," he said, "I believe you have white-washed the galvanising bay. I don't object so much to the cost of the whitewash, but the labour could have been used for something else. Please try to avoid unnecessary expenditure in future." Allan could hardly believe his ears—the cost of the material *and* the labour amounted to less than $20US. "It was then that I knew that I could not work for the rest of my career in that place," said Allan to a friend after he had left the firm.

Allan made it his business after the end of the shutdown to visit all the production foremen and explain to them what work had been done to their plant during the shutdown, what work was incomplete due to lack of spares, etc. Don Singer of the main press shop seemed particularly impressed by this behaviour. None of the previous works engineers had ever bothered.

Allan also phoned the head of production optimisation and work study, Tony Morris, as soon as Tony returned from holiday. PO&WS was a recently established department, which Stanley Horch was sure could save Automotive a lot of money by optimising production processes, reducing labour, etc. About ten industrial engineers and technicians worked in the department. This investment in intellectual capital was uncharacteristic for the Horches, who had flown by the seat of their pants for so long. In time, Allan would have some interesting conversations with members of that department, usually young engineers from England or Israel. His relationship with their departmental head, Tony Morris, was not so good, however. In response to his initial phone call, in which he had suggested they get together as soon as possible, to see how they could cooperate, Allan felt that Tony had snubbed him by saying he was much too busy to consider a meeting in the near future. Tony clearly displayed a "don't call us, we'll call you" attitude. Allan did have a tendency to get a bit excited about things which he considered important, and his manner may have put Tony off. The result was that they treated each other with suspicion from then on. Tony's engineers admitted he was a moody type and told Allan not to worry about Tony's manner.

As the plant swung back into production, things became more and more hectic. The factory had about ten pieces of equipment break down every day. Because of multiple production lines, this did not usually result in the factory stopping production, except if the dreaded cut-to-length went down.

One of the first big breakdowns which did affect production of one of the plants critically was a fault in one of the big spinning lathes. Da Silva, one of the two "ace" fitters, along with Santos (no one seemed to know their first names), had dismantled the piston and cylinder housing for the thrust mechanism which drove the shaping tool against the spinning workpiece. The cylinder had had a serious oil leak. Allan, Fred, and Da Silva had been examining the cylinder bore when Eric Nash rushed up. "Why is this machine down?" he asked in an agitated voice. "Get it back together!" Allan pointed to the 150-mm-long score in the cylinder, about 3 mm deep. "It is not just a matter of replacing the seal," Allan said. "We'll have to have a new cylinder made, or this one re-sleeved. And the seal design is poor, being an O-ring. I suggest we should modify the piston as well, to incorporate a proper seal which is designed for this sort of axial movement. We must also check why dirt got into the oil to cause this groove in the first place."

"We don't have time for all that fancy stuff now—get the machine back together!"

"With this score in the cylinder?" asked Allan.

"Yes!" replied Eric Nash, and he rushed off to see to some other problem in the plant. Allan, Fred, and Da Silva just looked at each other incredulously. "Put it back together," said Allan to Fred and the fitter and walked away in disgust.

That evening, just as he was about to go home, Allan was called to Eric's office. "I've been thinking about that cylinder on the spinning machine," said Eric. "You must arrange to repair it as soon as the production people will let you have the machine."

"Yes," said Allan. What else was there to say?

One piece of equipment which gave continual trouble was the camelback furnace. This was used to dry and bake small painted items by passing them through a gas-heated tunnel on an overhead conveyor. The tunnel started at ground level, then rose about two metres in the air at the point where the gas burners were. It had originally dropped back to ground level at the end of the heated section, but during some modification, subsequent to the initial installation, the gas burners had been extended into the ground-level section. This resulted in a lot of hot air being wasted by spilling out of the open ground-level exit of the tunnel rather than being constrained in the raised portion by convection. Allan wondered about who had done this, but by this time, he was too busy coping with breakdowns to suggest any modification to the furnace.

In particular, the rollers on the conveyor were seizing. Of several hundred rollers in the chain, a few seized daily, even though they were greased daily as they came out of the furnace. Allan investigated what grease was being used and called in a representative from a specialist lubrication company to advise on alternatives. After purchasing a good high-temperature grease to replace the general-purpose grease which had been used, Allan requested the relevant production foreman to switch off the furnace and stop the conveyor when convenient. Having no orders to immediately fulfil, the foreman obliged. Allan had a team of fitters, under Fred Horstmann's direction, check all the rollers and replace those that were showing signs of damage. He then had all the rollers thoroughly greased with the new grease.

While this was being done, Allan received a phone call from an obviously agitated Eric Nash. "What's all this greasing going on down at the camelback? Why isn't it going?" Unable to hide his anger and contempt, Allan explained, speaking through his teeth, that he was trying to maintain something properly for a change. "Oh," said Eric and hung up.

At their next arranged meeting, Eric said, "Perhaps this place is getting us both down at the moment." Allan agreed by nodding without speaking. It seemed to him that Eric enjoyed causing crises and then solving them.

Allan's solution was justified, however, in practice. There were no seized rollers in the camelback conveyor thereafter, for at least eight months, after which time Allan had left the firm and lost contact with the situation.

Shortly before this, Eric had issued a memo to Allan concerning some regular inspections which had to be done. It was the only planned maintenance directive which he received in his 13 months at the plant. One action given was to inspect the galvanising baths once a month. Wormhole leaks tended to develop in the welded corners of the bath. The implications of a leak were severe—molten zinc poured out all over the galvanising shop. On receiving the memo, Allan left it in his in-tray for several weeks. Then there was a leak in a galvanising tank. Eric Nash was surprisingly understanding about the whole matter. "Allan," he said, "I did tell you to inspect the baths, didn't I?" Allan could only agree. "Well, we all have to try to get a bit more organised around here," said Eric. Not a bad chap, really, thought Allan.

At a subsequent meeting, Eric instructed Allan to purchase the flat drive belts for the large flywheel presses from a certain supplier. Allan was about to ask why, as the company manufactured hygienic-quality belts for food conveyors. Eric anticipated the question: "Don't bother to do any investigations into other possible sources—we know these belts work."

On another occasion, Fred asked Allan to come and inspect a pneumatic cylinder that had been dismantled. It was a large cylinder, some 250 mm in internal diameter by 700 mm long. The seals on the piston had a very short life and had to be replaced every few months. Allan put his hand up the bore. He felt a distinct circumferential ridge halfway up the cylinder. He then realised why the cylinder was flanged in the middle as well as at both ends. It was in fact made of two short cylinders joined together. He had previously thought the cylinder must have been sleeved, as no self-respecting engineer would have bolted two short plain cylinders together to make a longer one without a full-length sleeve. Knowing the swing and bed lengths of all the lathes in the machine shop, Allan put two and two together. At some time in the past, the original cylinder had been replaced by this in-house device, made in two parts because of the limitations of Automotive's machine shop. Someone had probably patted himself on the back, thinking how much money he had saved Automotive, thought Allan. He then instructed Fred to see that the ridge, which was about 0.5 mm high at its highest points, was removed. This was done by spinning the cylinder slowly in a lathe while holding increasingly fine grades of emery paper on the ridge by hand. A laborious job, but it solved the seal failure problem.

Another concern Allan had was the terrible state of the plant's compressed air. There was no chiller in the system, so wet air was to be expected. To cater for this, water traps with silica gel cylinders were installed at various points in the air lines. Most of them were choked with oil. One day down at the compressor house, Allan was surprised to see the cylinders and heads for one of the large reciprocating compressors lying on the concrete apron in front of the door. Shorty Ford had taken it upon himself to do a compressor overhaul. Without going too deeply into the reasons, Allan deduced that the old man was out to impress the Horches, one of whom had complained to him personally about a "lack of air." No work order had been taken out. Shorty was a law unto himself in the boiler and compressor houses. "Shorty, you had better break that glaze on the cylinders, or the compressor is going to pump oil past the new rings. They will never bed in like that," said Allan.

"Mr. Riley," said Shorty, "I promised Mr. Stanley Horch that I would have the compressor back together this afternoon."

Shorty went on working, and Allan walked back to his office. The air supply was as oily as ever after the overhaul. Allan did nothing about bringing Shorty into line. Fred had left him alone for so long Allan did not feel he could do anything about it.

This was not the only bad practice of Shorty's. One day, Allan caught him welding a bracket onto a fully pressurised air receiver. "Shorty," said Allan, "you aren't allowed to do that. It's against the code."

"What code?" asked Shorty.

"The pressure vessel code to which the vessel is built—here, it gives it on the nameplate," said Allan. "The Factories Act requires that all parts of the code be complied with throughout the life cycle of the vessel."

Shorty did not seem to see the point but stopped welding.

Back in his office, a shed next to the main road through the plant, Allan spent the rest of that same afternoon collating the manuals for the various items of plant and compiling a list of machines for which no manuals were available. He had been around to all the production foremen and maintenance foremen, collecting what manuals he could. He had also approached the original suppliers of equipment to replace lost manuals. One item which required a manual was the electric forklift truck. All the other forklifts were diesel-powered, and the forklift mechanic could cope with them. The electric truck had been purchased as a bargain by the family at a sale and had never run properly. The electricians had tried to fix it, to no avail. Allan had ordered a manual for it, at a cost of some R50. On arrival, it had been intercepted by Stanley Horch. "What does he want to spend all this for?" he asked with a sigh.

Another programme of Allan's which caused concern with Stanley was his training programme for the air tool repair fitter. Fred had had an argument with the elderly fitter who had done the job before, because of his "cowboy" approach to his job—for example, the man would grind a taper on a shaft with a handheld air tool to fit a taper-bore flywheel governor, simply because he had those two incompatible parts in stock, rather than order a tapered shaft or parallel-bore flywheel. The man resigned. Allan sent his replacement, a young Portuguese fitter called Roy, to the manufacturer down the road where Automotive bought all its air tools. He took a lot of the plant's air tools with him, with instructions from Allan to get them back to as-new condition. The bill was several thousand rand, but the average monthly bill for air tool spares was almost that, and Allan now felt that they had a reliable set of tools that would work well for several months or even years. When Stanley got the bill, he hit the roof, Allan was later told, but he did not say anything to Allan at that time.

Plant was continually breaking down and being repaired. And there was always pressure from the management that everyone had to be busy all the time. This problem solved itself with the large backlog of half-finished jobs which developed. Artisans could always leave one job and work on another if they were waiting for spares on the first. Allan failed to see that, in carrying on with this policy of using a pool of uncompleted maintenance tasks, to balance his artisan workload, he was guilty of the same error as the production people with their piles of work-in-progress all over the factory. The obvious advantage of the system, which Allan inherited rather than caused, was that it did keep all the artisans busy all the time. This was particularly important to Fred Horstmann. He told Allan that Stanley had a spyhole up on the roof of the building. He would sometimes leave his office via the window and climb onto the fitting shop roof via an adjoining shed to spy on the activities in the fitting shop. He had similar spyholes in various parts of the factory, Fred said.

There were difficulties as well as advantages for Allan with the pool-of-broken-machines policy, however. The scheduling of restarts was practically impossible. This caused a problem for Allan one day. Aaron, Babe's son, in production, once asked Allan when a certain machine would be recommissioned. He said there was no particular urgency, that he just wanted to do some planning. Allan had seen parts of the machine strewn all over the fitting shop that morning, and it seemed likely that it would remain in pieces for several days, if not weeks, judging by the fitting shop's past performance with this sort of job. He told Aaron that the job could take up to two weeks. That same afternoon, the "ace" fitter, Da Silva, had the machine back together and working. Just what influence the family might have brought to bear on Da Silva to perform this extraordinary feat was not known to Allan. Soon the news was all over the factory. Riley had told Aaron Skoda that the job would take two weeks—he had made a fool of Skoda. Allan, in his typical naive manner, could not see what the fuss was about—what's wrong with giving someone something earlier than you had promised him?

One part of the plant where Allan felt he achieved a measure of success was the hydraulic press shop. The machinery here was newer than in the rest of the factory and, by and large, worked well. When there was a failure, the fitters could not always cope, as none of them had much hydraulic experience. It was left to the foreman electrician, Bernhard Dodge, to sometimes locate faults in the hydraulic or pneumatic circuits. Allan worked well with Bernhard. Sometimes the two of them could solve a problem that either working by himself could not. Allan marvelled at the synergy created.

Once they had traced the fault to a particular pneumatic spool valve, a fitter was brought in to effect a temporary repair. He replaced the multifunction spool valve with a set of four single-function valves which now hung from pieces of neoprene tubing out of the side of the machine. It worked but looked like nothing on earth, thought Allan. No one else seemed to mind. The machine was back online; that was all that counted. Allan took the spool valve and, after locating a grinding shop in the vicinity, asked if they could machine and grind a new spool. The owner of the shop grudgingly took on the job. "We don't like doing work for your company, you know," said the owner, without ever explaining why. In any event, the valve was refurbished and re-installed, to Allan's satisfaction, if no one else's.

Allan did not often work with Bernhard. Most of Allan's dealings were with Fred Horstmann. Fred had the unfortunate trait of praising his superiors to their faces and criticising them behind their back to his men and, Allan suspected, to the family as well. Despite that, Allan, naive and trusting, got on quite well with him.

The relationship first became strained when a new factory building had to be erected. "Development work" like this, as it was called by Eric Nash, was usually to be managed by Nash. This is what he had said to Allan when he first joined the company.

But if Nash was too busy or was not interested in the project, the work was passed down to Allan. Such was the case with the new building. It had been designed by the firm's own draughtsmen, and a materials list had been prepared by them. "Piece of cake," thought Allan, and he left the ordering of the steelwork to Fred. Fred ordered too much—about 25% too much. This was clearly Allan's responsibility, as Fred worked for him, but Allan made no attempt to defend himself. And he suspected later that Fred had told Stanley Horch that Allan had ordered the steel. Stanley had had enough and confronted Allan. Allan knew that in time the surplus steel joists and I beams would be used for other work around the factory, and so he was not particularly concerned, even in Horch's presence. Horch instructed Nash to keep a close eye on Allan's purchases from then on.

Eric Nash called Allan in and gave him a dressing-down. "I am becoming increasingly concerned about your attitude," he said. "I've told you before, you have to learn to do things the Horch way. I don't like everything that happens here, but I believe I serve the company as I should. You have a bad attitude—bad towards the family—and you are aloof with the workforce. It's time you changed your behaviour."

Startled by the emotion and intensity of Eric's attack, Allan did not know what to say.

Eric then changed the subject and gave Allan a ten-minute lecture on how he, Eric, had been able to design a replacement gear for a broken machine. "A good engineer can respond to a problem quickly and solve it," he said. Allan could not agree more and could not understand why this subject had come up.

The straw that broke the camel's back, from the family's point of view, followed a few months later. A new belt conveyor had to be installed to take completed, boxed products from one building, over a roadway, to a storeroom. Allan asked Nash if this, as a development job, was to be done by Santos, the man who had erected previous conveyors under Nash's supervision. "No," replied Eric Nash irritably. "Why should Santos have to do it? You and your men must do it."

Allan discussed the job with Fred Horstmann. Fred said he would put Da Silva on it, as he had never seen him run away from any job. But after a few days, it was plain to see Da Silva was in trouble. Reading drawings was not his strong point, and these drawings, prepared by a young engineer in the factory drawing office, were not very readable to start with. Allan decided he would have to complete the job using the young welder who had erected the steelwork for the building for which Fred had ordered the steel.

This was not such a bad idea. The young welder, called Tom, was also good at rigging and had made a good job of the building. The conveyor was supplied in modular sections and was intended to be bolted together like a Meccano set. Allan left Tom alone for much of the job, and it was soon erected. The trouble began when it was commissioned, by the plant optimisation people.

"What's wrong with your conveyor, Riley? The boxes tumble down it and fall off of it. The descent angle is much too steep at 50.675 degrees," said Tony Morris.

"How did you manage to measure the angle that accurately?" asked Allan, which did not help Tony's mood.

The conveyor suppliers were friends of Tony's, Allan later suspected. They had been in to see Stanley Horch and stipulated that the descending section of the conveyor could not exceed 45 degrees in slope. Nothing on the drawings had indicated this, and Tom had done his best to fit the poorly made conveyor sections into the confined space between the buildings. But the angle was too steep. Eric Nash called Santos in to fix up the mess. Fred Horstmann also gave Stanley Horch his personal assurance that he would do everything possible to sort out the problem.

The next day, Allan was asked by Nash to hand in his notice. "What do you mean?" he thought. "I was just leaving, anyway." He put on a brave front, but the scars of the failure to succeed in this job were to remain with him for years to come and would show in his lack of self-confidence at his next job, which was in the project management function that he had operated in before he joined Automotive. "At least," he rationalised, "I now have my government ticket." That was true. He had been allowed to sit the exam on the basis of his experience at Automotive a month before he left the company. He had been with the company thirteen months.

- **Postscript 1**: About a year later, Allan was introduced to a new employee where he was working. "This is Frank," Allan's boss said. "He has just joined us from Automotive Pressings."
  "What was your job there, Frank?" asked Allan.
  "Works engineer," replied Frank.
- **Postscript 2**: Several other employees of Automotive followed Allan to his new firm. A few years later, Allan heard that Automotive had been sold to a South African conglomerate and that the Horch family had all moved to the USA, except for the Old Man. He had been murdered in his house by a burglar.

TOPICS FOR DISCUSSION

Analyse the previous case from a maintenance management viewpoint. Be sure to include answers to the following questions:

1. What did Allan do right?
2. What did Allan do wrong?
3. Why do you think that while much of the staff at Automotive were stable, the position of works engineer was continually having to be refilled?
4. What should Allan (or Frank, or any of the others, for that matter) have done to ensure success in the works engineer position?
5. Do you think it was possible to succeed in that position, or are some jobs simply impossible?
6. Assume you are Tony Morris. Having made a very good impression with Stanley Horch in the production optimisation section, he has now appointed you as maintenance consultant, with the brief to advise on a complete revamp of the maintenance function in Automotive. What will you do? Your report will be required in two months' time. A new works engineer is to be appointed in the meantime in the line function. Give a description of what your report would contain.
7. Why was Allan's solution to the problem of the circumferential ridge in the split cylinder not necessarily foolproof?

CASE 3.2 RCM AT DRAGON PEAK

Author's Note: This case tells us how not to do it. For the veterans of the early days of RCM, I am sure this case will ring a few bells.

Mark Francis, a maintenance consultant, had been called down to the Dragon Peak pumped storage station from the head office of Pacific Hydro, who had recently purchased the facility from its previous owners. He had been approached by Steven Wong, the young engineer in the mechanical engineering section at Dragon Peak, to present a two-day course on RCM, the maintenance system which Steven had been told to implement on the plant.

A pumped storage scheme is a system of turbines and generators that can also be operated as pumps and motors. The problem with electricity is that it is a difficult product to store. In this respect, electricity has always been a "just-in-time" industry. Make it and sell it—there is no shelf life. This is generally the case, but there are some ingenious exceptions. For example, one method of storing electricity is to pump water up into a reservoir, then using the pump as a turbine, run the water back through the machine to generate electricity. This is the method used at Dragon Peak.

Mark was shown around the plant by Steven to familiarise himself with all the equipment. Mark was then taken to the lecture facility, and before the lectures began, Steven filled in the background for him: "There have been at least two attempts to implement RCM here before, but we don't have much to show for it. We have in fact compiled this manual of recommendations for how to modify our scheduled maintenance on most of the large items, but we haven't implemented it yet. As it's now my responsibility, I feel I need to know more about the process before trusting someone else's recommendations. That's why you are here—to advise me on this."

The audience for the lectures was composed of eight persons—a mixture of engineers, technicians, and senior artisans assembled by Steven for the occasion. During the lectures, Mark noticed that one of the senior artisans, Peter Kaplan, sat at the back of the class, not participating and looking bored. "Have you attended an RCM seminar before?" asked Mark.

"Yes, this is my third time, actually," replied Peter.

Mark thought this was a bit strange, but he pressed on with the lectures. At the end of two days, everyone thanked him for his excellent presentation. Steven and Mark then sat down to discuss the strategy for the coming weeks—whether they would convene an RCM committee and when they should start analysing the plant.

Steven spoke first. "You know, I'm not sure how to continue. Before my time here, the station tried to implement RCM, but nothing came of it. Can we study the worksheets in the manual I showed you earlier?"

Mark agreed.

The first system they investigated was the compressed air system. The station was equipped with 30 bar compressors driven by 132 kW motors. These compressors were used to pressurise large air receivers which were used to blow the water out of the turbine-pump units when they were to be used in synchronous condenser mode. (In this mode of operation, the units were motored in air and used for power factor correction on the system.) Mark and Steven studied the worksheets for the compressor and the air receiver. All scheduled maintenance for the compressor had been written out—the worksheets all stated, "No scheduled maintenance necessary." They had been signed as compiled by Peter Kaplan.

Mark and Steven next consulted the maintenance manual for the compressor as supplied by the manufacturer. The compressors were horizontally opposed twin-cylinder types, oil-free, with nylon rings. The manual stated, in large letters, in a special panel on the page, to draw attention to this fact: "It is most important that the cylinder head be removed after every 4,000 operating hours to check the clearance between the cylinder and the piston." Steven pointed this out to Mark, who replied, "If you let the rings wear to the point that you get metal-to-metal contact between the piston and the cylinder, you could have a very expensive repair on your hands. Judging by the design of the machine, if you seize the piston, you might also break the crankshaft. Are you sure none of the previous RCM recommendations have been implemented yet?"

They next studied the worksheets for the receiver. There were quite a few of them, all dealing with the drain cock on the bottom of the receiver. Extensive inspections and replacements of sealing glands were recommended. The author of the sheets was Peter Kaplan.

"I think I'll have to get back to you on this," said Steven, "after I've discussed the matter with my manager, Fred Pelton. He's been here for years, and I know he attended the course given previously by another RCM consultant. I must also find out why a fitter was left in charge of the process previously."

TOPICS FOR DISCUSSION

"RCM can go horribly wrong if not properly managed": Discuss this quote by Mark Francis given at one of his subsequent seminars at another location, in light of the previous case.

CASE 3.3 RCM AT TIGER SUBSTATION

Author's Note: If RCM AT DRAGON PEAK was the bad news about RCM, RCM AT TIGER SUBSTATION is (mostly) good news. What is the difference? I think it is all a question of how the RCM process is managed.

Maintenance consultant Mark Francis had been approached by George Galbraith, an industrial engineer at Pacific Hydro, to help him with an RCM analysis at a large substation in one of their regions. The substation, called Tiger, was in a remote location, and the job would require commuting by air on a regular basis.

On their first visit, George introduced Mark to the district manager, Bill Heaviside, at the district offices in the town of New Bedford. After a visit to the substation and a study of the layout drawings for the station, Mark proposed a schedule for the job.

"We will need the manager of the sub and maintenance and operating experts to sit on our committee. I would suggest six to eight people, certainly no more than that. And as none of the staff have had any exposure to RCM, it is essential that we begin with a training programme," said Mark.

"We can fly down for two days a week," he continued. "Say, every Monday and Tuesday, if that suits your staff. The first week will be for training in the technique, and the analysis will commence the following week. I estimate the job will take about eight sessions of two days each."

The district manager said, "I think I will ask the managers of the two other large substations in my area to join your team. They are both experienced men with a wealth of operating and maintenance experience. I will also invite the local engineering manager from the New London office to contribute. He has an oversight role over all maintenance in the region and is in charge of the updating of all maintenance manuals and that sort of thing."

"From our side we will bring down with us one of the more experienced substation engineers, Rudi Steinmetz, from the transmission department. You're out in the sticks here and perhaps don't have ready access to the latest expertise. Rudi's very capable, and what he doesn't know, he can find out from the manufacturers or their representatives, all of whom are located near our head office," said Mark.

"I will also bring down an engineer that has recently joined our department," said Mark. He does not have much experience on RCM, and I would like to train him up. He will serve as clerk of the committee, filling in the worksheets and seeing that they are typed up before the next meeting. He will also generate minutes and an action list for your people and for Rudi, so that outstanding items are attended to for each subsequent meeting."

"Don't you have software to help you?" asked Heaviside.

"Not at this stage," said Mark. "Many of the software packages available are just glorified word processors that do not add more value than they cost."

Mark continued, "We will also co-opt expertise on to the committee as required. For example, I will bring our current transformer expert down when we discuss the CTs. He also has a video recording to show you what can happen if an oil-filled CT is badly maintained . . ."

"I've seen that video," said Heaviside. "The thing blows up."

"Quite an impressive sight, isn't it?" said Mark with dry humour. But Mark's intent was serious. He realised what a tedious process RCM could be and wanted to make sure the committee retained its enthusiasm. This was another reason to co-opt experts in certain fields, to show videos, etc. It all helped to keep the interest up.

A few weeks later, the first two-day session was set up in a conference room in a hotel in New Bedford. Mark was introduced to the manager of the Tiger sub, Arnold Kirchoff. The managers of the other two substations were also there: Brian Watt of Panda sub, and Bill Faraday of Leopard sub. Also present was the engineering manager from New London, Don Hertz. Mark, George, Rudi, and the "engineer-clerk," Ron Lee, were present from head office.

Bill Heaviside welcomed the visitors and wished them well. He sat in at the first training session and then returned to his office. The two days of training went off well. Everyone on the team seemed eager to learn, except perhaps for the regional engineering manager, Don Hertz, who seemed a bit bored at times. "Perhaps he thinks he is too senior for this type of thing," thought Mark, "or perhaps he feels we are on his turf. Anyway, George and I have a mandate from the district manager to do this work. You can't please all the people all the time."

The following week, the four head office engineers returned for the first analysis session. The Tiger substation consisted of various voltage transformers, current transformers, breakers, and other electrical equipment. It had been decided to analyse the breakers first, as they were the most maintenance-intensive items. This analysis had to cover four different makes of breaker. Mark had asked for the substation managers to bring in examples of failed components from the breakers, as well as the maintenance manuals and drawings for each type. These items were all laid out on tables at the back of the conference room where they were meeting.

As Mark commenced with the FMEA of the first type of breaker, he noticed that very soon there was a good deal of argument between the three substation managers. They did not seem to be cooperating in the analysis but rather trying to show each other how much they each knew. At the end of the day, when he had time to reflect on the situation, he came to the following conclusion: All three men had come up through the ranks, and none had any tertiary education. The last time they were probably in a group situation like the present one, at desks with pencil and paper, being taught something as a group, was in high school. And in high school the aim is "impress the teacher."

Brian Watt, in particular, became quite intense when views contrary to his were expressed. This was particularly so if Arnold Kirchoff contradicted him. Mark hoped they would settle down, but at the end of the second day, he felt he had to ask Bill Heaviside to intervene. "Brian is being a bit disruptive," Mark said on the phone to the district manager. "I would appreciate it if you could take him aside before the session next week and tell him that."

At the following session, a suitably chastened Brian Watt contributed far more meaningfully. But the shakedown process was still not complete. Numerous actions had been placed on the various group members—Rudi had to consult with a breaker supplier regarding a possible failure mode, the substation managers had to complete cost calculations for recent breaker failures, and the district engineering manager, Don Hertz, had to bring in a set of revised maintenance manuals which had not yet been issued to the substations. Everyone came to the third Monday meeting except Don Hertz. Not only did he not arrive himself, but he also did not send the manuals or a replacement or an apology.

Mark now regretted that he had appealed to higher authority for help with his previous problem, which he now felt he could have resolved himself. He now had to have Bill Heaviside's help again. Don Hertz was more senior than Mark in Pacific Hydro and was in fact at the same level as Heaviside. In his approach to Bill Heaviside, Mark emphasised that in order to meet the deadline for the job set between them in their initial discussions, it was essential that every committee member pull his weight. If Hertz was too busy to attend all the meetings, he should send a representative.

Heaviside's telephone call to the New London office apparently worked. That Tuesday, an engineer from the New London office appeared at the meeting. After a quick overview of the RCM process, given to him by Mark, he entered into the spirit of the thing, always fulfilling whatever actions were placed on him timeously.

The group proceeded efficiently with its work over the remaining weeks. With the videos and specialist input that Mark had arranged, interest remained high. On some occasions, the substation managers would bring complete pieces of equipment to the meeting on the back of a pickup truck. The group would gather around the item in the car park and discuss failure modes and rectification methods.

In the evenings, there was much for out-of-towners to do. New Bedford is a town of much historical interest, and with the good rapport that had developed with the local Pacific Hydro staff, the head office staff were well hosted whenever they came down.

Another thing that Mark noticed during the committee sessions was the number of "spin-offs" that resulted from the RCM process but were not strictly part of it. For example, on one occasion, the group was discussing a plastic trigger in one of the breakers that continuously failed in the breakers at Tiger Sub.

Brian Watt said, "But we stopped using those years ago at Panda."

Bill Faraday of Leopard Sub concurred. "We changed to a brass design of our own and sent details of the mod to the New London office. They should have sent a notification of the mod to all the sites."

"We received it and have been using brass ever since at Panda," said Brian.

The notification of the modification had apparently never been received at Tiger. This was just one of several examples of configuration management anomalies which the RCM exercise helped to solve.

At the end of eight weeks, all the equipment included in the study had been analysed. As is usual with the recommendation coming out of RCM studies, there were considerably more scheduled inspections to be programmed into the CMMS and considerably less reworks and replacements.

What now remained was to incorporate all the new maintenance tasks and frequencies into the CMMS. This was left to Ron Lee, who had served as committee clerk, to do after Mark, George, and Rudi left New Bedford for the last time. Bill Heaviside was extremely pleased with what had been done. Mark suggested to him that the team should return after twelve months to revise the maintenance task schedule in the light of a year's operating experience.

POSTSCRIPT

Sometime after returning to head office, Mark had occasion to study a report finally prepared by the "engineer-clerk," Ron Lee, at the end of the job. To his dismay, he found that several of the recommendations proposed by the RCM committee concerning maintenance tasks and frequencies had inexplicably been omitted from the report and also from the CMMS. Lee had left Pacific Hydro in the meantime, after having failed to complete an RCM exercise at another site which Mark had entrusted him with. He had in fact been asked to leave the site by the site management.

"I should have been more careful to study the schedules in the report in detail before signing it off," said Mark. "I let my attention slip towards the end of the job, because of pressure of other upcoming work."

No request to return to New Bedford was ever received. The champion of the RCM process there, Bill Heaviside, was promoted to the New London office and was no longer able to influence the process.

GLOSSARY

CMMS: computerised maintenance management system

CASE 3.4 CROWNING A HANGAR QUEEN TAKEOUT: UNTITLED PICTURE OF PHANTOM AIRCRAFT (HANGAR QUEEN)

Author's Note: This is a case involving maintenance in the military. There is something amazing about it viz that one would expect maintenance "by-the-book" in the military, but that is hardly the case here. It is set during the Vietnam War and is reproduced from Chapter 10 of the book *Phantom Over Vietnam: Fighter Pilot, USMC*, by John Trotti, © 1984 by John Hall Trotti. Used by permission of Presidio Press, an imprint of Random House, a division of Penguin Random House LLC. All rights reserved.

Waiting for me at the line shack when I landed was Sergeant Olsen, a look of real concern on his face. "It's a bad day all around, Major. Thirteen just came in with a fragged stabilator. It's one of the new boron filament composites, so the tech rep is looking it over to see whether we can epoxy it here or send it to Da Nang."

"If it has to go to Da Nang, how long will it be tied up, do you think?"

"Be at least three weeks . . . maybe a month or more. They're backlogged to the hilt. Seven will be going back together tomorrow, so we might as well plan on Thirteen becoming our hangar queen. Seem reasonable to you?"

I was the squadron maintenance officer, which might sound important, but there was really not much for me to do except explain things to the commanding officer and pat people on the back. Sergeant Olsen humored me every once in a while by asking for a decision that we both knew he had already made, but at least he knew that I'd back him up, which was better than some maintenance officers I'd seen and about the only thing I was good for. The troops had things under control, which was not only the way it should have been but the reason why the squadron had always outflown the other units and would continue to do so, regardless of how Wing tried to even things out.

"When will Seven be ready for test?"

"It could be ready by late tomorrow afternoon if we get lucky, but mid-morning the day after is a better bet. You want me to go with you?"

There was some history behind this question. I had taken Sergeant Olsen flying during the previous week. It was highly irregular to say the least—probably marginally illegal. The occasion had been a mandatory test flight of a battle-damaged aircraft that had been down for nearly two months awaiting the arrival of a leading edge flap to replace the one that had lost an argument to a tree during a bombing run. Because it was obvious that the aircraft was going to be down for quite a while, maintenance used it as the "hangar queen," which is to say that any time we needed a part to get another aircraft "up," we would rob it from good old Number Six to avoid waiting for the material people to locate the part and provide it through the correct channels.

From the standpoint of availability, the spare parts situation had improved dramatically from the early days, but in terms of response time, the system was if anything worse. In 1965, squadrons had been much more responsible for their own parts procurement, and even when more Phantom squadrons arrived, the supply effort reflected its initiation from the operating units rather than (as later) its imposition from above. Before, you could call over to Supply with a requirement and even though the letter of the law said that you had to turn in the old part before Supply could pull and issue the new one, they would respond in the knowledge that every delay could cause you to bust your flight schedule. Now we played it according to the book, and even where a part was in ready supply, the system imposed a minimum delay of twenty-four hours.

As ridiculous as this was, we had learned to live with it by the creation of "goody lockers" and hangar queens, providing us with our own internal source of supply. Among the three Phantom squadrons in Chu Lai, there were enough illegal spares to totally circumvent the best efforts of the supply system to put us out of business. In desperation, the group supply officer went to the group commander and got him to issue an order requiring us to get rid of our goody lockers and return all illegal spares to Supply. The maintenance officers of the other two squadrons balked at first, but finally, they capitulated over my pleas for resolve in the face of the enemy. The weak link was the maintenance officer of our neighbor squadron, who told me it was my solemn duty as an officer to obey orders and turn in my spares. I told him that knew nothing about spares. Being literal about it, I told him that "I had not laid eyes on the spare parts locker for several weeks"—which I hadn't because I had received warning that this was in the offing and had ordered Sergeant Olsen to make sure that I would henceforth have no knowledge as to the existence or whereabouts of the spares. Sergeant Olsen had the spares removed to the living area so even he wouldn't know where they were. For the first several days after they had turned in their spares, the other two squadrons were able to manage because of their hangar queens, but gradually they fell farther and farther behind until they needed two queens instead of one, and then three. Within a week they couldn't make their flight schedules and we were beginning to pick up some of their Hot Pad launches. Finally, the other maintenance chiefs came to Sergeant Olsen asking for access to our spares.

"How do their bosses feel about this? Are they going to turn us in for cheating?"

"They say 'to hell with them.' They'll pay us back and build their own goodie lockers back up without telling anyone."

So the crisis was over, but it still took them another six months to get things back to normal.

In order to prevent the creation of hangar queens, there was a maintenance directive which required (under the pain of court martial) that any assigned aircraft *must* be flown at least- once every sixty calendar days. After fifty days, and with Number Six's gizzards having spread to every airplane in the squadron, we had to face the fact that we had a potential catastrophe on our hands. The new leading edge flap had arrived and been fitted, but there were at least 500 man-hours of labor staring us in the face, before getting the airplane into any condition to fly. It had already been decided that the aircraft was so badly damaged that it would be barged to Japan for repair, but we needed to get it to Da Nang, where it could be mothballed and embarked. Right in the middle of all of this, another aircraft had come back from a mission with a broken canopy, replacement of which could take as much as six weeks. True to its present mission, Six yielded up the required part, elevating the crisis to an almost insurmountable level.

That night Sergeant Olsen and I met with all of the shop chiefs to discuss the situation. We had nine days to virtually rebuild the airplane, and even then it would be nothing more than a shell. We could weld the flaps up and wire off the boundary layer control; we could make an air dam for the ejection seat mechanism for the front cockpit to prevent the wind from firing the seat in the absence of a canopy; we could add lead to replace the missing radar gear in the nose. The thing that worried us the most was that in having to race to put an almost totally dismantled bird back together, some little facet might be overlooked, leading to a needless loss of an aircraft merely to meet a silly regulation. Still, we had no choice, but it occurred to me that there might be a way to increase the odds of a flyable airplane, so I gave it a try:

"How long have you been after me to take you for a ride, Sergeant Olsen?" He lit up with a huge grin.

"Great. We'll weld the gear and flaps down and take it up to Da Nang on the test hop. No sense making two flights out of it."

As it turned put, the flight was a piece of cake. Every maintenance man in the squadron came out to watch as we fired up and went. Through our preflight checks. All the new parts that had been requisitioned for Six had been put on other aircraft, substituting them for 'high-time" components to send back to Japan. We exercised systems on the ground until the shop chiefs were satisfied that everything we needed to get to Da Nang was working and off we went. Stripped of every single nonessential piece of gear to augment our goody locker, and only partially fuelled to allow us to land immediately if need be, Six thundered into the air in a 60-degree climb attitude as we departed the pattern.

After we arrived at Da Nang, Sergeant Olsen borrowed a ground crew from a resident Phantom squadron and removed every part that he could get his hands on, going so far as to put on, where missing components should have been, dummy covers stencilled CLASSIFIED—NO ACCESS. When the Hummer showed up to take us back" we had to bump all the other passengers to make room for the parts. It isn't hard to imagine the reaction of the people at Nippi, the rework facility for F -4s at Atsugi, Japan, when this cannibalized piece of junk rolled off the barge. They probably yelled and screamed for a month, but they'd get over it and go to work making it as good as new.

~

So here we were with Seven coming back together again after more than a month. Like its predecessors, its innards had been in and out of it so many times to keep the others going, it was bound to have a few leaks and glitches, but hopefully nothing too serious. Thirteen was the obvious candidate to take its place, which was too bad, because she had always been one of the top performers. I was getting depressed just thinking about it. We never seemed to make any headway against the procession of ills that affected the airplanes. Besides, I needed to stretch my legs, so suggested to Sergeant Olsen that we snoop around the area.

"Come on, Top, quit trying to snivel your way into another flight. Let's take a walk around the joint and see what's moving and shaking."

My domain was the middle hangar and flight line area at the north end of the field between two other F4 squadrons. The hangar contained all but the flight line and ordnance shops and was able to accommodate three aircraft at a time. At the moment, Seven and Two were up on jacks with the engine bay doors hanging down, and Eleven was being rolled out following a hydraulic pump change. Once outside, it would sit on the apron while the various shops replaced the myriad pyrotechnic devices used for ejection and to ensure the destruction of classified pieces of equipment.

"She'll make the twenty—hundred Steel Tiger, Major," Sergeant Ross from the Maintenance Control branch reported, his tone showing pride in the rapid turnaround. "She will be out tomorrow afternoon sometime if the flap linkage pieces get here tonight like Supply promised." Frowning deeply, he consulted his clipboard for a moment. "Two's a mess. We fixed the leak the number one fuel cell and buttoned her up, only to come up with another leak during the high—powered turnup." He started to say something, thought better of it and slumped in resignation. "I don't know when it will be done." he mumbled, erasing and rewriting notes on his clipboard; then he brightened again. "Six is getting a compass swing, but she should be ready for an acceptance flight in another hour or two. It's been a total mission sorting out the wiring, but the electricians think they have it whipped."

Maintenance Control was the nerve centre of the effort and, next to morale, the key to our having a higher availability than our sister squadrons. Sergeant Olsen had instituted a system of close interval scheduling nearly a year before that revolved around the five- minute around-the-clock updates that allowed Sergeant Ross and his Maintenance Control people to shift men and materiel to where they could do the most good. Like most things, the maintenance manual for the F -4 had been written during peacetime for peacetime operations. It was aimed at the most cost-effective utilisation of resources in accomplishing the job. All well and good for the boys back in Memphis (home of Naval Air Maintenance), but in Vietnam, our job was to turn around aircraft to fly missions. When time was of the essence (as it nearly always was), it was often imperative that we pull people from one job and set them on another, even when it meant that we had five people doing the job that the manual said took only two. . . . As soon as Eleven had its pyrotechnics reinstalled, it would be towed clear of the apron, and Five would be in to have its radar replaced. At the same time, we planned to put the commelect people on her to see if they could get the radios to work a little better. Braun, who was our best trouble-shooter, thought that there might be a bad connection behind the air-conditioning pack, so it made sense to give it a shot.

Outside the hangar, the scramble claxon blared: Long . . . short . . . long. It was the signal for the B Pad-napalm and snakeyes, so we trotted after the stream of troops who were heading out to add their assistance. Eight—still hot from her earlier flight—was already loaded and sitting in the middle of the flight line ready to fill one of the soon-to-be vacated Hot Pad spots. "She'll take Nine's spot on the A Pad, and Nine will go out on Colonel Jeffries' TPQ." This was a tacit acknowledgment of what everyone knew—that Nine was a dog, better suited to straight-and-level work than heaving around a bombing pattern.

The first pad bird pulled out of the chocks, swinging its tail around to face us. The exhaust blast was warm even at a distance of fifty yards, and the billowing dust was overwhelming. It's amazing that the equipment stayed together as well as it did, considering the conditions under which it was forced to perform. What would Messrs. McDonnell and Douglas think of this?

Choking from the dust, I yelled out to the world in general, "Damn, I'll be glad when the revetments are finished and we can clean up the flight line." And then to Olsen, "Let's duck through here and talk to McDivitt."

McDivitt was the MAG tech rep attached to our squadron and the person most qualified to decide what was to happen to Thirteen. In the group, there were a dozen civilian tech reps providing us with technical assistance in various systems. There were people from Westinghouse (radar and fire control), Sanders (ECM), GE (engines), Hughes (gun pod), Raytheon (missiles), and half a dozen more. Without them, we would have been in deep trouble.

"No, I can't tell you about Thirteen until I hear back from St. Louis, and that won't be for another hour at least!" That was McDivitt's way. His office was a ten-by-twelve enclosure in the back of a partially collapsed SA TS hangar that served as a none-too-effective shelter for flight line dunnage and supplies. In front of him on his desk there was a Xerox Telecopier and a telephone, which together pointed up some more of the ridiculousness of the situation. Chu Lai—in fact almost all of Vietnam—had a better telephone system than I had back home. McDivitt could direct-dial anywhere in the world from his rinky-dink little office; and with his telecopier, he could even send photographs or documents halfway around the world in real time. (Yet over in the American area there was this communications centre with olive-drab boxes and antennae out the yin-yang that couldn't raise Saigon half the time, much less Washington. Their favourite reply to a transmission request was a five-minute spiel on sunspots, and even. under the best circumstances, their turnaround time was something in the neighbourhood of twenty-four hours.)

McDivitt started to say something else, but he was totally drowned out by the racket coming from the construction battalion crew, which made further conversation impossible. McD as he was known to the troops, waved his hand in disgust. Olsen and I beat a hasty retreat back outside.

There were at the moment two concrete pumps at work on the revets, requiring the services of twice the number of cement trucks as normal. They were lined up single-file blocking the entrance to our hangar, preventing our line people from dragging Five inside. It would never do. "Let's go find Steve," I shouted to Sergeant Olsen through cupped hands, and we took off at a trot. Lieutenant Commander Stevens was the head of the Seabee detachment building the revets, and we found him fuming over one of the concrete pumps that was currently out of commission.

"Hey, Steve, we've got to do something about the trucks in front of the hangar."

"It sure would be easier if we could use the taxiway, John. We could lay some aluminum matting down there between the end of the runway and the main service road that would cut off a mile of driving for our trucks and keep them away from your hangar to boot." We'd been over this one a dozen times before. After a few more grumbles, he signalled his resignation with a shrug. "Yeah, I'll pass the word to the drivers to keep the front of the hangar clear. I'll sure be happy when this job is over. We'd be finished by noon the day after tomorrow if nothing else breaks down, and then we can get started on the air wells and three-phase power."

"Steve, why don't you and your troops take about a month's vacation? That way, when you get back, Top and I will be long gone."

Sergeant Olsen and I poked our heads into the nooks and crannies of our domain for another half-hour, talking to people to get a feeling for how things were going. I have always been amazed by how good the morale was considering the circumstances. It was a real tribute to Sergeant Olsen and his NCOs.

Walking back to the hangar, we were intercepted by Sergeant Ross, who was brandishing his clipboard aloft like atrophy. You could always tell when Sergeant Ross had good news, because he'd barge right into the middle of anything and let fly. "Six is ready for test, Maj. You got anyone in mind?"

"Yep, me. Has anyone seen Lieutenant King around? I'd like him to go." Bud King was our senior flight test RIO and the person I generally took when I was doing the test hop myself.

"He's on his way down from Ops." There were five maintenance test pilots in the squadron, and I tried to see that everyone got about the same number of flights, but it was funny how people were able to anticipate me. I hadn't said anything to anyone about taking the hop myself, but, on that afternoon, felt the need to fly one. Our new Number Six had arrived two days before from Nippi, and the electricians had spent nearly 150 man-hours sorting out the weapons-select Wiring. About the only thing they couldn't handle was when some squadron had made its own field fix without telling anyone or sending along a schematic. A previous F-4 squadron commander, looking to get more flexibility in bomb coverage, had messed with the electrical systems of his squadron's aircraft. All was well and good until after he was long gone and his modified aircraft began showing up from overhaul where an ASC had been incorporated to accomplish exactly what he had set out to do, but in a different manner. Nippi's job was to install the new wiring, which they did in a literal manner, attaching wire A to pin B and "so forth without regard to what changes had been performed. The result was chaos, and it was up to the recipient squadron to sort out the mess that cost close to a hundred man-hours an airplane to get things right.

Then there had been the "great potting compound disaster." Potting compound is a polymer sealant used to isolate and protect from crosscurrent the individual pins of an electrical connector plug. The Phantom contains some

2,000 of these connectors known as cannon plugs, so when the potting compound began oozing and our all-weather airplanes began to go bonkers every time one ran into a cloud, everyone panicked. As it turned out, there was a spore residing in Southeast Asia that had an affinity for the coagulant in the sealant, so the fix was merely a matter of creative chemistry. But this shows what a profound effect seemingly insignificant things can have on the most sophisticated mechanical equipment, raising some doubt as to the level of confidence one ought to have in any machine.

GLOSSARY

Boundary layer: The slow-moving airstream close to the aircraft's surface during flight.

Dunnage: Lighter and less-valuable items of cargo on a vessel or aircraft, stowed between the more valuable items to prevent chafing and damage.

ECM: Electronic countermeasures.

Fragged: Military slang for damaged or destroyed. From *fragmented.*

Flap: Movable surface along the rear edge of the wing, deployed at low speed to increase lift.

Gear: Short for *landing gear* or *undercarriage.*

Hot Pad: A hard-standing area where an aircraft waits in readiness to be scrambled on a mission.

Hummer: Airforce slang for a modified C-47 cargo plane, known in civilian life as a DC-3 or in UK military service as a Dakota.

MAG: Marine Air Group. A support group assisting with maintenance and logistic support.

Napalm: A mixture of fuel and soapsuds in a container, delivered as a bomb. The **mixture explodes and catches fire on impact**.

Revet: Short for *revetment.*

Revetment: A barrier against explosives. A facing (usually of masonry) that supports an embankment. The embankment itself.

Phantom: The McDonnell-Douglas F-4 fighter bomber, mainstay attack aircraft of the US Marine Corps, US Navy, and many Air Forces during the Vietnam era.

RIO: Radar intercept officer. The second crewman on a Phantom, the first being the pilot.

SATS: Short Airfield for Tactical Support. An airfield which can be constructed within 48 hours, with hangars, offices, etc.

Scramble: Slang for a rapid start to a mission.

Seabee: A member of the US Navy's Construction Battalion (CB).

Snakeye: A fin-retarded bomb which allows the aircraft time to escape the bomb blast after a low-level bomb drop.

Stabilator: A combination of stabiliser and elevator, also known as an all-flying tailplane. The aft horizontal control surfaces of an aircraft.

Steel Tiger: An assault mission against enemy supply convoys in the jungle. A Phantom aircraft when on such a mission.

Tech rep: Short for *technical representative.*

TPQ: Ground-controlled bombing. A remotely located radar link sends the aircraft to the target and delivers the weapon.

ASSIGNMENT

Read the case and comment on it as follows:

1. What is a *hangar queen*?
2. Various bad practices, according to the textbooks, are apparent in the way the persons in the case maintained the aircraft. List what they are and why you think they were resorted to.
3. What was the result of these unusual maintenance practices?
4. Comment on the squadron maintenance officer's management style.
5. Comment on the organisational structure that places a flying officer in overall charge of maintenance.
6. Add any other notes and comments on any item of the case that you think is particularly important.

CASE 3.5 A STRANGE CASE OF TPM

Author's Note: This is probably the most bizarre case I have ever encountered.

During lunch at one of the author's training sessions in maintenance, a woman came forward to discuss how TPM was implemented in the company she worked for. It was a fast-moving consumer goods company.

"You know," she said, "since we introduced TPM a few years ago, we don't just clean the plant once. We clean it every month."

"Oh, why is that?" I asked

"I don't know, but I don't like it."

"Why?"

"Because I work in the buying department, and I don't see why all of us ladies from Buying and HR have to wear overalls and spend a morning a month cleaning the factory."

"What do the plant operatives do while you are doing this?"

"They think it's great—they go to the canteen, and we call them when we are finished."

ASSIGNMENT

Comment on this company's practice. What could be in the minds of the management that instituted such a scheme?

CASE 3.6 *CHALLENGER*

Author's Note: One of the most infamous disasters of the last century—seen by millions on television as it happened. Largely, but not entirely, maintenance-related. Hence its inclusion in this chapter. Also significant as an example of the project/program management triangle of schedule, budget, and specification. When pressure comes on the schedule or the budget, the specification is debased.

INTRODUCTION

January 28, 1986, 11:38 a.m., Cape Canaveral

The launch of *Challenger* came at 11:38 a.m. on a bitterly cold morning in January. The 25th space shuttle flight, which President Reagan was preparing to refer to in his State of the Union address, had begun, with a civilian schoolteacher, Crista McAuliffe, on board. President Reagan was preparing to say in his address that space flight for civilians was now easily possible.

For each second of its ascent, ten tons of liquid hydrogen, liquid oxygen, and solid fuel were consumed. The crew were euphoric. Pre-launch scrubs were a thing of the past—*Challenger* was on its way at last.

At 19,000 ft, *Challenger* exceeded Mach 1 and the computers throttled back to 65%, anticipating what the stress engineers called maximum aerodynamic pressure. In eight more minutes, *Challenger* would be in space.

For 14 seconds, the crew swayed and jolted silently in their seats while the shuttle arched through the altitude of maximum wind shear.

Clear of the zone of maximum stress, the "throttle up" command resulted in a violent surge of power, as the *Challenger*'s main engines returned to maximum thrust. As *Challenger* climbed, its computer processed millions of bits of data. A dozen telemetry channels beamed this data down to antennas at Mission Control. Mission Control personnel, sitting in front of computer screens, saw that *Challenger*'s engines had returned normally to full thrust. The ascent was proceeding perfectly.

It was exactly 70 seconds after lift-off, with the shuttle at just under 50,000 ft, that *Challenger* began increasingly violent manoeuvres. For three seconds, there was violent buffeting, and then the digits on the mission control screens abruptly went dead.

Hundreds of press cameras continued to click as hundreds of tons of propellant exploded in a huge steam cloud. *Challenger* tumbled wildly and was ripped apart by g-forces that were beyond its airframe to bear. Both booster rockets continued to careen violently upwards, only half their fuel expended. The right-hand booster burned through *Challenger*'s wing, and the volatile, hypergolic fuel for the manoeuvring engines also ignited, giving an orange glow to the steam cloud. The fatal haemorrhage feared by the Morton Thiokol engineers had occurred.

PRELUDE TO DISASTER

To most knowledgeable people, including the ill-fated astronauts on board *Challenger*, the solid rocket boosters were amongst the safest and the most reliable components of the space shuttle. The solid rocket boosters consisted of four massive cylindrical sections joined by tang and clevis segments, known as field joints. These joints were designed to seal with fail-safe redundancy. To this end, they depended on two synthetic rubber O-rings in each joint. There was, however, a fatal design flaw in these field joints, which had already been diagnosed by the responsible officials at NASA's Marshall Space Flight Centre in Huntsville, Alabama. The engineers at Morton Thiokol, the company that designed and built the boosters in Wasatch, Utah, were also concerned about the design. However, although the flaw had been known about for more than a year, the danger was minimised, and outside investigation quashed. Almost half of the first 23 space shuttle flights had experienced partial failure of the boosters' field joints. In fact, the previous five flights before *Challenger*'s had been launched under a formal constraint stemming from partial joint failures. In each case, the constraint had been waived to allow take-off.

In January 1986, Morton Thiokol was involved in an urgent redesign of the field joints. The Marshall Space Flight Centre managers tried to minimise the problem to maintain their reputation and for fear of losing scarce funding. Thiokol, on the other hand, had long experienced a monopoly on the booster contract and did not wish to see that monopoly attacked. The company stood to lose the contract, worth about a billion dollars, if space shuttle operations were halted long enough to correct the design.

To millions of people, the space shuttle was a product of scientific prowess, free enterprise innovation, and insightful political leadership. That was the myth. The truth was less wholesome. The claim that the space shuttle provided safe, economical, and routing access to space would have been seen to be untrue to any impartial person observing the space programme before the accident.

In fact, the shuttle's safety and reliability were seriously deficient. Over 700 pieces of flight hardware, including the boosters' field joints, were rated "Criticality 1," which meant that the failure of any one element would cause destruction of the vehicle and crew. NASA engineers were later to testify that they did not actually understand the function or nature of several of such key components.

In addition, the operational economy of the shuttle was terrible. Less than half of the launch costs could be recouped from paying customers. Far from providing routine access to space and the profitability of a major airline, as NASA had predicted, flying the shuttle taxed NASA and the contractors to the limit. *Columbia*'s launch prior to *Challenger*'s had seen a record number of delays and launch pad scrubs. There were, in fact, three schedule slips and four launch pad scrubs. The consequent delays put pressure on the launch date of *Challenger* as NASA attempted to maintain its launch schedule.

By the 15th of January, with *Columbia* safely in orbit, the Level 1 Flight Readiness Review, or FRR, for *Challenger* could take place. This was between senior NASA officials in Washington, the launch site at Cape Canaveral, Mission Control in Houston, and the Marshall Space Flight Centre in Huntsville, Alabama. The FRR was the management and engineering conference that should guarantee that *Challenger* was ready for flight.

The FRR procedure had been designed at the start of the shuttle programme as a formal disciplined ritual of certification. In this multistage process, all government officials and aerospace contractors were required to present evidence that the vehicle, its payload, and the ground support equipment were all in a state of readiness for the flight.

The original rationale for these reviews was to create a logical, smooth flow of information from the hardware engineers of the various contractors (level IV) to the project offices (level III), and from there to the centre directors and the National Space Transportation System Offices (level II). Finally, information would flow to NASA's Washington headquarters (level I).

The meeting of January 15 was the culmination of such a process in which *Challenger*'s flight readiness was to be guaranteed. Simply stated, the engines, external fuel tank, solid rocket boosters, satellite, and scientific payloads should all be ready for launch.

This elaborate procedure was deficient, however. The Rogers Commission, investigating the shuttle accident, found that there were at least 15 separate meetings or telephone conversations concerning the effects of low temperatures on the O-ring seals. None of these were communicated to levels I and II. And level II was ultimately responsible for the shuttle's launch.

Even as *Columbia* landed on January 18, *Challenger* was still not ready to fly. The increased flight rates of the previous six months had completely overpowered the inadequate logistic systems.

A T-38 jet trainer was required to bring in vital parts, cannibalised from *Columbia*, for installation on *Challenger*. These parts included a propulsion system temperature sensor, an important nosewheel steering box, an air sensor for the crew cabin, and a computer (one of five needed for the flight). However, even these actions were not sufficient to bring *Challenger* to launch readiness, and the initial launch date of January 23 had to be delayed.

As *Challenger* stood on its launch day, its wing flaps had been removed from *Atlantis*. Its orbital manoeuvring system pods had come from *Columbia*, which had in turn borrowed them from *Discovery*. Finally, *Challenger* had received three vital main engine heat shields and several avionics boxes direct from *Discovery*.

Hence, while NASA's press releases talked of a four-orbiter fleet, no mention was made of their operational readiness. As was later testified at the Presidential Commission investigating the accident, had the *Challenger* accident not occurred in January, shuttle missions would have had to stop by April due to shortages of spare parts.

DESIGN OF THE SHUTTLE

After the orbiter *Columbia* landed on the 4th of July 1982, having completed its fourth and final test flight, President Reagan announced that the space shuttle was now ready for operational tasks. NASA announced a very optimistic schedule of 12 flights for 1984, 14 in 1985, and 17 each in 1986 and 1987. The actual flights achieved were only 5 in 1984 and 8 in 1985.

As a result, NASA came under heavy criticism—the "routine" flights that the press had been inspecting had failed to materialise. The cost of a launch was between $150 million and $200 million, of which only $71 million could be recovered in commercial fees. America's Congress was becoming disenchanted with NASA, and its budget was consequently cut. The message was clear: NASA's shuttle had to begin paying its way. To accomplish that goal, the shuttle had to start delivering the routine cost-effectiveness of dependably regular operations.

However, questions on the fundamental reliability of the shuttle were increasing. Contrary to the popular myth of the shuttle design being the innovative excellence of America's best scientists and engineers, the shuttle was a compromise shaped more by economic considerations and hidden political manoeuvring than technical considerations. A $5 billion development cost and a payback time of ten years had been the NASA promise to Congress. Initial estimates of $14 billion were reduced to ensure at least some funding.

NASA's design problems had been considerable. The shuttle had to perform both as an aircraft and as a spacecraft; it had to stand up to vast extremes of transonic atmospheric flight as well as the harsh orbital environment and had to prove robust enough for routine operations with short turnaround times between missions.

Faced with payload weight requirements, normally aspirated engines had to be eliminated to save weight. Therefore, the aircraft would have to glide in to land. The vehicle was a heavy, fast, low-lift glider that had to descend from orbit with absolute precision and land on its first and only approach on every flight. No margin of error existed in the case of failure or orbital emergencies requiring unplanned re-entry. The original design had provided for jet engines to be used for re-entry and landing. The safety implications of leaving them out were glossed over by NASA.

The payload weight requirement also forced the designers to remove the launch abort escape system from the original design. All previous manned American spacecraft had carried crew escape capsules for extraction during the critically dangerous early moments of the mission.

The next serious design compromise arose in connection with the booster rockets. To achieve the rapid turnaround requirements of a commercial space freighter, NASA became committed to strap-on, cost-effective solid rocket boosters. Liquid rocket boosters would have required intricate plumbing, which would require extensive renovation when recovered from the sea. The use of solid fuel boosters was a departure from proven technology and introduced the added danger that they were not throttleable.

The compromised design carried three cryogenic engines and two orbital manoeuvring engines. The weight of this on-board propulsion system required that the engines had to be made lighter and deliver more thrust than any cryogenic engines built up to that time. The propellant pumps had to operate reliably at 30,000 rpm for up to 50 consecutive flights, a requirement that caused NASA considerable problems.

The fragility of the final design was evident to many inside and outside of government, but President Nixon gave the shuttle his unqualified support because he saw it as a programme for generating employment and hence support in the Republican Sun Belt, where most aerospace contractors were located.

Prime contracts were awarded to Rockwell International for the shuttle and main engines and to Morton Thiokol for the solid rocket boosters. These were, to some extent, at least, political appointments. It is ironic to note that a competitor's design included crew escape and a burn-through detection device which would have triggered booster rocket trust termination. This may have saved the shuttle and crew. NASA was not prepared to trade the increased weight for increased safety.

The shuttle design that emerged from this political battle pleased nobody. As Professor Roland Thomas said, "Instead of a horse, NASA got a camel." This statement is perhaps excessively critical; however, NASA certainly did not get the Rolls-Royce that it wanted.

The combination of the politicking process to win funding and NASA's commitment to reliable reusable flights had, by 1985, forced the agency into a corner where it was battling to maintain its schedule. This was to have a direct impact on the decision to launch on January 28, 1986.

THE EFFECT ON EMPLOYEES

By May 1985, NASA was in an extremely difficult position. In order to keep the confidence of commercial and military customers, as well as the civil government, NASA would have to fly the shuttle with an increasingly ambitious launch schedule. This created conditions threatening reliability and launch safety. NASA nevertheless had developed the credo "Fly out that manifest," and a key to maintaining that manifest was fast turnaround times. Shift workers, engineers, and project managers were forced to work increasingly longer hours. The effect was inevitable—fatigue-produced errors in judgement which led to accidents and to an eventual breakdown of safety standards.

The thousands of systems and subsystems on the orbiter, the external fuel tank, and the solid rocket boosters had to achieve a near-perfect level of reliability under extreme conditions. NASA's Flight Critical Rating system identified 700 items as Criticality 1, which meant, if such a component failed, the result was "catastrophic loss of shuttle and crew." For these 700 items, there were no backup systems, so a stringent system of maintenance and safety accountability was set up, involving voluminous documentation.

This system functioned well with the easy flight schedule prior to 1984. But as the schedule sped up, team leaders very soon found themselves in a paper chase to expedite detailed work certification documents that were a key part of the maintenance safety system. Replacing a thermal tile, for example, may have taken a few hours, but acquiring the necessary inspection stamps could take days.

In January 1986, there were a total of five launch pad scrubs and only two launches for the *Columbia* and *Challenger* missions. Three of these scrubs occurred on weekends. No important shuttle technician or engineer at Kennedy had a day off in the 27 days preceding *Challenger*'s launch. Serious mistakes were occurring, symptomatic of the effect the work schedule was having on the employees at the Kennedy Centre. For example, 18,000 lbs of liquid oxygen were inadvertently drained off four minutes before launch. This mistake occurred in the 11th hour of a 12-hour shift.

JANUARY 27–28: RECOMMENDATION TO LAUNCH

After the initial launch date of January 23 had been delayed due to *Challenger* not being fully operational, the date of January 27 was also scrubbed due to a fault on a lock-and-latch apparatus for the access hatch. The thick circular hatch was one of two regularly used openings in the pressurised crew compartment (the other being the airlock to the payload bay). The hatch was covered with thermal tiles and provided both pressure integrity in the vacuum of space and thermal protection on re-entry. To verify that the hatch was sealed, the technicians relied on two microswitch indicators.

The closeout crew, following the Operational Maintenance Manual (OM1), found that only one microswitch was reading correctly. Inspection showed that both latches were apparently correctly latched, but as a precaution, the technicians were ordered to open and close the hatch so that astronaut Ron McNain could verify that the locking pins were falling into place. This was duly completed and verified, and the closeout crew continued with their checklist. A problem then developed in removing the hatch locking handle, due to a captive fastener bolt having stripped. This caused this launch to be scrubbed.

The consequence was that the next day, Tuesday, was the last day available to launch if another mission on the 6th of March was not to be delayed. This March mission was critical—if successful, it would enable the USA to beat the USSR in securing coverage of Halley's Comet, then rapidly approaching Earth.

A second consequence of the delay on Monday was, the launch, if taking place on Tuesday, would be in much colder weather. Predicted temperatures for Tuesday were lower than the design temperature for launch. However, the launch managers decided to proceed with the countdown in the hope that the temperatures would climb sufficiently before noon. In addition, the operations support personnel were to investigate in detail any possible launch constraints that the cold weather may trigger. For the Morton Thiokol engineers who had designed the rocket boosters and who had intimate knowledge of the problematic field joints, this was an opportunity to impress on the managers how serious the situation really was.

To understand why the engineers felt that the question of the field joints was so important, it is necessary to understand the background to the solid rocket boosters and their construction and maintenance.

The field joints of the solid rocket booster had performed anything but flawlessly and had been the cause of mounting anxiety since the beginning of the programme. There was a long history of potential catastrophic defects. As part of the budgetary constraints and consequent design compromises imposed during the shuttle's development, the boosters were designed to be reusable and built in sections small enough to be transported from the plant in Utah to the launch site in Florida.

The rockets, when recovered from the sea, were dismantled and transported back to Utah. The segments were then filled with propellant at Morton Thiokol's plant and transported back to the Kennedy Centre in Florida. Here the segments were assembled vertically in the vehicle assembly building, like sections of drainpipe, fitting into each other. The field joints that joined the segments together consisted of a circular groove in the one segment, called the clevis, which mated with a circular tongue section called the tang, in the other segment; 177 thick steel pins connected the tang to the clevis at intervals around the booster circumference.

The joint was sealed against the pressure developed by the burning propellant by synthetic rubber O-rings, housed in machined grooves in the clevis section. Pressure from gas to O-ring was transmitted via a thin coating of flameproof zinc chromate putty, which was applied to the surfaces during assembly. When the rocket ignited, the pressure would force the putty down towards the O-rings. This would create a pressure pulse between the putty and the O-rings, forcing the O-rings into place.

While this system had been proposed by Morton Thiokol, they stated that their design was based on the seals used in the highly reliable solid rocket booster employed in the Titan missile. There were, however, crucial differences. On the Titan booster, the clevis faced down and the single O-ring was secured in the proper position by a technician before the joint was closed during final assembly. The clevis in the orbiter's rocket booster segments faced upwards and had to be activated by the ignition pressure pulse. This modification to proven technology was to have disastrous consequences. The Titan booster joint was simple and reliable. The joint used in the space shuttle was complex and unreliable. Tests reported in Aviation Week subsequent to the blast indicate that water collected in the upward-facing clevis turned into ice on launch day. This contributed to the delayed action of the seals.

There were other fundamental flaws in the joint design. A 1959 test of the booster casings, which used pressurised water to simulate the pressure of the booster combustion process, produced surprising results. Instead of the tang and clevis joint being forced more tightly closed by the ignition force, the pressure caused the unexpected phenomenon that became known as "joint rotation." This occurred as the metal parts bent away from each other, opening a narrow gap through which hot gases could leak. The steel booster cases ballooned outward during the first fractions of a second after ignition, bending the tang away from the inner clevis lip. This rotation caused a momentary drop in pressure around the O-rings instead of the anticipated increase in pressure. The implications of this discovery were ominous; without the pressure seal, combustion gases could pass through the putty, burn the O-rings, and erode the O-ring grooves. If this erosion was widespread, an actual flame path could develop, destroying the booster and the shuttle. Although concerns were raised, they did not result in a redesign, and in 1980, the Shuttle Verification and Certification Committee recommended that the field joints be accepted for flight.

After the first four missions, NASA introduced a new lightweight booster rocket, with a thinner casing to save weight. Joint rotation was thereby exacerbated. In 1982, the field joints and their O-ring seals were reclassified Criticality 1 from their previous status of Criticality 1R (redundant). No one at this stage called for a halt to the shuttle flights. The engineers felt that they understood the joint rotation phenomenon and that it did not have serious consequences.

Nevertheless, over half of the subsequent shuttle flights experienced O-ring damage and even actual blowby of hot gases. By 1985, it was recognised that a potentially catastrophic situation existed. Analysis of the April 1985 *Challenger* mission to the spacelab showed that hot gas had blown by the primary O-ring and had scorched the secondary O-ring badly. The project manager, Larry Mulloy, was obliged to a formal launch constraint. Over the next six months, Mulloy imposed and then waived five further launch constraints.

The anxiety over the boosters was, however, soon eclipsed by the unpredictable main engines. The July 12 *Challenger* launch was aborted at T–3 seconds with all engines firing. The July 29 launch was also aborted due to a faulty engine temperature sensor.

The climate at the Marshall Space Flight Centre under the autocratic Dr William Lucas was such that "under no circumstances is the Marshall Centre to be the cause of delaying a launch." Marshall's engineers did, however, consider the O-ring problem serious enough to order new booster casing segments that had a "capture feature," a circular lip on the inside of the field joint tang which would prevent rotation of the joint. But it would be a year before this feature could be incorporated. In the meantime, flying with the present design was deemed to be an acceptable risk.

Between the end of August 1985 and the *Challenger* launch five months later, there were five space shuttle missions. Four of these flights experienced O-ring damage; three had actual blowby.

On the 27th of January, the Morton Thiokol engineers amassed their test data about the effects of cold temperature on O-ring performance and therefore on the functioning of the field joints. Engineers McDonald and Boisjoly were convinced that the unusually cold conditions exacerbated the O-ring problem. Cold temperatures made the O-rings less elastic, and they took longer to spring into place. As a result, a teleconference was arranged between all the relevant parties, at the launch site in Cape Canaveral in Florida, at the Marshall Space Flight Centre in Huntsville, Alabama, and at the Morton Thiokol plant in Utah. The Morton Thiokol engineers recommended that the launch be delayed. Their database only went down to 53°F, which was the estimated temperature for flight 51-C the previous January, in which the hitherto most serious O-ring degradation had taken place. This information proved difficult to relay on the telephone, and a further teleconference was scheduled for the evening, while important data was telefaxed to NASA to aid their officials in understanding Morton Thiokol's point of view. NASA argued that development boosters fired in the horizontal position during cold weather did not show blowby. The Morton Thiokol engineers explained that this was because the putty between the joints had been carefully packed before ignition to eliminate blowholes. The only valid test data was therefore that of Flight 51-C at 53°F. As the predicted temperature for the morning launch was outside the database, the engineers at Morton Thiokol recommended delaying the launch. NASA's reaction was to dispute this recommendation, as it was based on only one data point.

The NASA project manager's response to the engineers' recommendations was, "When do you expect me to launch—next April?" After four hours in the teleconference, the Morton Thiokol engineers withdrew their objection. The concern over the O-rings was not relayed to the launch management at "the Cape."

THE DAY OF LAUNCH, JANUARY 28, 1986

As feared, the temperatures on the 28th of January were very low, and the launch pad inspection crew found ice buildup on the structure surrounding the shuttle. This buildup was of such a magnitude that it posed a danger to the orbiter. In the period of intense vibration after ignition, the shower of dislodged ice could be blown onto the shuttle, damaging the protective tiles. The concerns of the inspection team were not shared by the launch directors, however.

Further problems arose on the day of launch, only discovered subsequent to the accident. An operator controlling the delicate hydrogen disconnect valve, which separates the main engine fuel system from the external delivery hose, or umbilical, issued a computer command to open the valve while a large pressure differential existed on either side of the valve. For 20 seconds, the valve was held closed against this pressure before it slammed open. Had this valve been damaged, the hydrogen tank could have been less than full at take-off. This would have led to premature main engine shutdown after launch, with subsequent loss of the vehicle.

This accident was not reported.

Engineers measured the main nose cone temperature. This was the temperature point which was used to determine whether a launch should proceed or be aborted. The coldest temperature recorded was 24°F at 7:00 a.m. This was below the limit, and a formal launch constraint should have been issued. The solution was to readjust the acceptable temperature limit down to 10°F. The temperature measured at the right rocket booster rear section, where the initial failure occurred, was later confirmed to be 19°F. This did not trigger any action as there was nothing in the Launch Committee's criteria about actual booster temperatures.

A MAJOR MALFUNCTION

At 11:25 a.m., the launch decision was taken and the final nine-minute countdown began.

At the moment of ignition, large chunks of ice broke free. Some chunks struck the left-hand booster, but no damage due to this cause was later found.

The primary O-ring was too cold to seat within the 300 milliseconds required. The putty did not serve its purpose either. Subsequent investigation of the recovered booster showed that combustion gases had vaporised both O-rings over 70 degrees of arc.

800 milliseconds after ignition, *Challenger*'s computers fired the pyrotechnic bolts that held the vehicle to the launch tower. The shuttle started its climb into space. Cameras focused on the right side of the craft showed a series of ominous puffs of black smoke emanating from the right-hand booster's lowest field joint. These appeared at a frequency of three puffs per second, closely matching the booster's natural vibration frequency. This puffing action

continued for 12 seconds, after which time glassy aluminium oxides from the combustion process sealed the joint, temporarily replacing the original O-ring.

This fortuitous event was not to last, however. After 59 seconds of flight, the most violent wind shear ever encountered on a space shuttle mission caused shattering of the oxides. A plume of bright orange flame erupted from the field joint. Photography shows this plume growing for about 15 seconds until it was several feet wide and about 40 feet long. Chamber pressure from the booster indicated that it was 4% below the usual level, indicative of a massive leak. At 71 seconds, the 100,000 lb reduction in thrust was beginning to affect the vehicle's flight path. The control system noticed this and began to compensate, causing the vehicle to weave from side to side.

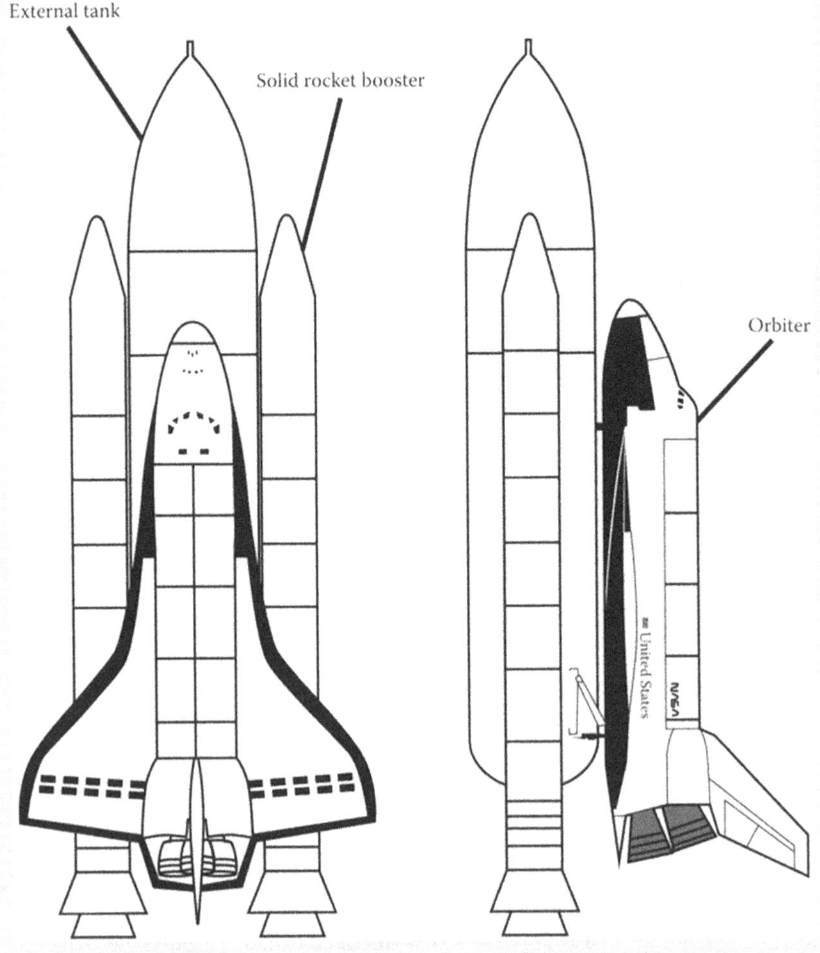

FIGURE 3.4 Space shuttle.

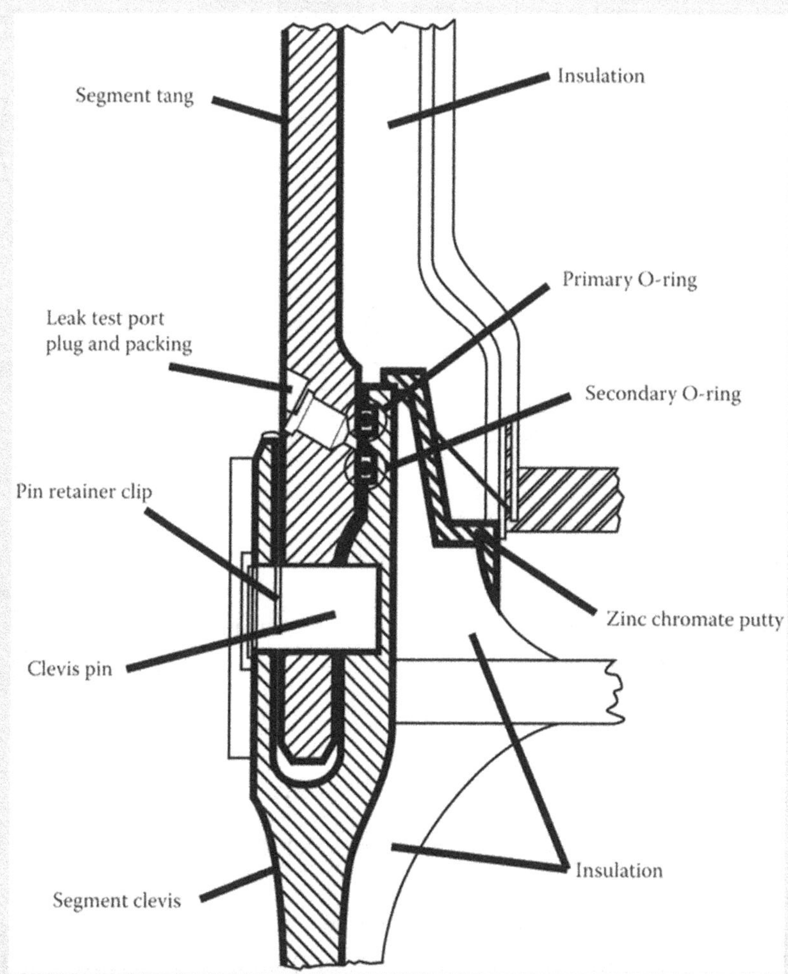

FIGURE 3.5 Rocket motor joint in cross section.

The fire plume grew bigger and was deflected downwards towards the hydrogen fuel tank. The flame acted as a blowtorch, burning through the tank. The hydrogen ignited, and the tank collapsed. The struts holding the rocket to the lower part of the tank broke, allowing the rocket to pivot into the liquid oxygen tank and break it. The oxygen and hydrogen now combined to make the steam cloud seen in the camera shots of the disaster. The orbiter was ripped apart.

The space shuttle is shown in Figure 3.4, the troublesome joint in Figure 3.5, and the behaviour of the joint under load in Figure 3.6.

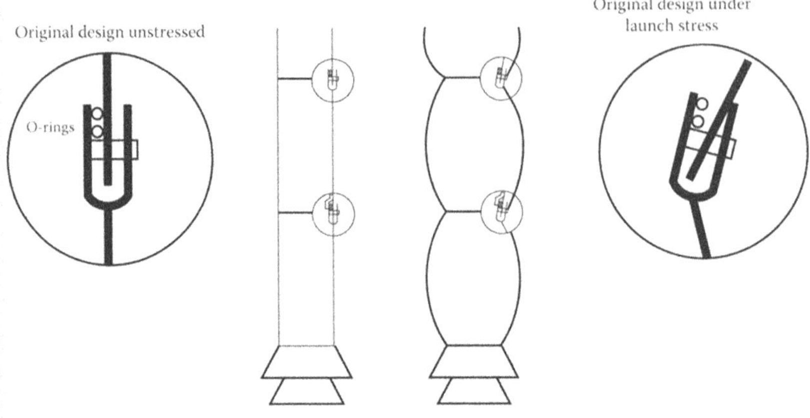

FIGURE 3.6 Field joint rotation.

ASSIGNMENT

Prepare a one-page report summarising the disaster in your own words and give recommendations as to how to prevent such accidents in the future.

CASE 3.7 ELECTRIC MOTOR MAINTENANCE REGIMES

Author's Note: This case is adapted from *Case Studies in Maintenance and Reliability*, by Narayan, Wardhaugh, and Das, Industrial Press, New York, 2012. Although written from an oil refinery perspective, the generic and ubiquitous nature of electric motors makes the case of universal application.

Jim Wardhaugh is an electrical engineer extensively involved in the implementation of CMMS schemes at oil refineries in the Far East. On being recalled to his head office in Europe, he was tasked to assist locations that were underperforming. He noticed that electric motor repairs were a significant ongoing cost. He decided to put some effort into collecting data on the subject so that he could make informed decisions.

MOTOR FAILURE DATA

A few of his company's bigger locations as well as the Institute of Electrical and Electronic Engineers (IEEE) had been collecting data for many years. This data was captured and collated, as shown in the Tables 3.6 and 3.7. It was seen that there were large differences between the company's numbers

TABLE 3.6
Motor Failure Rate in Failures per Motor-Year

Location	Offshore			Onshore			
Data Source	**IEEE**	**Wardhaugh**		**IEEE**		**Wardhaugh**	
Voltage	HV	HV	LV	HV	LV	HV	LV
Failures per motor-year	0.1	0.1	0.07	0.07	0.08	0.07	0.05

TABLE 3.7
Percentage of Failures per Failed Component

Location	Offshore		Onshore	
Data Source	**IEEE**	**Wardhaugh**	**IEEE**	**Wardhaugh**
Bearings	40	30	50	50
Stator	12	23	23	23
Rotor	12	7	8	4
Other	36	40	19	23

and those of the IEEE. They believed that the IEEE's numbers were too conservative and that better results could be achieved at a well-managed site.

The information shows that motors are generally very reliable pieces of equipment. There are few electrical failures, and most failures are bearing-related. The most significant underlying causes of failure are:

- Defective components
- Poor installation
- Poor maintenance
- Poor lubrication
- Water ingress

ELECTRICAL FAILURES

Scrutiny of the data showed three prime causes of electrical failure:

- Catastrophic failure due to bearing collapse, leading to rotor rubbing on stator
- Water ingress due to cooler leaks or cleaning with high-pressure water hoses
- Breakage of connections (due often to inadequate bracing)

There did seem to be some deterioration of motors with age. This was most apparent in large motors with high starting frequency. The larger the machine, the higher the stress level. Manufacturers had algorithms which could predict the end of useful life with reasonable accuracy. For this they needed service and operational data, such as frequency of starts. Such information is valuable with older machines in critical services.

BEARING FAILURES

Scrutiny of the data showed four main causes of premature bearing failure:

- Wrong bearing being installed
- Poor installation causing initial damage
- Poor lubrication
- Poor alignment between driver and driven machine

PERFORMANCE OF SEVEN MOTORS AT DIFFERENT LOCATIONS

Seven of the company's sites in six different countries were studied. The results are shown in Table 3.8. Each was a company plant, built to corporate standards, with most rotating equipment having an installed spare. However, there were a variety of maintenance strategies in place. Not all sites used the same strategy. At all sites, there was a mixture of time-based, condition-based, and run-to-failure maintenance, but in different proportions.

Considering the previous table, it was seen that:

- The large percentage of time-based overhauls of motors at Site 7 did not reduce breakdowns significantly.
- Site 6 seemed more effective than Site 7, but perhaps still not cost-effective.
- Site 5 had several breakdowns even though several motors were "caught in time" by condition monitoring. Therefore, the condition

TABLE 3.8
Motors Repaired Annually as a Percentage of All Motors on the Site

	Site 1	Site 2	Site 3	Site 4	Site 5	Site 6	Site 7
Time-based	1	1	1	3	2	10	14
Condition -based	2	4	5	5	5	3	4
Run-to-failure	4	3	3	3	5	2	3
Total failures per site	7	8	9	11	12	15	21

monitoring was not very effective at predicting failures. This site also had an extreme "blame culture."

- Sites 4, 3, and 2 seemed best at using condition monitoring.
- Site 1 had minimised repair efforts by using run-to-failure as the default strategy.
  - The small amount of time-based maintenance was for a very few "unspared" furnace fans which were overhauled when the plant was shut down every four or five years.
  - This site had a fairly sceptical view of the merits of condition monitoring. They would keep motors running until imminent failure was very apparent.
  - What they had also found was that running motors of less than about 20 kW to failure did not result in significant additional consequential damage and cost compared to any pre-emptive action.
- In summary, then, the proportion of breakdowns is fairly constant whether one does condition monitoring, time-based maintenance, or simply runs motors to failure.

SUMMARISED FINDINGS

- Electric motors are basically very reliable.
- Smaller motors, that is, 20 kW or less, make up the bulk of the population and can be run-to-failure without significant risk of consequential damage to the shaft and the windings.
- In many locations, there is a high level of redundancy, so the consequences of failure are low.
- Condition monitoring has to be inexpensive, or else, it is not an economical strategy. The following conclusions are drawn:
  - Vibration monitoring could not be justified for most motors.
  - Ultrasound was useful in cases where there are closely packed fractional HP motors.
  - Winding monitoring could not be justified for most motors, but the technique is improving and may become more viable in future.
- Bearings do wear out, but long life is a function of a few simple things, viz:
  - Correct bearing selection (sometimes different from the OEM's original supply).
  - Correct installation. Bearing heaters are a big help here.
  - Correct type and quantity of lubricant.
  - Correct alignment.
- Windings do not exhibit significant wearout unless:
  - They are started very frequently.
  - They are large and highly stressed.

- The winding connections are inadequately braced, leading to movement and breakage.

A MAINTENANCE STRATEGY FOR REFINERY MOTORS

A number of possible strategies were evaluated. It was concluded that for very large motors:

- A proprietary monitoring installation that continually monitors vibration and axial displacement should be the norm.
- Information should be centrally monitored.
- Alarm and trip information should be set after agreement with the manufacturers.
- Regular annual inspections must be performed using a borescope.

TABLE 3.9
Periodic Inspection and Testing of Motors

Type of Inspection	Description	Frequency
External, visual, motor running	• No visible unauthorised modifications. • Bolts, cable entries, gaskets, earths, cables OK. • Clean. • No unusual vibration, sound, or heat. • Circuit identification OK.	Every four years.
Internal (need to shut down)	• Tightness of electrical connections. • Test the operation of all standard industrial protection devices. • Do vibration checks before shutting down.	• Six to eight years or at planned shutdown.

TABLE 3.10
Monitoring of Motors while in Operation

Category of Motor	Techniques Suggested
LV Motor—noncritical	No routine monitoring unless in association with driven unit. Monitor vibration before plant shutdowns.
LV Motor—critical	Vibration monitoring.
HV Motor < 1,000 kW—noncritical	Vibration monitoring and bearing temperature monitoring.
HV Motor < 1,000 kW—critical	Vibration monitoring, bearing temperature monitoring, and current spectrum.
HV Motor > 1,000 kW—critical	Vibration monitoring, bearing temperature monitoring, and current spectrum.
HV Motor > 1,000 kW—critical	Vibration monitoring, bearing temperature monitoring, and current spectrum.

For all other industrial motors, Tables 3.9 and 3.10 offer acceptable maintenance strategies for periodic inspection and testing and for condition monitoring while in operation.

If ultrasound and thermographic tools are available, they can be useful in the case of a large number of small motors, closely packed.

LESSONS LEARNED AND PRINCIPLES DEMONSTRATED

Assignment

Prepare a bullet list of lessons learned and principles demonstrated in this case. Your answer should cover about ten points.

CASE 3.8 THE SCRAPER WINCH CLUTCH

Author's Note: Scraper winches are discussed in Chapter 2 of this text. For some makes, the design has not changed in years. Several still use the old-fashioned technology of external contracting brakes or clutches. This case shows the need for mechanic's skills in maintenance—using unskilled labour to repair items of equipment has its consequences.

Mark Francis, while a student studying for his mechanical engineering degree, was employed during the university vacation at a mining company near his hometown. He commuted to the mine premises each morning on his motorcycle. The mine's chief mechanical engineer, Dr Gordon Grey, gave two students, Mark and another student, the task of investigating scraper winch unreliability.

The main problem was short clutch lining life. Dr Grey stated that operating conditions were not good. He had personally seen an operator pour sand into a clutch to get it to grip better, for example.

The two students went their separate ways, visiting various underground installations and observing scraper winch operation.

Mark also visited the maintenance depot to see how the winches were maintained. He observed on elderly man tasked with lining replacement. "Well, young feller, I recently came out of retirement when offered the chance of this job. I used to work in the stores here. Now I repair clutch bands all day—not very interesting, but the pay is good, considering."

Mark watched the man cut linings to length, clamp them at three or four points to the clutch band, and then drill the holes through from the back of the band. He then riveted the linings in place.

That night, Mark explained to his father what he had seen. His dad, a mechanic himself with a local construction company, said, "It is no wonder that they experience so many clutch failures. When I worked on the mines myself, I had to reline clutches on similar machinery. As you have described it to me, they just lay the lining flat in the band and rivet it. That way, all the friction gets transmitted through the rivets, which stretch, and the lining tears off. You must spring the lining into place by riveting the ends first and creating a bow in the middle, which you push down until it springs into place against the band. Now there is friction between the lining and the band which holds it in place when the clutch is operated. That's the way we always did it. Band brake linings have been fitted this way since the days of the Model T. That's what an apprenticeship gives you, son—skills for life." Mark saw that clearly his father had the intelligence to have benefitted from further education. Sadly, because of his family's economic circumstances, he had had to leave school at the age of 16 and take up a trade.

The attached figure shows how the linings should be fitted (Figure 3.7).

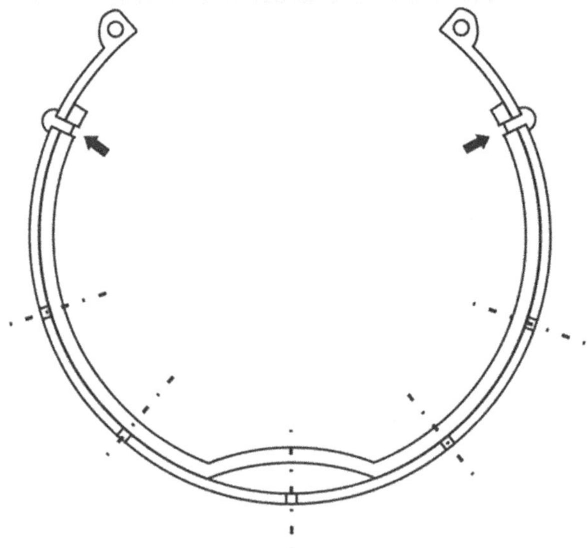

To reline external band brakes

1. Cut lining to length
2. Rivet to the band at the two ends as shown. Lining must be slightly too long, so that it buckles at the bottom as shown
3. Push the buckle into place against the band
4. Drill the rivet hole from the outside of the band through the lining
5. Rivet the band in place
6. Drill the other rivet hole from outside the band and rivet in place

The lining is now secured by friction against the band, not just at the rivets but along its entire length.

FIGURE 3.7 Relining clutch bands.

Mark included this information in his report to Dr Grey, detailing the number of winches inspected, etc. Dr Grey commended him for his effort. When the other student was asked for his report, he simply said that he did not think a report was necessary—his answer was simply that the rivets were too ductile.

ASSIGNMENT

1. Comment on the approaches of these two students to the problem.
2. Do you agree with Mark's father that there is a need for skill in maintenance, or is unskilled labour sufficient in this age, anyway (i.e. the twenty-first century)?

NOTE

1. Readers might also note a characteristic of these cases in this book—sometimes the names of the persons reflect the industry they work in. See how many car company names you can spot!

4 Condition Monitoring

For want of a nail the shoe was lost;
For want of a shoe the horse was lost;
For want of a horse the battle was lost;
For the failure of battle the kingdom was lost—
All for the want of a horse-shoe nail.

Ancient Proverb

THE FOUR KINDS OF MAINTENANCE

As has been mentioned previously, there are four kinds of maintenance:

1. Run-to-failure. This is also known as reactive maintenance.
2. Scheduled maintenance. Also known as planned maintenance, predictive maintenance, or time-based maintenance, which uses past history and manufacturers' recommendations to determine when an item should be shut down for maintenance. This is to pre-empt in-service failure.
3. Condition-based maintenance. Also known as proactive maintenance, which uses proxies for failure, such as vibration levels, to warn of impending failure and so allow the plant to be shut down when convenient to allow for maintenance. As with scheduled maintenance, the aim is to prevent in-service failure, but because of the statistical variation in failure history, condition-based maintenance is likely to prove more accurate and prevent both premature removal from service and failures in service resulting from inadequate historical data. This chapter will concentrate on condition-based maintenance.
4. Maintenance is maintenance elimination. This depends on an industrial engineering study of the efficacy of all the types of maintenance previously mentioned. Even with excellent condition-monitoring facilities, some items will prove uneconomic to run and maintain, with low reliability and low maintainability. Such items should be eliminated and replaced with more reliable systems.

The previous information can be summarised in the form of Table 4.1. It should be noted that none of the four types of maintenance is the only correct type. A well-managed organisation might simultaneously apply all four, as applicable, to different systems and components in the plant. The maintenance optimisation technique reliability-centred maintenance makes allowance for the simultaneous use of all four, for example.

The second type of maintenance, the time-based variety, has been covered earlier in this book. This chapter will now concentrate on the third type, namely, condition-based maintenance.

DOI: 10.1201/9781003326489-4

TABLE 4.1
Different Types of Maintenance

Type of Maintenance	Run-to-Failure	Scheduled Maintenance	Condition-Based Maintenance	Maintenance Elimination
Other names	Breakdown maintenance, reactive maintenance	Planned maintenance, predictive maintenance, time-based maintenance	Proactive maintenance	Design-out maintenance
Notes	Sometimes applicable for minor items where production does not stop with failure of the item and there are no secondary effects.	Depends on good maintenance history and small standard deviations from the MTBF. In Weibull terms, requires high β and high γ.	Applicable where the cost of monitoring is lower than the cost of failure. Moves items from the forced outage regime into the planned outage regime. Eliminates knock-on or secondary damage. Good maintenance history is still a requirement to set up a condition.	An industrial engineering exercise that examines the effectiveness of the other three maintenance regimes and then eliminates items with low reliability and low maintainability.

THE MAJOR TYPES OF CONDITION MONITORING

Condition monitoring has existed since the beginning of the Industrial Revolution at least, in the form of the human being's natural senses. One can, for example, hear if a bearing is starting to become noisy. One can also feel a bearing housing to see if the temperature is starting to climb too high. One can sometimes smell if something is running too hot. And by visual inspection, one may discover many incipient problems. The only human sense that is not used as a condition monitor is probably taste, although in the case of an escaping gas, one might be able to taste it in the air, but normally taste and smell are connected anyway.

In a modern sense, another form of condition monitoring is the control desk in the plant or the dashboard in a car. The petrol gauge and the temperature gauge on a car's dashboard are also condition-monitoring instruments.

With regard to more specific dedicated forms, there are three main technologies:

1. Vibration analysis (including acoustic monitoring)
2. Oil analysis
3. Thermography

These types will now be discussed in detail.

VIBRATION MONITORING

Vibration is inherent in all rotating machinery, as it is impossible to balance any rotating object 100%. Various defects will result in additional vibration, sometimes at the rotational frequency, but also at other frequencies.

THE TIME DOMAIN

There are two vibration traces that are of interest: the trace in the time domain and the trace in the frequency domain. The latter is derived from the former by means of a fast Fourier transform, which deconstructs the complex time trace into its constituent sine waves, each of a unique frequency and amplitude. It is these constituent sine waves that indicate, singly or in some combination, various defects within the machine being analysed. Before the advent of powerful computers, the Fourier transform had to be computed manually, a laborious and time-consuming process. Now it is done, literally, at the touch of a button.

THE FREQUENCY DOMAIN

An example of a deconstructed time trace is given in Figure 4.1. Although the time trace often yields useful information, the frequency spectrum is where most of the information lies. In the hypothetical trace in Figure 4.2, there is the usual 1× peak (i.e. at the machine's rotational frequency), which will always exist because of residual unbalance. In this case, there are two other peaks at 2× and 3×, from some vibrational problem or problems within the machine.

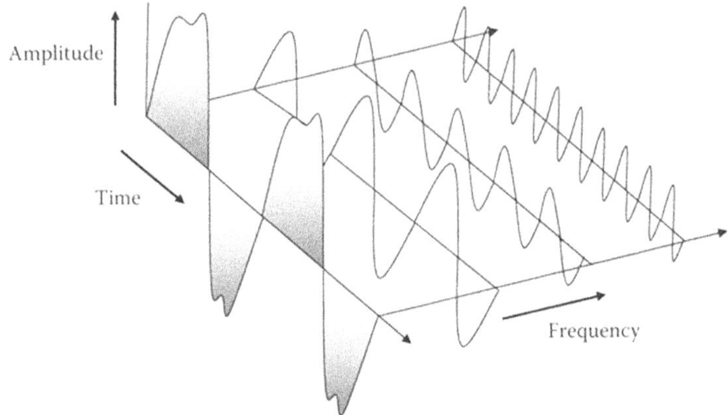

FIGURE 4.1 Time trace (shaded) is made up of the three sine waves of varying amplitude and frequency.

FIGURE 4.2 Frequency spectrum derived from Figure 4.1 (idealised).

The Measurement of Vibration

The transducers that measure vibration are basically of three types:

1. Displacement
2. Velocity
3. Acceleration

Each type of transducer has its own field of application—displacement transducers are used with plain bearings. Velocity and acceleration transducers are used to detect defects in rolling element bearings, gears, and many other machine components. The three types of transducers are shown in Figures 4.3 through 4.5.

Variations in the magnetic field between the transducer probe and the shaft are detected by the transducer and amplified, as shown in Figure 4.3, into the displacement

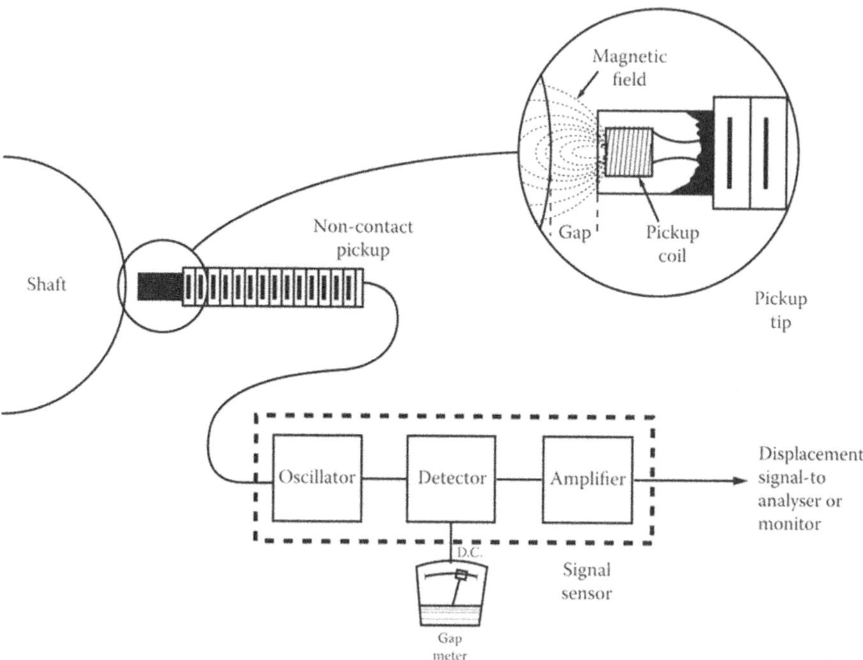

FIGURE 4.3 Displacement transducer. (Source: From Higgins, L. R., and Brautigam, D. P. *Maintenance Engineering Handbook*. McGraw-Hill, 1994.)

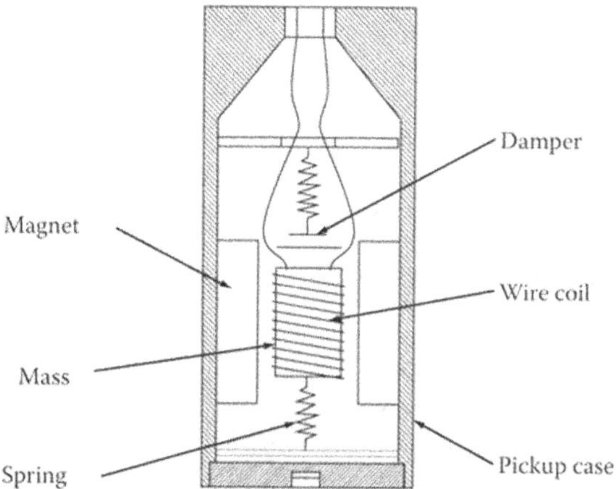

FIGURE 4.4 Velocity transducer.

signal. It should be noted that displacement transducers are intrusive, going through the machine casing. The other two types of transducers are not intrusive.

The voltage output from a velocity transducer is proportional to the velocity of the vibration. It will be seen from Figure 4.4 that the velocity transducer is a

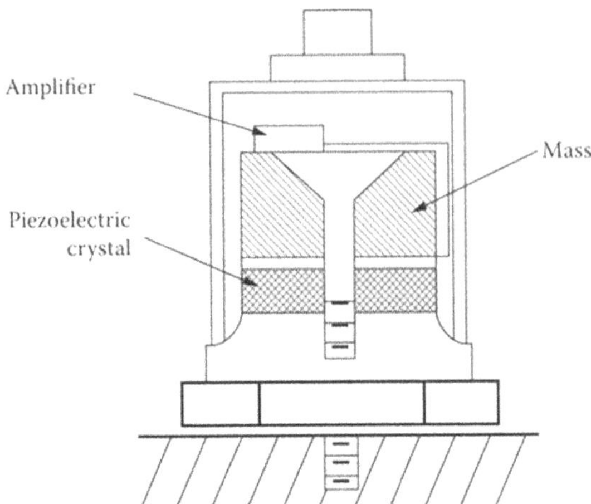

FIGURE 4.5 Accelerometer.

small vibration labo-ratory in itself, with a moving mass, a coil, a spring, and a damper. It is therefore quite a delicate device, subject to going out of calibration if dropped or bumped. The displacement transducer and the accelerometer are more robust. The vibration and acceleration transducers tend to be used in different parts of the vibration frequency spectrum—the accelerometer is used for higher frequencies.

The accelerometer consists of a mass anchored to a piezoelectric crystal. The crystal generates a voltage when deformed by a force, and since force = mass × acceleration, the output voltage is a proxy for acceleration.

With hardwired systems, best practice is to have the transducers screwed to the machine surfaces in the relevant positions, but many installations still use handheld devices where the transducer is in the handheld device and is just pressed against a prepared, cleaned surface and a reading taken at regular intervals.

Positioning of Transducers

As a first and general rule, transducers should be positioned in two radial positions, 90° apart on each external bearing. Space permitting, this translates into transducers in vertical and horizontal positions. A third transducer should be added to measure axial vibration.

Detection of Various Vibration Problems

Table 4.2 gives a high-level summary of the types of vibration problems that can be encountered. It should only be taken as a rough guide—the indications given may vary depending on many factors. Different authorities make different claims, based on their own experience.

TABLE 4.2
Basic Vibration Identification

Cause	Amplitude	Frequency	Phase	Remarks
Resonance				Effects of resonance must be known before checking for other faults.
Unbalance	Proportional to unbalance—radial	1× rpm		Most common cause of vibration—machine must be balanced before any other diagnosis is made.
Misalignment—parallel	Radial		In phase either coupling	
Misalignment—angular	Axial		In phase either side of the coupling	
Misalignment—parallel and angular	Radial and axial	1×, 2×, or 3× rpm	Different phase either side of the coupling	
• Bearing cocked on shaft • Bent shaft	Large axial amplitude—50% or more of radial amplitude	1× rpm		Eliminate misalignment—if vibration still present, check for cocked bearing or bent shaft.
Rolling element bearing problems		Very high		Depends on number of rolling elements in the bearing.
Oil whirl in plain bearings		Less than 50% of rpm—unique signal to left of rpm		Very characteristic signal—must be corrected or bearing and shaft will fail.
Gear problems	Low amplitude	Very high (no. of gear teeth) × rpm		
Mechanical looseness		1× to 2× rpm or 0.5×, 1.5×, and 2.5× rpm		
Mechanical rub		Very low		Sometimes below limit of sensor.
Drive belt problems		1× to 4× rpm		
Electric problems	Signal disappears when power is turned off			
• Aerodynamic forces • Hydraulic forces		1× rpm, or no. of blades × rpm		
Reciprocating forces		1× and higher		Inherent—cannot be eliminated.

Resonance

It is necessary to be able to calculate resonant frequencies of components like shafts. If the resonant frequencies are not known, misinterpretation of the vibration spectra may result.

Unbalance

This is the most common cause of machinery unbalance and can never be completely eradicated. Serious unbalance, however, will show up as a high 1× peak.

Misalignment

Shaft misalignment can be radial, angular, or a combination of the two, as shown in Figure 4.6. Most texts state that radial misalignment will produce a large 2× radial trace, angular misalignment will produce large 1×, 2×, or 3× axial traces, and the combination misalignment will produce both. Furthermore, for the first and second case, the vibrations on each side of the coupling will be in phase; in the third case, they will be out of phase. However, in a reliability web article on the internet, it is shown to be more complicated than that. In an article by Ganeriwala, Patel, and Hartung entitled "The Truth behind Misalignment Vibration Spectra of Rotating Machinery," the following is stated:

- Simple vibration spectral analysis for a given operating condition does not provide a good, reliable tool for detecting machinery misalignment.
- A proper analysis of the shaft misalignment–induced vibration is rather complex. It may require modelling machinery non-linear dynamics.
- Misalignment vibration is a strong function of machine speed and coupling stiffness.
- Softer couplings seem to be more forgiving and tend to produce less vibration than stiffer couplings.
- The rules provided in training courses and wall charts are doubtful at best.
- More work is needed to develop simple rules, if possible, for diagnosing machinery shaft misalignment.

As with all vibration interpretation, the dictum remains important: *Know your plant*. The specific characteristics of any installation (e.g. the type of coupling, etc.) will determine what the traces will look like.

At this juncture, an important statement must be made. Since unbalance and misalignment are the major causes of machine vibration, *it is essential that all machinery must be balanced and aligned before the institution of a vibration monitoring programme. Furthermore, any resonant effects must be understood.*

Rolling Element Bearings

Rolling element bearings are especially amenable to vibration analysis. They are items of precision engineering, which are often abused in storage, installation, or

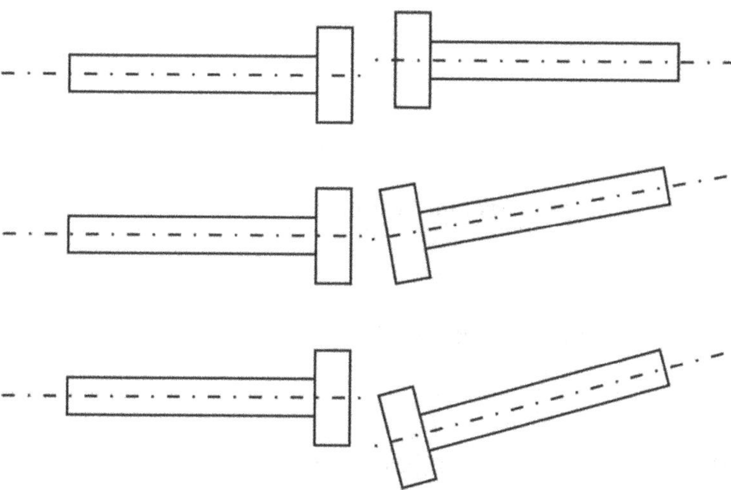

FIGURE 4.6 Misalignment.

operation, and they readily show that they have been abused if their condition is monitored. In fact, they will indicate their distress by two forms of vibration signal—the first acoustic, followed later by conventional accelerometer detection. The acoustic signal is ultrasonic, and special microphones are used to pick it up. Much further information is obtainable from the vibration signal, which can show whether the problem is in the outer race, the inner race, the rolling elements, or the cage.

For this information to be apparent, all the following information is required:

1. Bearing code number
2. Inner diameter
3. Outer diameter
4. Number of rolling elements
5. Speed of the various shafts

Knowing the aforementioned information, we can generate the following information from the following equations:

- Ball spin frequency, BSF
- Fundamental train frequency, FTF
- Ball pass frequency, outer race, BPFO
- Ball pas frequency, inner race, BPFI

The ball spin frequency, BSF:

$$\text{BSF} = 0.5N \times (D/d) \times [1 - (d/D)^2] \tag{4.1}$$

where D is the pitch diameter of the bearing, d is the diameter of the ball, and N is shaft speed in rpm or hertz.

This calculation is important to identify frequencies at which ball damage will manifest.

Cage rotation frequency or fundamental train frequency = FTF:

$$FTF = 0.5N \times [1 - (d/D)] \quad (4.2)$$

where D is the pitch diameter of the bearing, d is the diameter of the ball, and N is shaft speed in rpm or hertz.

This calculation is important to identify frequencies at which cage damage will manifest.

Ball pass frequency, outer race = BPFO:

$$BPFO = 0.5Nn \times [1 - (d/D)], \quad (4.3)$$

where D is the pitch diameter of the bearing, d is the diameter of the ball, N is shaft speed in rpm or hertz, and n is the number of balls.

This calculation is important to identify frequencies at which outer race damage will manifest.

Ball pass frequency, inner race = BPFI:

$$BPFI = 0.5Nn \times [1 + (d/D)], \quad (4.4)$$

where D is the pitch diameter of the bearing, d is the diameter of the ball, N is shaft speed in rpm or hertz, and n is the number of balls.

This calculation is important to identify frequencies at which inner race damage will manifest.

Ultrasonics

The first symptom of distress given by a rolling element bearing cannot be measured by vibration monitoring equipment—it is an ultrasonic scream, far higher in frequency than the range of the human ear. An ultrasonic microphone can be a very good investment. Not only does it give early warning of bearing problems, but it can also be used to trace air and gas leaks, as well as electrical sparking. It can also be used to monitor steam traps, where it is often difficult to determine, by sight or ear, whether they are passing steam or not. There are also specialised probes on the market which are placed against the machinery to detect ultrasonic bearing sounds. Such probes are dedicated to this service and cannot be used for leak detection.

Plain Bearings

Plain bearings can exhibit unique problems compared to rolling element bearings. One such problem is *oil whirl*, leading to *oil whip*. This is shown in Figure 4.7. Figure 4.8 shows the characteristic displacement transducer trace of oil whip, which is a peak at less than 50% of 1× rpm.

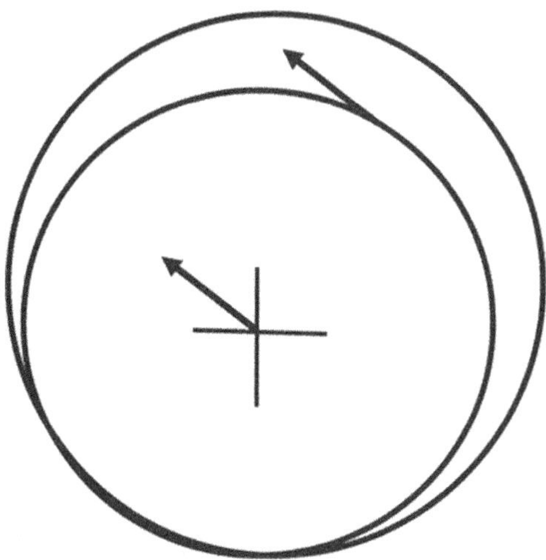

FIGURE 4.7 Oil whirl: shaft precesses as well as rotates.

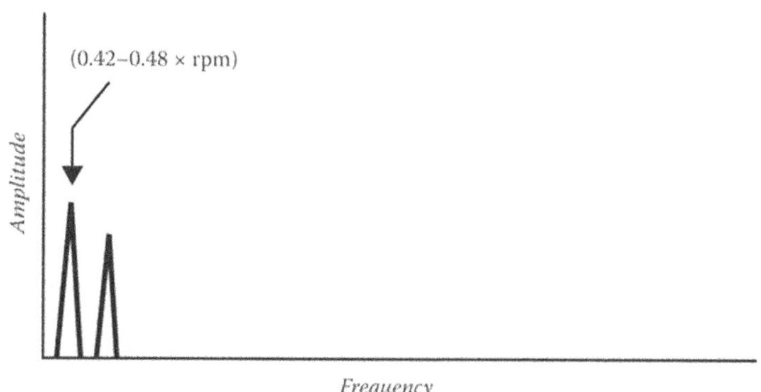

FIGURE 4.8 The oil whirl phenomenon.

Oil whirl is caused by the precession of the shaft in the bearing and often occurs when the bearing is lightly loaded. Oil whirl instability occurs at 0.42–0.48× rpm and is often quite severe. It is considered excessive when the amplitude exceeds 40%–50% of the bearing clearance. If not corrected, it can lead to the trace as shown in Figure 4.9, where later stages of sleeve bearing wear are normally evidenced by the presence of whole series of running speed harmonics (up to 10 or 20). Wiped sleeve bearings often allow high vertical amplitudes compared to horizontal ones. Sleeve bearings with excessive clearance may allow a minor unbalance or misalignment to cause high vibration, which would be much lower if bearing clearances were to specification.

Oil whirl is an oil film–excited vibration where deviations in normal operating conditions cause the oil wedge to push the shaft around within the bearing. A

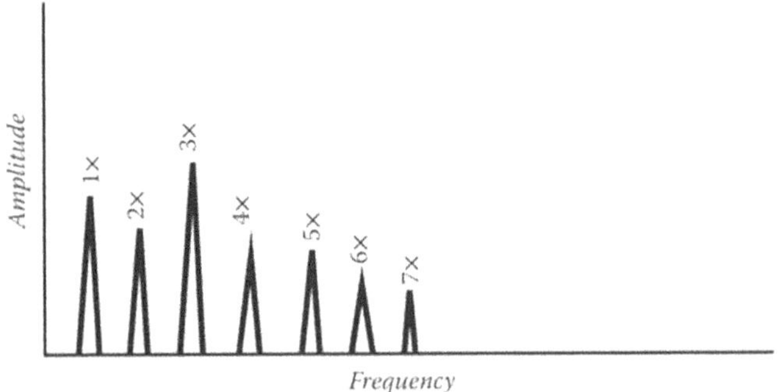

FIGURE 4.9 A "near death" trace for a bearing with a whirling shaft.

destabilising force in the direction of rotation results in whirl. Whirl is inherently unstable since it increases centrifugal forces, which, in turn, increase whirl forces. In most cases, there will be a breakdown of the oil film. Changes in oil viscosity, lube pressure, and external preloads can affect oil whirl, for good or ill.

Gearboxes

Gearboxes require radial monitoring of each bearing cap with two sensors at 90° to each other. If there are single helical gears used, axial measurements are needed as well. This also applies for spiral helical gears. For each shaft, the following should be monitored: unbalance, misalignment, gear meshing frequencies, bearing defects, and mechanical rub.

Sidebands

Gears, in particular, present failure frequencies in which *sidebands* are shown. This phenomenon is analogous to a carrier wave in electronics. There is a basic frequency associated with the carrier wave, and then the sidebands represent lower-amplitude waves at frequencies on each side of the main wave. Any defect in the gear, such as a cracked tooth, will result in sidebands in the spectrum. This is caused by modulation, where *modulation* is defined as any change in the amplitude, frequency, or phase of a wave. Consider Figure 4.10, which gives one example of how sidebands may occur. The eccentricity of the gearwheel will cause the pressure on the pinion to increase and decrease as the pair rotates. This will produce the *amplitude modulation* shown in Figure 4.11, which will appear as sidebands in the frequency spectrum as shown in Figure 4.13. The effect of gear teeth is shown in Figure 4.12, where *frequency modulation* is now added to the amplitude modulation. This results in extra sidebands, as shown in Figure 4.13.

The interpretation of sideband phenomena is not easy, as very often the sidebands are obscured because of very low resolution on either the horizontal (frequency) axis or the vertical (amplitude) axis. It is essential, therefore, to have the capability in the

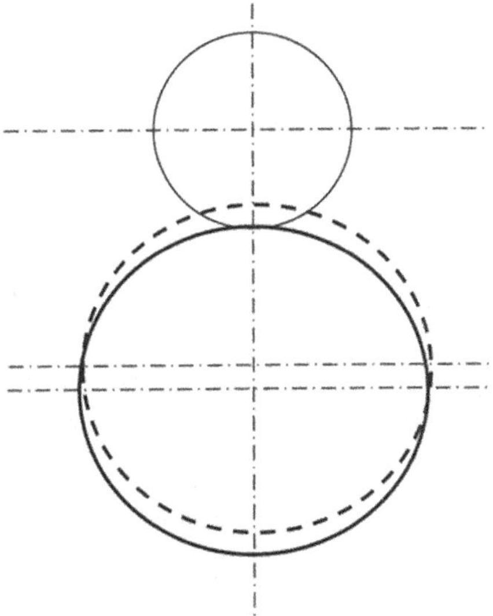

FIGURE 4.10 Eccentric gearwheel (exaggerated).

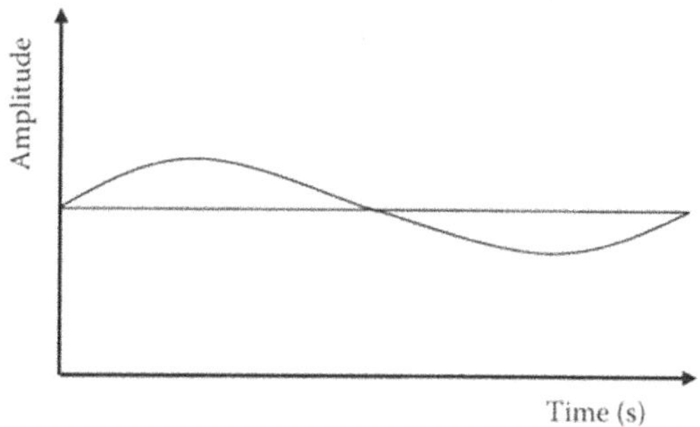

FIGURE 4.11 Amplitude modulated waveform.

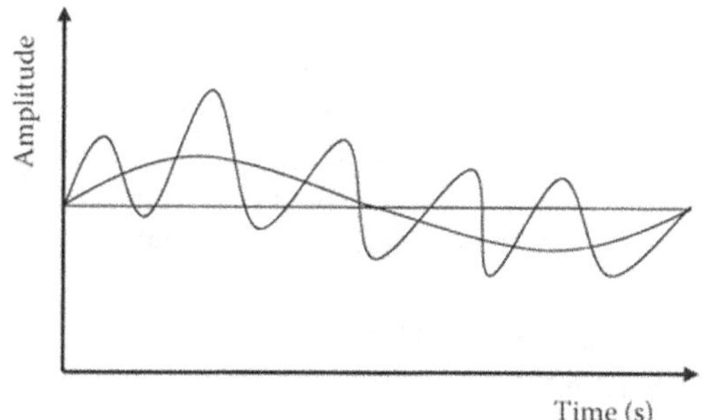

FIGURE 4.12 Amplitude and frequency modulation.

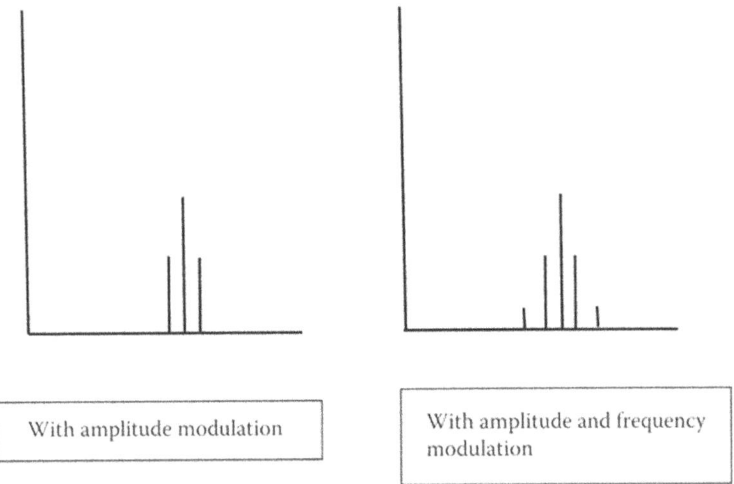

FIGURE 4.13 Sidebands (idealised).

software to change the scales on the axes and to provide "narrowband" windows, where the specific frequencies and amplitudes in a specific part of the spectrum can be studied. Even with these facilities, sideband interpretation is, for experts, very often.

A further practical complication may arise. The numbers of teeth on each gear need to be known. Like bearing details, this information is not always given on the relevant drawings. In such cases, a visit to the maintenance stores may save the day, as components such as gears and bearings may be inspected in the stores. (And then repackaged in the proper way, of course!)

Compressors

Compressors cover the whole gamut of vibration monitoring needs, depending on the type, of which there are many. These include single-stage centrifugal, multi-stage centrifugal, axial, screw type, and reciprocating. Multistage centrifugals are further divided into in-line and bull gear types. Rolling element, sleeve, and tilting pad thrust bearings may be encountered, depending on the machine.

Single-stage centrifugals should be monitored as follows: Vane pass frequency should be a primary indicator of condition. Imbalance, misalignment, bearings, and mechanical rub should also be monitored. Aerodynamic instability or surging should also be catered for as well and automatically corrected.

Multistage in-line centrifugals and axials should be monitored as for single-stage centrifugals, but with vane pass frequency for every stage.

Multistage bull gear centrifugals should be treated as a combination of gearbox and centrifugal compressor. These machines have a large helical gear (the bull gear) driving several smaller gears mounted on impellor pinion shafts. Very high speeds, in the regions of 50,000 rpm, are encountered; hence, the impellors and gears should be monitored closely—things, if they go wrong, will go wrong very quickly. Except in the smallest sizes, these machines use plain bearings and usually have displacement transducers permanently mounted to monitor the pinion shafts. Because the helical gears are used, axial measurements also need to be taken at each bearing. Each shaft requires monitoring as follows: imbalance, misalignment, gear mesh, vane pass, bearings, thrust bearings, surge, and mechanical rub.

Screw compressors use either rolling element or plain bearings, and the relevant transducers need to be installed in either case. Such machines have extremely close axial tolerances that will allow no more than 0.01 mm of movement before the rotors contact. Two radial and one axial measurement are required at each bearing cap. The following vibration parameters should be measured: imbalance, misalignment (unless the motor is flange-mounted, integral with the compressor), rotor meshing, bearings, surge, and mechanical rub.

Reciprocating compressors are now restricted to some specialist high-pressure applications, such as plastic bottle moulding. Even with cylinders mounted at 90° to each other, there will still be unbalanced forces. The resulting vibrations are different to rotating machines, often with a 2× peak being predominant instead of the true crankshaft rpm.

Fans

Fans are either centrifugal or axial, and the centrifugals are either centreline types (one bearing each side of the fan) or overhung. They are normally designed to operate just below their first critical speed. This means that there will be extreme vibration if operated above their running speed. The following should be monitored: imbalance, misalignment, bearing defects, blade pass, and mechanical rub.

OIL ANALYSIS

Lubrication is discussed at length elsewhere in this text. Here, we will only be dealing with the oil of oil analysis as a condition-monitoring technique. It is claimed to be the oldest technique, having been in limited use since the 1940s.

History

In 1946, the Denver and Rio Grande Railroad research laboratory successfully detected diesel engine problems through wear metal analysis of used oils. A key factor in their success was the development of the *spectrograph*, a single instrument that replaced several wet chemical methods for detecting and measuring individual chemical elements, such as iron or copper. This practice was soon accepted and used extensively throughout the railroad industry.

By 1955, oil analysis had matured to the point that the United States Naval Bureau of Weapons began a major research programme to adopt wear metal analysis for use in aircraft component failure prediction. These studies formed the basis for a Joint Oil Analysis Program (JOAP) involving all branches of the US Armed Forces.

The JOAP results proved conclusively that increases in component wear could be confirmed by detecting corresponding increases in the wear metal content of the lubricating oil.

In 1958, Pacific Intermountain Express was the first trucking company to set up an in-house used oil analysis laboratory to control vehicle maintenance costs. This extensive history makes oil analysis the oldest of the proactive maintenance technologies.

In 1960, the first independent laboratory was founded to provide a complete oil analysis diagnostic service to all areas of business and industry.

It is claimed by the oil analysis industry that 50% of bearing failures are lubrication related and that oil analysis can give warning of bearing failures two months before vibration analysis. In practice, it is not a case of "either or." The two techniques should be used in conjunction.

Requirements for Oil Analysis

A good oil analysis package looks at four main criteria to provide a unique way of indicating what is going on inside the equipment:

1. Oil condition (oxidation, viscosity, additives degradation, etc.)
2. Equipment condition: type of wear (normal, abnormal)
3. Contaminants: type and quantity (dirt, water, debris, other lubricants, etc.)
4. Equipment operating conditions (load, temperature, alignment, balance, speed, etc.)

It will be seen that some of the equipment conditions monitored previously are also monitored in a vibration-monitoring programme. This duplication is valuable—if condition is verified by two independent tests, then costly mistakes can sometimes be avoided.

In oil monitoring itself, the importance of using a combination of physical and spectrochemical tests to monitor lubricant and component condition is now universally accepted. Oil analysis test procedures are established and reviewed by such agencies as the International Organization for Standardization (ISO), the American Society for Testing and Materials, and the Society of Automotive Engineers.

In complex or large machinery, the first step is to define sampling points to determine what component or condition is important to monitor and how often to monitor it. Consistency is important as well. Once the sampling point is selected, this should always be the point of sampling for the equipment involved. Samples can be taken from a dedicated sampling point or by the use of a *thief pump*, also known as a *vampire pump* (Figure 4.14).

For gearboxes and other equipment, a thief pump draws oil through the dipstick of certain models of gearboxes or through the fill port (if it can be done safely while the gearbox is operating). A sample can also be obtained through the drain if there is a valve and the piping is flushed. The ideal place to obtain a representative sample is from a gearbox that has a circulating system with a sample tap.

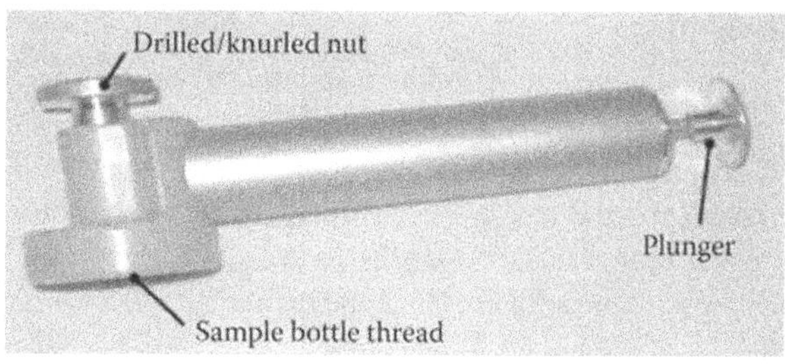

FIGURE 4.14 Sampling pump.

A Basic Test Programme

There are a few key condition-monitoring tests for industrial lubricants:

- Viscosity
- Water content
- Acid number
- Oxidation
- Elemental analysis

Viscosity

Viscosity is a measure of a fluid's resistance to flow. It is the single most important property of a lubricant. The most common units for reporting kinematic viscosities are centistokes (cSt) and Saybolt Universal Seconds (SUS—now dying out internationally), which are reported at 40°C for industrial lubricants. Where a lubricant's viscosity is defined in ISO-VG grades, a change of approximately ±25% from typical levels indicates a need for corrective action. To some, this seems to be too much of a change from specification, but it must be remembered that oil viscosity rapidly degrades in use. If a more stringent requirement were imposed, oil changes would be very frequent.

If the lubricant's viscosity shows an *increase* above the typical value, it may be attributed to high soot or insoluble content, water contamination, contamination with higher-viscosity lubricant, oxidation, nitration, or low load operation. *Corrective action* in this case could be as follows:

- Apply methods such as centrifuging, vacuum dehydration, and filtration to remove particulate or water contamination from the oil, thus extending the oil's life.
- If this is not practicable, change the whole charge.

If the lubricant's viscosity shows a decrease, it may be attributed to water contamination, fuel dilution, dilution with a lighter-viscosity oil, or "shear down" of a viscosity index improver. It is unfortunately not possible to restore viscosity in this case. The solution, therefore, is to change the oil, either partially or completely.

Water Content

Water content is a particular problem for gear oils. A gear oil is only considered unequivocally fit for further use if its water content is less than 0.2 wt.%. Water can enter the gear oil from contaminated oil, a cooling system leak, or bad sealing. The nature of water (fresh or salt) can be determined from the sodium and magnesium levels detected. Water reduces film strengths and reacts with aggressive components in the oil to form acids. These acids degrade the oil and attack the gears and bearings.

Acid Number

The acid number is the amount of acidic substances present in the lubricating oil. A high value implies that the oil is degraded by high acidity. If the oil has a high acid number and remains in service, the equipment elements will be attacked and create corrosion. The corrective action necessary is to change the oil, either partially (if emergencies dictate) or preferably completely.

Oxidation

Oxidation is measured using infrared test methods. This test is run for all samples, with the exception of some synthetic-based oils. Oxidised oil normally increases its viscosity and turns dark in colour. The oxidation process can be attributed to high operating temperatures or extended drain periods. Corrective action can only be to change the oil, either partially or preferably completely.

Element Analysis

Element analysis measures the presence of elements in the oil, in parts per million. This test is run for the purpose of diagnosing wear. Elements are identified and measured on an inductively coupled plasma emission spectrometer. Typically, 12 elements, such as iron, antimony, copper, and so on, are detected, but more elements can be measured, depending on the application. Limit values for each element depend on the original equipment manufacturer and the application. If the values are above the attention limits, the test indicates changes in machinery wear and lubricant contamination. Corrective action will be to:

- Review the tendency.
- Correct any problem, if possible, according to the element(s) present.
- Change the oil, either partially or completely.

If the values are above the alert limits, the test indicates severe changes in the machinery regime wear. Corrective action must then be to:

- Shut down the equipment and inspect it for failure or damage.
- Correct the failure or damage.
- Change the oil completely.

Additional Oil Analysis Tests

In addition to the key tests, there are also optional tests for gear and industrial lubricants. These include particle count, ferrography, and flash point.

Particle Count

The particle count test is a method of monitoring the cleanliness of a system. The size range of particles considered to be most damaging are those between 6 and 14 µm. ISO Standard 4406 provides size ranges for the particle count test. The shape of the particles is also important—some look like machinings from a lathe, others are rounded, and so on.

Ferrography

Ferrography is the microscopic evaluation of a used oil sample, examining particle sizes from 10 µm to more than 1,000 µm. This test should only be conducted by well-trained, experienced personnel. Ferrography yields two results:

The PQ (particle quantifier) index, generated when magnetic and non-magnetic contaminating particles in the sample change the magnetic field in the particle quantifier or induce an electrical magnetic field. The result shows the PQ index without any units.

The second result from a ferrographic investigation is the rotary particle depositor (RPD). Sample fluid is dropped on a rotating glass plate, which has three concentric magnets beneath it. Magnetic particles form rings on the glass plate, while non-magnetic particles move toward the rings, as a result of centrifugal force.

Flash Point

Flash point is the temperature at which oil releases enough vapor at its surface to ignite when an open flame is applied. Oil lubricants with a flash point below 150°C should not be used. A decrease in the flash point indicates fuel dilution or contamination with another substance with a lower flash point. Corrective action could include stopping the source of fuel or other contaminant ingress and thereafter changing the oil.

THERMOGRAPHY

Thermography is the other main form of condition monitoring that we will be discussing. What is required is an infrared camera and a software program covering the following:

- Route description and tracking
- Data collection
- Field analysis
- Report generation
- Data storage

The main use of thermography is in detecting electrical faults. The camera is equipped with a screen that displays a false-colour image of the equipment being monitored. The colour scale can be changed depending on the temperature range being scanned. Bad connections and other sources of unnatural heat are easily found.

The other advantage of the thermographic camera is that the sensing is remote—there is no need to approach close to dangerous equipment when monitoring its condition. Common uses are for the monitoring of breakers, transformers, and high-voltage lines.

Thermography does have applications for mechanical components as well and can be used to detect high temperatures in bearings, drive belts, steam traps, and piping insulation. It can also detect hot spots in refractories, heat exchangers, and building insulation.

The infrastructure needed for thermography is much less than that needed for vibration monitoring or oil monitoring. It is simply a case of buying the camera, installing the software on a laptop or tablet, switching it on, and going for it!

5 Incident Investigation or Root Cause Analysis

I am still convinced that the ability to read a failure is one of the most valuable attributes of an engineer.

Rudd, T, 2000

Tony Rudd (1923–2003) was an English engineer who worked for Rolls-Royce, BRM, and Lotus Cars. Of his many achievements, one was getting the previously maligned, very complex V16 BRM racing car reliable and winning races, through a painstaking process of failure analysis and redesign.

INTRODUCTION

The investigation of incidents is an important task in industry. Serious incidents that can lead to fatalities, injuries, or serious plant damage occur. In the first place, the investigation must determine what actually went wrong and then recommend what must be done to rectify the situation and to ensure that the incident never recurs.

In this chapter, we will discuss various incident investigation methodologies and then study some cases of well-known incidents, namely, Chernobyl, Piper Alpha,

FIGURE 5.1 Tony Rudd. Source: Photograph by Raimund Kommer, photosubmission@wikimedia.org, 1967.

DOI: 10.1201/9781003326489-5

Flixborough, and the Hyatt Regency Hotel. These accidents are all from what the author describes as the Era of the Big Bangs—the 1970s and 1980s of the last century. We seem to have learned from such accidents, as various initiatives have been put in place to prevent the recurrence of similar events.

First, Chernobyl stands alone as the worst industrial accident ever, with the effects of released radiation continuing to this day. It is the author's assertion, as a nuclear-trained mechanical engineer and an amateur historian, that this accident had geopolitical significance, being the last nail in the coffin of the crumbling Soviet Empire.

Piper Alpha, on the other hand, is a study of poor communication, poor maintenance, and the need to keep the paperwork in order. Who would have thought that errors in a permit-to-work system could cause such a disaster?

Flixborough and the Hyatt Regency Hotel are similar in that very simple, very obvious errors led to catastrophic damage. But then, hindsight is an exact science.

This chapter ends with the description of an aircraft accident in the United States. The reader here is required to analyse the accident given information from interviews with the maintenance staff.

SCOPE OF ANY INVESTIGATION

All incidents that result in injuries or fatalities should be investigated. Incidents in which the direct damage is estimated to be greater than a certain amount, say, $10,000, should also be investigated. Lesser incidents may also be investigated at the discretion of the manager of the facility involved. Apart from the incidents mentioned previously that must be investigated, any employee should be able to instigate an investigation if he believes it to be in the company's interest.

DEFINITIONS AND ABBREVIATIONS

Definitions

Root Cause Analysis

The technique or set of techniques that determines the reason for an incident having occurred. The practice is predicated on the belief that problems are best solved by attempting to correct or eliminate root causes, as opposed to merely addressing immediately obvious symptoms. The term is unfortunate in that, to some, it implies that there is only one main root. Some methodologies actually state this as a fact. Others allow for multiple "roots."

Abbreviations and Acronyms

FTA fault tree analysis
IAEA International Atomic Energy Agency
RCA root cause analysis

Root Cause Analysis

A term frequently encountered in conjunction with incident investigation is root cause analysis (RCA). In some people's minds, the two processes are identical. There are problems with this view. The mere term *root cause analysis*, taken at face grammatical value, implies that there is one root cause. For this reason, it is best to use the term *incident investigation* instead.

INCIDENT INVESTIGATION TECHNIQUES

Incident investigations have been performed for many years, using a variety of techniques. The most common of these have been the following:

- The "five whys" technique
- The Kepner-Tregoe technique, or its derivatives
- The ASSET methodology
- Fault trees

Each of these techniques will now be described. This procedure is not prescriptive as to what technique is to be used. It will be found that, in some instances, none of the aforementioned techniques are particularly suitable. However, the author has found that the last technique listed, namely, fault trees, is both easy to understand and yet powerful enough to handle most incidents efficiently. It is therefore recommended. Some investigators, already trained up in one of the other techniques, may prefer to continue using it.

The Five Whys

This technique was originally developed by Sakichi Toyoda and was used within the Toyota Motor Corporation. It is also used in the Six Sigma suite of processes developed by Motorola. An example is as follows.

Problem statement: My car will not start.

1. Why? The battery is dead.
2. Why? The alternator has ceased functioning.
3. Why? The alternator belt is broken.
4. Why? The belt was beyond its service life limit and had not been replaced.
5. Why? I have not been maintaining my car according to the recommended service schedule.

There is no magic in having five "whys." As seen in the previous example, the list could go on:

6. Why? I have not been able to afford the servicing costs.
7. And so on . . .

It is found in practice, however, that "five whys" is usually sufficient.

Criticisms of the technique include the fact that it is not rigorous enough. Different people applying the technique will come up with different causes for the same problem. The "five whys" is a good technique to use at the start of an investigation but is seldom sufficient on its own to solve complex incidents.

Kepner-Tregoe and Its Derivatives

The Kepner-Tregoe (KT) method of problem-solving is long established, having arisen out of initial research the authors did while at the Rand Corporation in the 1950s. It has been taught by KT globally for many years and is positioned as a series of rational thinking processes that can be utilised across a wide range of issues (see the References). The KT processes include situation appraisal, problem analysis, decision analysis, and potential problem/potential opportunity analysis. The KT problem analysis process is used to identify the cause of a defect or deviation and can be applied in almost any industry, including sales, manufacturing, technology, aerospace, and so on. Although the underlying concepts are simple, training in the use of the KT problem analysis approach is required before it can be used with confidence. As with the previously described technique, KT is best described by means of an example. This one is taken directly from *The New Rational Manager* (Kepner and Tregoe, 2006).

The techniques of problem analysis are divided into these activities:

- State the problem.
- Specify the problem.
- Develop possible causes from knowledge and experience.
- Test possible causes against the specification.
- Determine the most probable cause.
- Verify assumptions, observe, experiment or try a fix, and monitor it.

Step 1: Problem Definition

This should be as precise as possible. In this case, the problem concerns leaking oil filters, and the problem definition is, "Filter leaking oil on the floor."

Step 2: The Problem Described in Detail: Identity, Location, Timing, Magnitude

This can also be described as what, where, when, and to what extent, as follows:

- *What* is the malfunction? No. 1 Filter leaking oil.
- *Where* is the malfunction? Northeast corner of filter house.
- *When* did it start? Three days ago, at the start of the shift.
- *To What Extent*? Five gallons per shift.

Step 3: Where Does the Malfunction NOT Occur?

- *What?* No other filters.
- *Where?* Nowhere else.
- *When?* Never before.
- *To What Extent?* None.

Step 4: What Differences Could Have Caused the Malfunction?

- *What?* No. 1 filter has square-cornered gasket.
- *Where?* There is less vibration elsewhere in the plant.
- *When?* Monthly maintenance done three days ago.
- *To What Extent?* Not applicable.

Step 5: Analyse the Differences

- *What?* Square-cornered gasket installed three days ago.
- *Where?* The extra vibration has always been there, so this is not an issue.
- *When?* (See *"What?"* prior.)
- *To What Extent?* Not applicable.

Step 6: Verify Proposed Solution

In this case, this would mean installing a new gasket of the old type to see if the leak stops. We will leave the problem there.

The ASSET Methodology

Background to the Process

This methodology was developed by the International Atomic Energy Agency (IAEA), and a summary of the technique is reproduced here with the permission of the IAEA, which retain the copyright (ASSET Guidelines Revised 1991 Edition, IAEA TECDOC No. 632). One of the best things about the technique is the way that it divides the failure up into three parts, which we will call by the mnemonic the "three Ps"—personnel, procedures, and plant. The technique also specifies what it calls the root cause(s) and the direct cause(s). This is sometimes not easy to follow. An example is given here:

Direct Cause → Root Cause → Occurrence → Event

Table 5.1 can be understood by reading from right to left as follows. Any event is attributed to a failure in one of the three Ps or in more than one of them. Any such failure is in turn attributed to a failure in the configuration management (CM) system, because it did not forestall and repair the latent weakness in one or more of the three Ps. In Table 5.1, therefore, the CM system is the ultimate barrier against incidents. Therefore, there must be a CM programme in place. This programme must

TABLE 5.1
The ASSET Methodology of Incident Investigation

Direct Cause	Root Cause	Occurrence	Event
Existence of a latent weakness in:	Failure of the configuration management system to eliminate the latent weakness in:	Failure in:	Event
Personnel		Personnel	
Procedures	Personnel	Procedures	
Plant	Procedures	Plant	
	Plant		
Due to:	Due to:		
Degradation during operation, or	Poor detection		
Poor commissioning	Poor restoration		

thoroughly and continuously assess the proficiency of personnel, usability of procedures, and operability of equipment.

If CM is not in place, the ASSET methodology does not work. This limits the use of the ASSET methodology because, as far as ASSET is concerned, some aspect of poor CM is always a root cause.

Nevertheless, the ASSET guidelines contain much useful advice in how to conduct an incident investigation. In particular, there are sets of suggested questions that may be incorporated into any form of incident investigation. These question sets cover the identification of latent weaknesses in the "three Ps": personnel, procedures, and plant.

The Sequence of Actions in the ASSET Process

The first stages of the process are similar to other techniques:

- Define the incident.
- Collect information.
- Prepare a logical event sequence to describe what occurred before the incident.
- Analyse the incident according to personnel, procedures, and plant.
- Determine the corrective actions required.

Personnel Errors

The basic characteristics of personnel proficiency according to ASSET are given under three headings:

- Reliability at work
- External influences
- Qualifications to perform the task

These are further divided as follows:

- Reliability at work:
  Vigilance
  Physical and mental fitness
  Understanding one's own abilities
- External influences:
  Communications
  Physical environment
  Safety culture
- Qualification to perform the task:
  Experience
  Training
  Educational background

When conducting an investigation, one should therefore check the following:

Reliability at Work

Vigilance

- Check that appropriate attention was given by the operator to the task and that he or she was not distracted by other activities.
- Verify that the operator was conscious of the importance of the task and employed self-checking practices.
- Check that the individual did have plant safety in mind.
- Verify that the operator's vigilance was not impaired because of very frequent execution of the task.

Physical and Mental Fitness

- Was the individual's ability impaired by any of the following factors?
  - Sickness, injury, or substance abuse
  - Stress as a result of personal problems
  - Stress created by the job in hand, including fear of failure
- Check that the individual's attitude towards his job was correct and that his interest in the task was adequate.
- Was the work schedule of any influence (overtime, shift change, late/early shift, etc.)?
- Was the workload on the individual too high?

Self-Capability and Understanding

Verify that the performance of the task was not affected by the individual:

- Not being conscious of his capabilities or limits.
- Reliance on co-workers and supervision.
- Underestimating the complexity of the task.
- Preparation for the task—was it adequate?
- Overconfidence.
- Complacency.
- Disdain for procedures.

External Factors

Communications Was the individual's performance affected by a breakdown in communications, particularly the following?

- What was his supervisor's understanding of the job?
- Did his supervisor pass on adequate information to the individual concerning the job?
- Were there any other communication difficulties between any other persons associated with the job?
- Was all electronic communication equipment in good order?

- Was all the necessary communication provided (labelling of equipment, instrumentation, paperwork, etc.)?

Physical Environment

- Was the work area cramped or untidy?
- Was the work area crowded or noisy?
- Was there prolonged exposure to high temperature or humidity?
- Was there prolonged exposure to low temperature?
- Was protective clothing or respiratory equipment required?

Safety Culture

- Are management goals clearly established and clearly understood?
- Are policies on safety clearly defined, disseminated, and enforced?
- Does management respond to problems and take account of input from the staff?
- Is there adequate recognition of the resource level required?
- What is the general level of staff morale?

Qualification to Perform the Task

Experience

- How long has the individual done the job?
- Has the individual done the job before?
- How often is the individual required to perform the task?

Training

- Does the basic training that the company provides give a good general knowledge of the plant, systems, and physical phenomena involved? In other words, did the individual know the plant?
- Has the individual received training for the specific task in question?
- Did this training include assessment of the individual's competence to perform the task?

Educational Background

- Is the educational level for the individual's assigned post clearly defined?
- Does the individual conform to this required level?
- *Note:* The records relating to the individual's educational level must be checked.

The reason(s) for personnel errors should now be apparent and will be noted down. There may also be errors in the procedures and plant, however, so these aspects of the incident must still be investigated.

Contributors to the Identified Errors

Even when the errors produced by the previous analysis have been isolated, there is still another step in the ASSET process. *Contributors* to the latent weakness now have to be identified. These will be attributed to one, or both, of two causes:

- Inadequate preparation of personnel before the job assignment:
  - Check whether there was inadequate verification of personnel proficiency
- Degradation of proficiency of personnel during their employment:
  - Check for evidence of degradation in proficiency

Procedural Errors

Having covered personnel errors, the next area to be investigated is procedural errors. The same threefold approach as with personnel errors is used:

- Reliability of the procedures
  - Availability to the user
  - Assurance of currency
  - Scope limitations
- Adaptation to working conditions
  - Utilisation
  - Ergonomics
  - Environment
- Task qualification
  - Task orientation
  - Adequacy of content
  - Background support

Reliability of the Procedures

Availability to the Intended User

- Is the procedure detailed enough and appropriate to the complexity of the task?
- Was the procedure accessible to the user, and was it clearly identified by the work authorisation permit or some other document?
- Was the procedure used?
- Is the procedure appropriately identified on each page?
- Are all pages included?

Assurance of Currency

- Check if the procedure is a permanent one or temporary, issued for the specific job at hand.
- Check that the procedure is not outdated.
- Check when the procedure was last reviewed.

Scope Limitations

- Is there a clear statement of the purpose for which the procedure is intended?
- Are scope, terminal points, and access and safety requirements clearly specified?
- Check if the qualifications required of the person performing the task are indicated.

- Review the document control policies at the plant.
- Review the quality assurance manual for alignment with this procedure.

Adaptation to Working Conditions

Utilisation

- Check that the procedure covers what preparations for the task are required, what precautions must be observed, and what equipment is required.
- Check that the instrumentation and tools required for the task were available.
- Verify the legibility of the procedure.

Ergonomics

- Is the language precise and unambiguous?
- If there are symbols used in the procedure, are they understood by the reader?
- Are acceptance criteria clearly spelled out, with tolerances or minimum values where necessary?

Environmental Adaptation

- Was it possible to apply the procedure in the environment existing at the time of the incident (lighting levels, etc.)?

Task Qualification

Task Orientation

- If more than one person is required to perform the task, are separate instructions provided for each person?
- Are the instructions presented in the same sequence as their required performance?
- Are check-off features provided for the successive steps of the procedure?

Adequacy of Content

- Check for consistencies with reference documents.
- Check that acceptance criteria are all in quantitative terms.
- Check whether any required references to drawings, specifications, and so on are missing.
- Does the procedure cater for reasonable contingencies? (Actions have to be taken if specified requirements cannot be achieved.)

Background Support

- Review the documents on which the procedure depends, and determine whether they cover those areas not specifically covered by the procedure.

Contributors

- Check if inadequate proceduralisation and degradation in procedures could be contributing causes to the incident.

Deficiencies in Plant

As with personnel and procedures, these are given under three headings:

- Plant reliability
- External influences
- Qualification of function

Plant Reliability

Availability

- Check the history of the plant. Determine if actual equipment performance problems are identified during system start-up, shutdown, and normal and emergency conditions.

Endurance

- Check records to find trends that indicate degradation in performance as a result of aging, modifications, changes in operating practices, and so on.

Performance Limitations

- Check performance specifications against actual performance data to see whether the plant was operated outside of its design envelope.

External Influences

Auxiliary and Support Systems

- Review operating logs, maintenance records, and other data on systems that could interact with the failed system. Did a degraded support system contribute to the failure?

Physical Environment

- Did the operating environment contribute to the failure (high humidity, high temperature, etc.)?

Operating Practices

- Determine if good practices such as good housekeeping, good work order management, regular inspections, and so on are established and performed.

Qualification of Function

Commissioning, Maintenance, and Testing

- Was the plant properly maintained before commissioning/recommissioning?
- Were the necessary tests performed before commissioning/recommissioning?

Manufacture, Storage, and Installation

- Were the components used made to specification?
- Were any components used properly stored before being used?
- Were the components correctly installed?

Specification and Design

- Compare the design specifications with the actual operating requirements (humidity, pressure, voltage, etc.).
- Compare the design specifications for compatibility with requirements such as actual working conditions, for example, frequency of starts, and auxiliary system support (such as cooling temperature and flow rate).

Contributory Factors

- Check whether equipment preparation was adequate.
- Check whether there has been degradation of the equipment in service.
- Check whether discrepancies were detected.
- Check whether detected discrepancies were corrected.

Proposal of Solutions

The ASSET methodology is considerably less prescriptive and detailed when it comes to proposing post hoc solutions than it is to the analysis of the incident. In summary, the procedure is as follows:

- Review the direct and root causes of the incident. In other words, review the latent defects in the three Ps (plant, personnel, and procedures), as well as the deficiencies in the surveillance programmes.
- Determine the corrective actions needed to eliminate the direct causes.
- Correct inadequacies of preparation of the three Ps.
- Correct degradation in the three Ps.
- Eliminate deficiencies in the plant surveillance programme.
- Eliminate deficiencies in the contributing factors.

Tables

The following tables will assist anyone in performing an ASSET-type analysis:

- Table 5.2: Personnel Check Sheet for Incident Investigation
- Table 5.3: Procedures Check Sheet for Incident Investigation
- Table 5.4: Plant Check Sheet for Incident Investigation

Management of the Incident Investigation Process

A good way to manage an investigation is for a chairman and secretary to cooperate as follows: Using the check sheets as a general guide, the chairman addresses the various parties, both witnesses to the events and others. The secretary notes down on the check sheets any significant points. Sometimes, the chairman might instruct the secretary concerning a specific point. But the chairman does not do the note-taking. This allows him the freedom to conduct the investigation and to concentrate on the witnesses, their responses, and their body language. In this respect, the management of an incident investigation is similar to reliability-centred maintenance (RCM) or value engineering.

TABLE 5.2
Personnel Check Sheet for Incident Investigation

		Personnel		Y	N	Cause 1	Cause 2
						Poor preparation for task	Degradation in proficiency
Reliability	Vigilance	Paying attention?					
		Aware of importance?					
		Bored? (Does task often.)					
	Fitness	Physical	Sick, injured or drugged?				
		Mental	Stress (off-site or job) or bad attitude				
		Schedule influences	Shift change, overtime, or graveyard shift?				
	Understanding	Aware of own limitations?					
		Disdain for procedures?					
		Poor task preparation?					
External	Communications	Supervisor passed on info?					
		Software (drgs, etc.) available?					
		Plant ID OK (labels, etc.)?					
	Environment	Temperature too hot/cold?					
		Humidity?					
		Noise?					
	Safety	Policy understood?					
		Protective clothing?					
		General staff morale level?					

(Continued)

TABLE 5.2 *(Continued)*
Personnel Check Sheet for Incident Investigation

				Cause 1	Cause 2
Qualifications	Experience	Length of time on job.			
		Has he done this job before?			
		How often does he do the job?			
	Training	Did he or she know the plant?			
		Had training for this job?			
		Been assessed as competent?			
	Education	Education level defined?			
		Did he or she meet this level?			
		Level required?	Must check documents!		

TABLE 5.3
Procedures Check Sheet for an Incident Investigation

	Procedures		Y	N	Cause 1 Poor preparation for task	Cause 2 Degradation in proficiency
Reliability	Availability to user	Detailed?				
		Used?				
		Complete?				
	Current?	Up to date?				
		Permanent?				
		Last review?				
	Scope	Purpose stated?				
		Terminal points?				
		Qualifications required?				
Adaptation	Utilisation	Preparations required?				
		Precautions stated?				
		Equipment required stated?				
	Ergonomics	Language clear?				
		Symbols—understood?				
		Threshold values stated?				
	Environment	Possible to apply at time?				
		Lighting level to read?				
		Distractions?				
Task description	Orientation	Multiple-person task?				
		Instructions in sequence?				
		Tick boxes provided?				
	Content	Acceptance criteria?				
		All quantitative?				
		What if aim not met?				
	Support documents	Available?				
		Are areas not in our procedure covered?				
		Cross-referencing adequate?				

TABLE 5.4
Plant Check Sheet for Incident Investigation

					Cause 1	Cause 2
		Plant	Y	N	Poor commissioning or recommissioning	Plant degradation
Reliability	Availability	Unreliability	Normal running, start-up, or shutdown?			
	Endurance	Degradation due to:	Aging, modifications, or changes in operation?			
	Performance	Outside of specification	Plant thrashed or reduced output—why?			
External	Auxiliaries	Contributed to failure?				
		All working?				
		All needed?				
	Physical conditions	Temperature out of specification?				
		Humidity out of specification?				
		Other?				
	Operating	Housekeeping?				
		Inspections?				
		Work order management?				
Function	Maintenance	Optimised?				
		Adequate?				
		Commissioning?				
	Components	Pirate parts?				
		Storage OK?				
		Installation OK?				
	Design	Omissions in specification?				
		Specs honoured?				
		Mods configured?				

Fault Trees

Introduction

The final method to be discussed is that using fault trees. Fault trees are often used in probabilistic risk analysis to estimate the probability of an incident occurring. In this form, they use both AND gates (for events that must occur together) and OR gates (for events that, if any one occurs, it will lead to the next event higher in the tree). For incident investigation, OR gates are unnecessary as all the events that led to the incident have been discovered (if the investigation has been correctly conducted). Hence, it is not really necessary to even include the gates.

What is noteworthy about the fault tree presentation is how a fairly complex incident is communicated clearly and succinctly. This is one of the main advantages of the technique. The fault tree often describes the incident so well as to make narrative almost unnecessary.

Figure 5.2 is a fault tree drawn up from an actual incident.

Construction of the Fault Tree

The technique described following will include aspects of the various techniques discussed so far. Construction of the fault tree is by a combination of trial and error and common sense. It is nothing more than an exercise in logical thought. The fault tree process may be divided into the following steps:

- Define the incident.
- Obtain an understanding of the system by visiting the site and consulting drawings, maintenance manuals, manufacturer's literature, and so on.
- Form a committee to perform the investigation if necessary. If the incident is large and complicated, this is probably necessary as the system might be too complex for one person to easily comprehend. The committee should consist of technical experts knowledgeable in the system and of the various disciplines involved (mechanical, electrical, C&I, etc.). It might also be valuable to include a member of the company's legal department, particularly if there are safety implications to the investigation. A person with legal training could be useful in any investigation because of the rigorous thought process they bring to the investigation, which is similar, but not identical, to the engineering thought process.
- Call witnesses to the incident to testify to what happened, in formal sessions convened for the purpose. Ask questions to clarify what is being said. It is useful at this stage to structure the questions as per the ASSET methodology—for example, where there are defects in the three Ps (plant, personnel, and procedures), as well as defects in the surveillance system (if one is in place) that should have detected these defects? Reference may be made here to the lists of questions (see Tables 5.2 through 5.4) that will assist in the investigation. Other helpful questions are WHY, WHEN, WHERE, WHO, WHAT, and HOW. It might also be helpful to discuss the negatives of the prior, for example, where the incident did not occur, when it did not occur, and so on, as per the KT technique.
- Compile a narrative of the incident.

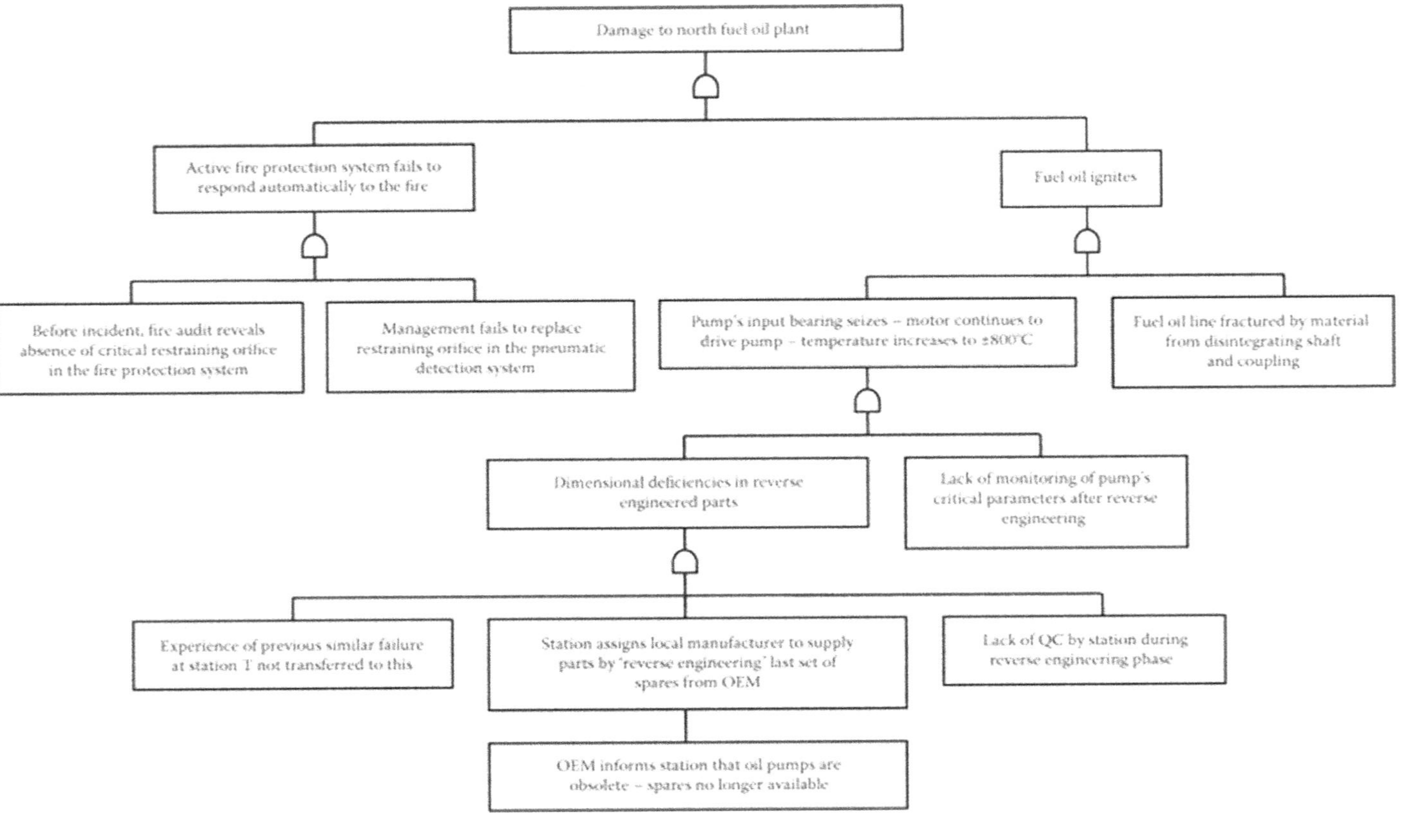

FIGURE 5.2 Fault tree of oil plant fire.

- Construct a fault tree describing the incident. This is usually an iterative process, construction beginning during the witness-calling session or before. As the investigation progresses, it will usually reveal errors in the logic of the fault tree, which is then modified to suit. Certain sections of the tree might be analysed separately, and the main tree might be built up later. The committee should review the fault tree and agree that it accurately represents what happened.
- Consider recommendations to eliminate the roots of the tree. It should be noted that simply cutting off one root will ensure that the incident can never occur again. However, similar incidents may occur unless all the roots are dealt with.

Finally, compile the report. This should include the following:

- A summary of the incident, the investigation, and the recommendations
- The lead investigator, who is also the main author of the report
- The participating committee members
- A description of the investigative process and the witnesses called
- A narrative of the incident
- The fault tree of the incident
- Conclusions
- Recommendations

Ishikawa Diagrams

Ishikawa diagrams (also called fishbone diagrams, herringbone diagrams, or Fishikawa) are diagrams that show the possible causes of a specific event. Such a diagram is shown in Figure 5.3. Common uses of the Ishikawa diagram are product design and quality defect prevention, to identify potential factors causing an overall effect. Each cause or reason for imperfection is a source of variation. Causes are usually grouped into major categories to identify these sources of variation. The categories typically include the following:

- People. Anyone involved with the process.
- Methods. How the process is performed and the specific requirements for doing it.
- Machines. Any equipment, computers, tools, and so on required to accomplish the job.
- Materials. Raw materials, parts, pens, paper, and so on used to produce the final product.
- Measurements. Data generated from the process that are used to evaluate its quality.
- Environment. The conditions, such as location, time, and temperature, to name a few.

Ishikawa diagrams were proposed in the 1960s by Kaoru Ishikawa, who pioneered quality management processes in the Kawasaki shipyards. However, the diagram was also used in American shipyards in the 1940s, before being popularised by Ishikawa. The fishbone diagram is considered one of the seven basic tools of quality

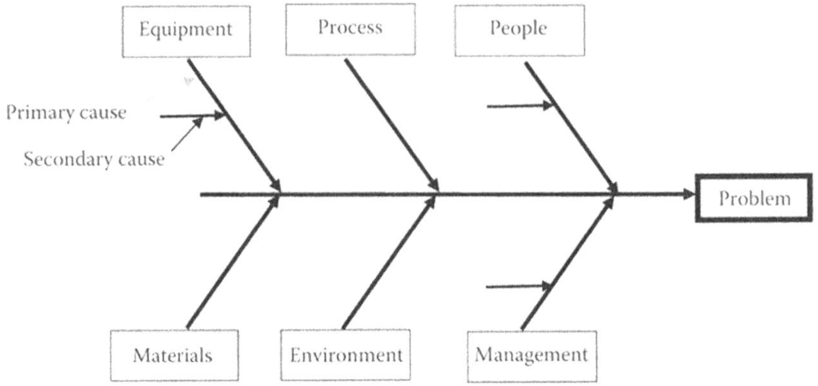

FIGURE 5.3 An Ishikawa or fishbone diagram showing factors for equipment, process, people, materials, environment, and management.

control. It is known as a fishbone diagram because of its shape, which is similar to the side view of a fish skeleton.

Mazda Motors famously used an Ishikawa diagram in the development of the Miata sports car. The main causes included aspects such as "touch" and "braking," with the minor causes including factors such as "50/50 weight distribution" and "able to rest elbow on top of the driver's door." Every factor identified in the diagram was included in the final design.

Causes

Causes in the diagram are often categorised, such as the 8 Ms, described next. Cause-and-effect diagrams can reveal key relationships among various variables, and the possible causes provide additional insight into process behaviour.

Causes can be derived from brainstorming sessions. These groups can then be labelled as categories of the fishbone. They will typically be one of the traditional categories mentioned prior, but there are other flavours of Ishikawa, as shown next:

The 8 Ms (used in manufacturing)

- Machine (technology)
- Method (process)
- Material (includes raw material, consumables, and information)
- Manpower (physical work)/Mind power (brainwork)
- Measurement (inspection)
- Mother Nature (environment)
- Management
- Maintenance

The 8 Ps (used in the service industry)

1. Product (or service)
2. Price

3. Place
4. Promotion
5. People
6. Process
7. Physical evidence
8. Productivity and quality

The 5 Ss (used in the service industry)

- Surroundings
- Suppliers
- Systems
- Skills
- Safety

Questions to Be Asked While Building a Fishbone Diagram

- **Man/Operator**: Was the document properly interpreted? Was the information properly circulated to all the functions? Did the recipient understand the information? Was the proper training to perform the task administered to the person? Was too much judgement required to perform the task? Were guidelines for judgement available? Did the environment influence the actions of the individual? Are there distractions in the workplace? Is fatigue a mitigating factor? Is the operator's work efficiency acceptable? Is he responsible/accountable? Is he qualified? Is he experienced? Is he medically fit and healthy? How much experience does the individual have in performing this task? Can he carry out the operation without error?
- **Machines**: Was the correct tool/tooling used? Does it meet production requirements? Does it meet process capabilities? Are files saved with the correct extension to the correct location? Is the equipment affected by the environment? Is the equipment being properly maintained (i.e. daily/weekly/monthly preventative maintenance schedule)? Does the software or hardware need to be updated? Does the equipment or software have the features to support our needs/usage? Was the machine properly maintained? Was the machine properly programmed? Is the tooling/fixturing adequate for the job? Does the machine have an adequate guard? Was the equipment used within its capabilities and limitations? Are all controls, including emergency stop buttons, clearly labelled and colour-coded or size-differentiated? Is the equipment the right application for the given job?
- **Measurement**: Does the gauge have a valid calibration date? Was the proper gauge used to measure the part, process, chemical, compound, and so on? Was a gauge capability study ever performed? Do measurements vary significantly from operator to operator? Do operators have a tough time using the prescribed gauge? Is the gauge fixturing adequate? Does the gauge have proper measurement resolution? Did the environment influence the measurements taken?
- **Material** (includes raw material, consumables, and information): Is all needed information available and accurate? Can information be verified

or cross-checked? Has any information changed recently/do we have a way of keeping the information up to date? What happens if we don't have all the information we need? Is a material safety data sheet readily available? Was the material properly tested? Was the material substituted? Is the supplier's process defined and controlled? Was the raw material defective? Was the raw material the wrong type for the job? Were quality requirements adequate for the part's function? Was the material contaminated? Was the material handled properly (stored, dispensed, used, and disposed)?

- **Method**: Was the canister, barrel, and so on labelled properly? Were the workers trained properly in the procedure? Was the testing performed statistically significant? Were data tested for true root cause? How many "if necessary" and "approximately" phrases are found in this process? Was this a process generated by an integrated product development (IPD) team? Did the IPD team employ design for environment principles? Has a capability study ever been performed for this process? Is the process under statistical process control? Are the work instructions clearly written? Are mistake-proofing devices/techniques employed? Are the work instructions complete? Is the work standard upgraded to current revision? Is the tooling adequately designed and controlled? Is handling/packaging adequately specified? Was the process changed? Was the design changed? Are the lighting and ventilation adequate? Was a process failure modes effects analysis ever performed? Was adequate sampling done? Are features of the process critical to safety clearly spelled out to the operator?
- **Environment**: Is the process affected by temperature changes over the course of a day? Is the process affected by humidity, vibration, noise, lighting, and so on? Does the process run in a controlled environment? Are associates distracted by noise, uncomfortable temperatures, fluorescent lighting, and so on?
- **Management**: Is management involvement seen? Inattention to task? Task hazards not guarded properly? Other (horseplay, inattention, etc.)? Stress demands? Lack of process? Training or education lacking? Poor employee involvement? Poor recognition of hazard? Were previously identified hazards not eliminated?

Criticism of the Ishikawa Technique

In a discussion of the nature of a cause, it is customary to distinguish between necessary and sufficient conditions for the occurrence of an event. A necessary condition for the occurrence of a specified event is a circumstance in whose absence the event cannot occur. A sufficient condition for the occurrence of an event is a circumstance in whose presence the event must occur. A sufficient condition naturally contains one or several necessary ones. Ishikawa diagrams are meant to use the necessary conditions and split the "sufficient" ones into the "necessary" parts.

Dean L Gano (2003), in his book *Apollo Root Cause Analysis*, is very critical of the Ishikawa technique, stating categorically that it does *not* identify causal relationships:

> To many quality engineers, a cause and effect chart is synonymous with the Ishikawa Fishbone Diagram. This method is a combination of categorical thinking and causal

relationships. It asks us to guess at possible causes of the event, usually in a brainstorming session, place them like a fish skeleton within predefined categories and then vote on which causes we think are the ones that caused the problem. The Ishikawa Fishbone Diagram is a good brainstorming tool, but it does not identify causal relationships *and it is not a cause and effect diagram.*

Causes are listed according to which category they belong to, not to their causal relationships. There are no causal relationships in the diagram, only guesses at various causes. I say guesses because this is brainstorming and no evidence is required to support possible causes.

One of the most common complaints I hear about the Ishikawa root cause analysis method is that it is difficult to come to an agreement on what the root cause is. The reason for this tension is the fact that it does not provide a true representation of causal relationships and this causes misunderstandings, arguments and ineffective solutions.

What is most intriguing to me is that despite the logical flaws in the Fishbone Diagram, it sometimes seems to provide the desired results. I believe this occurs not because of the process, but because the investigators are very experienced and the human mind is very good at understanding causal relationships at a subconscious level.

Kipling's Serving-Men

The verse from the poem that follows serves as a useful mnemonic for the words already mentioned prior—what, why, when, how, where, and who—all of which are useful words to use in an incident investigation:

I keep six honest serving-men
(They taught me all I knew);
Their names are What and Why and When
And How and Where and Who.

Apollo Root Cause Analysis—Dean L Gano

The method proposed by Gano is logically identical to the fault tree method but uses different terminology. For our purposes, Gano is useful in his categorisation of RCA methods, which he divides into categorisation schemes and causal relationship methods:

Categorisation schemes have been the most popular method in use over the past 25 years, but they are no longer used in many industries because of their ineffectiveness. Categorisation methodologies, such as checklists, provide a hierarchical outline, checklist or "cause tree" from which a root cause is chosen. Through discussion and consensus, a solution is attached to the defined root cause and is implemented. Focusing on *the root cause* is a highly ineffective problem-solving strategy because it prevents us from understanding all the (possible) causal relationships.

Causal Relationship Methods

Causal relationship methods are based on the premise that everything that happens has a set of very specific causes and that, by knowing these causes, we can control them and hence find effective solutions. While a very sound basis, these methods are only as effective as the tools provided. Fault trees and the Apollo Event tree method are two such techniques.

The Nature of Failure

The cases presented later in this chapter include the complex (Chernobyl), the simple (Flixborough and the Hyatt Regency Hotel), and the intermediate (*Challenger* and Piper Alpha). If we study these and other cases, we will notice certain patterns appearing. Faults can be attributed to any combination of design, maintenance, operational, and management errors. There are lessons to be learned for anyone engaged in the design, maintenance, operation, or management of engineering systems. In one of the cases, the Embraer aircraft crash, the student must investigate the case himself using the techniques presented earlier.

The approach taken by the author for all these cases is that all errors are human errors—the plant failed because someone designed it incorrectly, maintained it incorrectly, or whatever. The electrical, mechanical, chemical, civil, and electronic hardware is neutral; it is not emotional, it does not make mistakes, and it does not go out on strike.

The author has in fact analysed many cases of failure, and the results are presented in Table 5.5. The official reports of these accidents were studied, fault trees were constructed, and errors were apportioned as shown.

The error codes chosen were as follows:

- *B* for buying or specification error.
- *C* for commissioning error.
- *D* for design error.
- *F* for a functional failure, where the human agent could not be found. Such failures are of two types:
  - Type a, where the plant does not operate when it should (e.g. a failure of a circuit breaker to open)
  - Type b, where the plant operates when it should not (e.g. a circuit breaker that opens when it should not)
- *M* is for management error. Such errors are those that cannot be attributed to the other agents in the chain. Such errors include errors in proceduralisation, training, and so on.
- *O* is for operator error.
- *P* is for production or manufacturing error. As *M* has been reserved for management error, we cannot apply *M* here.
- *R* is for repair error. Maintenance error in other words.

An extra classification for any of the aforementioned error types was for errors committed within the person's knowledge base, type 1 errors, and errors committed outside the person's knowledge base, which the author calls type 2 errors. For example, the designer who specified steel with a high sulphur and phosphorus content for the *Titanic* ocean liner could not be blamed for this, contributing to the brittle fracture of the hull. Brittle fracture in steel was not understood in the early twentieth century. Therefore, although it was an error to specify such steel, it was a type 2 error, outside of the designer's knowledge base. It is hence specified as a D_2 error. When errors of the same kind are committed at the same site, they are multiplied together (e.g. the seven operator errors at Chernobyl are represented by O_1^7).

TABLE 5.5
Error Code Count for 12 Large-Scale System Failures

		B	C	D	F	M	O	P	R	Total
1	Bhopal					M_1^2	O_1		R_1^3	6
2	Challenger			D_1		M_1			R_1	3
3	Chernobyl			D_1^4		M_1	O_1^7			12
4	Clayton					M_1^2	$O_1^2O_2^2$		R_1	7
5	Flixborough			D_1			O_2			2
6	TMI				$F_a^3F_b^2$	M_1M_2	$O_1O_2^3$			11
7	Titanic			D_1D_2			O_1^2			4
8	Piper Alpha				F_a	M_1^3	O_1^4		R_1	9
9	R101			D_1^2	F_a	M_1^2	O_1^4		R_1^2	11
10	DC 10 –1					M_1^2	O_2^3	P_1		6
11	DC 10 – 2	B_1		D_1^3		M_1	O_1O_2		R_1	8
12	DC 10 –3			D_1	F_a	M_1	O_2^2		R_1	6
	Total	1		14	8	17	34	1	10	85

The incidents studied by the author were the following:

- Bhopal, a chemical plant disaster in India.
- *Challenger*, the American space shuttle disaster.
- Chernobyl, the worst disaster of all time, with geopolitical consequences. Chernobyl was a nuclear power plant in the Ukraine during the Soviet era. The consequences of that failure are still with us in the twenty-first century.
- Clayton Tunnel, an incident in nineteenth-century England when one train ran into the back of another in a long dark tunnel.
- Flixborough, a plant in the North of England, similar to Sasol in South Africa, in that it used coal as a feedstock to generate liquid hydrocarbons. It blew itself off the map in 1974.
- TMI, short for Three Mile Island, a nuclear plant accident in Pennsylvania in the United States.
- *Titanic*, the luxury ocean liner that sank on its maiden voyage in 1912.
- Piper Alpha, the British oil rig in the North Sea off the coast of Scotland.
- The R101, the British airship that crashed in France.
- Three incidents involving the Douglas DC-10 airliner.

These incidents are summarised in Table 5.5, with the error types totalled. It will be seen that operator error predominates, with 34 of the total of 85 identified errors being operator-based. Management errors come next with 17, followed by design with 14. Repair or maintenance comes fourth with 10 errors.

Why should operator error predominate? This author thinks it is because of the *average time available to make a decision.* For all the participants, the operator is the most constrained here. A maintenance person may install a thrust bearing the wrong way 'round, go home, ponder his possible error, and correct the situation the next morning. The operator has no such luxury at his disposal. When the alarm sounds and he presses the wrong button, he cannot possibly change the chain of events that follow.

What is certain is that operator training and operator selection are vital parts of reliability management.

While the pattern of Table 5.5 is interesting, it is not to be thought of as universally applicable. In a similar study that the author performed on six in-house incidents at the company he worked for at that time, it was found that management error predominated. In particular, poor proceduralisation was seen to be the cause of most problems. Procedures were seen to be missing, out of date, or that the personnel were not aware that a procedure existed.

Careful study of Table 5.5 is nevertheless profitable for the insights it shows. For example, it shows Bhopal as the worst incident for maintenance errors, while Chernobyl is the worst for operator errors, with seven culpable errors committed by the operators. The reasons for this provide for an interesting discussion.

What is apparent from Table 5.5 is that disaster can strike from any quarter and that there is very seldom only one cause, making the term *root cause (singular) analysis* an oxymoron. There is almost always a combination of causes, any one of which, if eliminated, would have ensured there was no incident. In reliability terms, the causes are *serial*, not parallel.

To minimise the possibility of incidents, a reliability manager must cover all the areas given in the table: specification, design, manufacture, commissioning, operation, maintenance, and management.

A final point can be made from Table 5.5. The number of very serious incidents occurred from the 1970s to the 1990s of the last century. As previously mentioned, since then, we seem to have learned from these failures, and so we have not seen such large disasters in this century as we have in the table. For example, in the chemical engineering field, the HAZOP procedure has been developed, which stands for hazard and operability analysis. All chemical plants today have the designs checked by performing a HAZOP. Modifications on existing plants are subject to HAZOPs as well. Several mining operations also make use of the technique.

CASE 5.1 CHERNOBYL

Chernobyl is the worst man-made disaster of all time. It happened in the eighties of the previous century, but its effects are still felt today. An extremely complex accident, it is included here for the following reasons: its importance as an industrial accident and to show the power of the fault tree technique. Once the technology is understood, a fault tree can be used to summarise this very complex accident on one page.

CLASS ASSIGNMENT

Read the case below and make sure you understand it. See if you agree with the fault tree. The case describes the events leading up to the Chernobyl accident as rendered in the available literature. The accident is analysed in the context of the main contributing factors that were involved and gives some suggestions on the way the lead-up to the accident could have been prevented.

After this analysis, a summary of the International Nuclear Safety Advisory Group's recommendations is provided for reference. Some of the effects that the accident had on the political climate in Europe and elsewhere, on nuclear programmes and perceptions of the Soviet programme, are also listed.

CHERNOBYL: BACKGROUND AND HISTORICAL INFORMATION

The Chernobyl power plant operated graphite-moderated nuclear reactors of the type known as RBMK. (Russian: RBMK = heterogeneous water-graphite channel–type reactor.) These are larger, 1,000-MW versions of the USSR's first nuclear plant's reactor, the 5-MW Obninsk, which started operating in 1954; 14 similar reactors were in operation in the Soviet Union before the Chernobyl accident, as well as one larger, 1,500-MW version of the same reactor, the latter having been commissioned in 1984. A sectioned view of the reactor is shown in Figure 5.4, and a diagrammatic presentation of the reactor/boiler/turbine is given in Figure 5.5. The more recent nuclear power plants in the USSR and the Eastern Bloc, however, employ pressurised water reactors known as VVERs.

Plans to build a nuclear power plant in the Ukraine, a Soviet republic, in proximity to major load centres, were initiated in 1970. Chernobyl was chosen as the first site because it was relatively remote and not part of the traditional "breadbasket" of the Ukraine Republic. It is situated approximately 130 km north of Kiev, the capital city of the Ukraine. The ceremonial foundation stone was laid in March 1970, and the first RBMK-1000 unit, which consists of one reactor that drives two turbines, came on stream in October 1977. Further units at the plant were commissioned in 1978, 1981, and 1983, giving a combined generating capability of 4,000 MW. Unit 4 was the last reactor to come online, although subsequent plans were to construct an additional two RBMK-1000s at Chernobyl.

RBMK CHARACTERISTICS

Each RBMK reactor core consists of an array of 0.25-m^2- and 7-m-long graphite blocks, set together to form a cylindrical shape approximately 12 m in diameter. There are approximately 1,600 vertical fuel channels and approximately 200 control rod channels in the core. The fuel rods consist of two fuel subassemblies consisting of 18 zircaloy-clad fuel pins of 2% enriched uranium dioxide. The reactor is cooled by means of boiling light water, which passes up through the fuel channels. A mass of stainless steel pipes carry the boiling water up from the top of the fuel channels to four steam separators. The separated steam feeds directly to the turbines, while the condensate is fed back to the bottom of the pressure tubes in the reactor by eight main circulation pumps, of which only six are needed during normal operation.

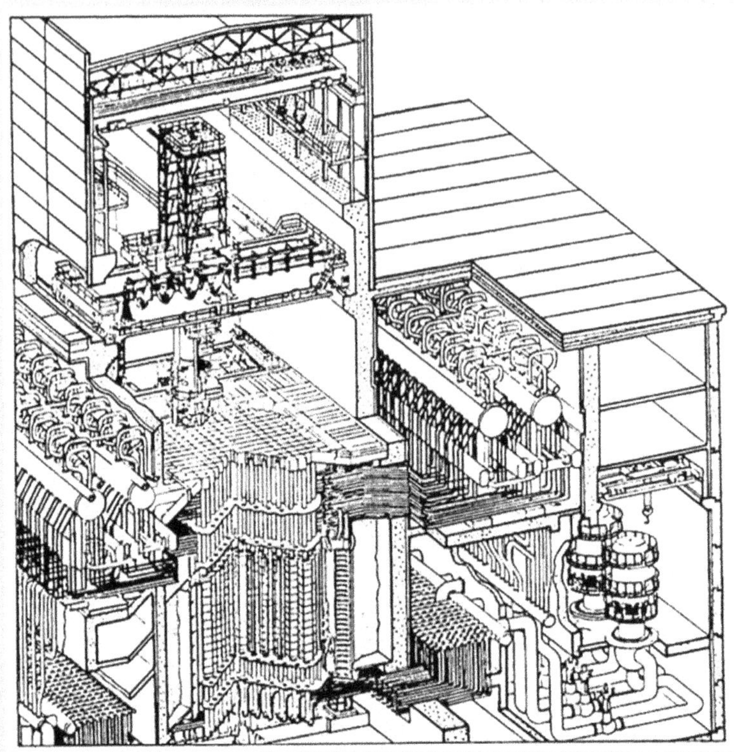

FIGURE 5.4 View of the RMBK-1000 reactor.

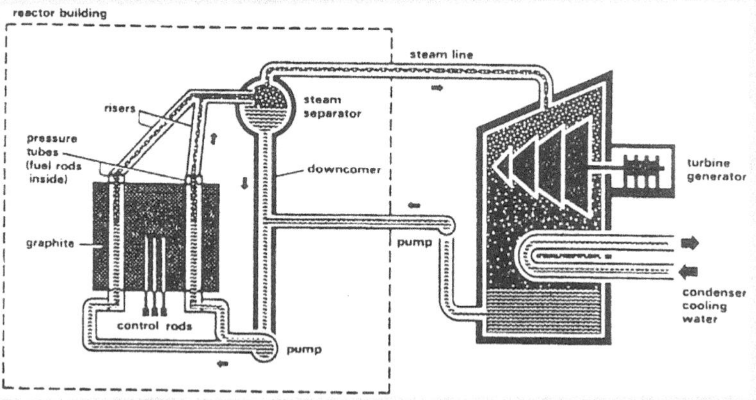

FIGURE 5.5 Diagrammatic representation of the Chernobyl reactor and turbine (in practice, the reactor feeds two turbines).

The RBMK-1000 reactor was equipped with an emergency core cooling system (ECCS) that is designed to deal with rupturing of the main circulation system. The core is surrounded by a thin cylindrical steel vessel. Refuelling can be performed when the reactor is in operation by means of the massive fuelling machine carried on a gantry above the reactor. Because the reactor building roof is more than 30 m above the core, there is enough headroom for the fuel rods to be removed and transferred to a separate storage pool for spent fuel at the other end of the reactor building.

Yet another characteristic of the RBMK reactor, one that it turned out was at least partly to blame for the Chernobyl disaster, is the positive reactivity coefficient. This implies that as the temperature within the core rises, the boiling of light water coolant increases, resulting in there being less neutron absorbing light water between the fuel and the moderator, which in turn results in rising reactor power and temperature. In contrast, light water–cooled *and* light water–moderated reactors have a negative void coefficient of reactivity, which implies that as temperature increases, reactivity drops. Controlling a reactor with a positive void coefficient of reactivity is not impossible, but good instrumentation and computer control systems are necessary to cope with any disturbances in the power level.

THE ACCIDENT AT UNIT 4

Unit number 4 at the Chernobyl power plant had been operated successfully for approximately three years. It was in fact considered to be one of the most successful nuclear-powered units of the RBMK type. It was planned that on April 25 and 26, 1986, while unit 4 was being shut down for routine maintenance, a special electrical systems test at the unit would be carried out. The purpose of this test was to prove that the turbine generators could support the essential plant systems, chiefly the ECCS, during their rundown period after the reactor had been shut. This was thought to be necessary because the backup diesel generators were slow in starting up during sudden power cuts. Previously, during similar testing procedures, the voltage could not be maintained at a constant correct level for any period. Now, however, a new voltage regulating system was in use on the generator, and the tests were to confirm that power at a constant voltage could be supplied from the turbine during its inertial rundown period.

The test was planned by the electrotechnical experts as it was considered to be a predominantly electrical test. Since it was thought that there would be no effect on the reactor as a result of the test, minimal consideration was given to nuclear safety during the planning stage, and nuclear experts were not involved or consulted. The procedures that resulted were very deficient in safety considerations, and the operators were given the go-ahead for the test without the necessary approval from the plant safety technology group.

The plan for the test was as follows: Initially, the power output of the reactor was to be reduced to between 700 and 1,000 MW (thermal), or approximately

30% of capacity. The ECCS was to be blocked off to prevent it from being started during the course of the test. The main circulating pumps were to be reconfigured so that four of the eight would be driven by the turbogenerator, which was being run down, providing the necessary load on it. Finally, the turbogenerator concerned would be isolated from the supply of steam from the reactor.

Preparation for the test began at 1:00 a.m. on April 25 with the start of reactor power reduction. Later that day, at 1:05 p.m., the power output reached 50% of capacity, and turbogenerator number 7 was shut down. The ECCS was isolated, according to plan, at 2:00 p.m. At this stage, however, the power reduction was halted on request of the system controller, who wanted to maintain a supply of power to the region for the remainder of the day, but the ECCS was kept isolated. Keeping the ECCS isolated at this level of reactor output was a violation of operating procedure and illustrates the nonchalant attitude towards the test on the part of the operators. It is also speculated that during the extended operation of the reactor at this level of power output, large quantities of xenon gas would have been built up, which, in turn, could have contributed later to the sudden power reduction and consequent reactor instability.

At 11:10 a.m., the power output reduction towards the target output of 700 to 1,000 MW (thermal) was resumed. This level had been chosen because it would keep the reactor output at just above the minimum allowed operating level of 700 MW (thermal). During the first seconds of April 26, however, as the operator was transferring the unit power control from the local to the global regulating system, he failed to enter a "hold power" request. As a result, the reactor rapidly lost power to below the allowable operating level of 700 MW (thermal). The reactor reached an output level of 30 MW (thermal), or roughly 1% of capacity, very quickly. The operator managed to get the output to 200 MW (thermal) by manually withdrawing most of the control rods. By 1:00 a.m., the power output had stabilised at 200 MW (thermal). At this output (below 700 MW), however, the power and feedwater control become very difficult as the relationships governing the reactor processes are highly nonlinear. Operation of the reactor at this level made the reactor very sensitive to plant or reactor perturbations, compromised the effectiveness of the protection system, and was in violation of a number of operating procedures. Despite this inherent instability and violation of procedure, the operator decided to continue with the test.

At this stage, the operator turned the fourth main circulation pump in the left and right loops of the cooling system. Owing to the low void condition of the reactor, however, the flow of water through the pumps was excessive, exceeding the allowed limits of operation. The operator attempted to compensate for this resulting condition by increasing the feedwater flow (of the condensate from the turbine). He was, however, unable to get the level of water in the steam drum to a desirable height. This was because the control of feedwater flow at this low power level was very sensitive.

At 1:19 a.m., the water level in the steam drum was still critically low. The operator increased the flow of feedwater yet further in an attempt to compensate. Although the level of water in the steam drum improved, the other effects of the increased flow through the reactor were becoming apparent.

The primary effect of the increased flow of water was void reduction (i.e. the water in the reactor began to boil less vigorously). This added negative reactivity to the system. The automatic control rods attempted to compensate, but this was insufficient. Withdrawal of the last remaining manual control rods was required. The operator responded by withdrawing them too.

Pressure, void (the level of boiling) and water flow within the reactor are all physically related. In order to prevent the reactor from tripping as a result of the unusual conditions indicated by the values of these parameters, the operators disabled further automatic control functions. They were determined to proceed with the experiment, come what may.

Having decided that the water in the steam drum had reached a satisfactory level, the operator drastically reduced the inflow of feedwater into the reactor. The effect of this, as expected, was opposite to that just described before this step. Reduced water flow immediately resulted in increased voidage, more positive reactivity, and increasing pressure. In order to maintain a constant reactor power, the automatic control rods were again driven into the core.

At this point, just before the beginning of the actual test, the operator obtained a printout of the neutron flux distribution within the core. The printout clearly showed that too many control rods were still withdrawn from the core and that there was not enough reactivity reserve to shut down the reactor.

Even at this late stage, the operators decided to carry out the test. A further reactor trip function was disabled. In order to be able to repeat the test, the operators did not want the reactor to trip when they shut down the second and last generator; this was against the explicit instructions laid down in the test procedure. At 1:23:04 a.m., the emergency stop valve to the turbine was closed and the experiment began.

Had the second turbine trip function not been disabled, the reactor would have tripped and still shut down safely even at this late stage. In this event, it didn't trip; rather, the flow of water through the reactor started to drop as the main circulation pumps were running down with the generator. The pressure and voidage within the reactor began to increase. It should be noted at this stage that the reactor was at this point in such a state that small changes in power would have considerably greater effects on the steam void, which, in turn, would lead to a further increase in power, because of the positive void coefficient of reactivity mentioned earlier.

By approximately 1:23:40 a.m., the power of the reactor began to increase dramatically. The shift manager at this point realised that the situation was critical and ordered that the reactor be shut down, but it was too late. The few rods in the core had insufficient reactivity left, and the remaining control rods could not be inserted fast enough to counteract the power increase being caused by the combination of factors previously described. The only automatic trip control, on excessively high power, which was not disabled, now came into play, and an automatic reactor shutdown was initiated. However, this came too late to be effective as well.

The positive void coefficient characteristic of the RBMK continued to play its part. All the monitoring and recording equipment could not cope with the state of the reactor. It is only through calculations and simulations carried out after the accident that it could be estimated that within four seconds of 1:23:40 a.m., the power of the reactor reached 100 times its rated full power. The fuel within the core fragmented, and this led to very rapid generation of steam. Pressures within the reactor space increased further until eventually an explosion ensued. This, what was to be the first of two successive explosions, damaged the reactor containment vessel and the reactor building and lifted the upper plate, above the core, weighing approximately 1,000 metric tons, into a vertical position.

Approximately three seconds later, the second explosion occurred. It is suspected that this was caused by a reaction between hydrogen in the air that had, by this time, leaked into the core. The hydrogen, it is believed, was produced as very hot steam came into direct contact with the zirconium alloy used to protect the uranium fuel. Especially at very high temperatures, zirconium is known to have a strong affinity for oxygen; at the temperatures that prevailed, it combined with the water, producing zirconium oxide and large amounts of hydrogen. As a result of the second explosion, some 25% of the graphite and fuel channel substances were ejected from the core, some of which were even found outside the reactor building. The water-filled shielding tanks around the reactor were also ruptured at this time.

Using the auxiliary feedwater pumps, water was injected into the reactor space for approximately the next 12 hours. This was stopped, however, when it was discovered that there was a danger of units 1 and 2 being flooded. The presence of this water, however, led to great amounts of steam being produced. This escaped from the damaged reactor building during the first day after the accident, contributing to the radiation spread that ensued. During the following days, large quantities of materials, including sand and boron, were added to the reactor vault by response teams, thereby decreasing the airflow, suffocating the burning graphite moderator, and helping to slow the release of radioactive smoke from the building. The response teams then injected nitrogen into the building in order to put out the fire completely.

The operating personnel of the reactor and electricity generating plant suffered forms of acute radiation syndrome. Two of the operating personnel died at the scene; their causes of death were cited as extensive steam burns and injury sustained from falling objects. The accident had, however, a much more extended effect because of the deadly radioactive plume that was blown from the plant throughout much of Europe.

ANALYSIS OF THE ACCIDENT

The following information was mostly gleaned from the Soviet experts and their working documents during the post-accident review meeting in Vienna during August 1986. From this, a plausible explanation of the accident was obtained, but no attempt has been made to verify this or look for alternative explanations.

During the aforementioned review meetings, the Soviets presented a large amount of information, but at no stage did they concede that the RBMK-1000 reactor was defective in its design. It can be speculated that even if deficiencies in the design did exist, as is maintained by some Western experts, it would not have been expedient for the Soviets to admit this fact because it could put a question mark over the continued operation of the remaining reactors of the same type. It is clear from the literature that the RBMK reactors were still strategically important to the Soviet Union, and it could not afford to question the safe operation of this significant proportion of its generating capability. Consequently, the blame is laid squarely at the feet of the operating staff, and therefore it is analysed here in the context of human reliability.

There is no doubt that actions by the operators contributed to the accident. From the initial planning stages, correct procedures were not followed. Some analysts speculate that the almost 100% trouble-free operation of unit 4 at Chernobyl for the preceding three years gave the operating and management staff a false feeling of security, pride, and infallibility. This is perhaps human nature; one likes to be recognised for one's achievements, but it should be addressed especially when the people involved are in control of plants, such as a nuclear reactor. They should not be allowed to feel that they can "drive this thing with their eyes closed." It is the feeling of the author that, on more than one occasion in the lead-up to this accident, the operators threw all caution to the wind and relied on their past successful experience and good luck to bring them through.

PLANNING THE EXPERIMENT

The notable shortcoming at this stage was that nuclear safety experts were not consulted, nor were the procedures submitted to them for their approval. The thinking, as was mentioned earlier, was that the test was to be of a purely electrotechnical nature and had nothing to do with the safety of the reactor. Hindsight proves that this was a wrong assumption, but it is again felt that with any digression in the operation of the nuclear plant as a whole, the safety aspects should always be re-evaluated.

In addition to this obvious error in judgement, from the available information, the level of detail that was included in the procedures given to the operators that were to be followed during the course of the test is not completely clear. Again, it is felt that nothing should have been left to chance. Every action should have been specified in detail, together with the consequences envisaged to result from that step. It should have been explicitly stated that any deviation from the expected consequences outlined in the procedure should initiate an immediate shutdown of the reactor and the test be terminated. It is apparent that because a similar test had been conducted successfully in the past, the test was not specified in its entirety and that a lot was left to the past experience of the operators.

From the literature, the level of training or qualifications that the operators and shift supervisors possessed was not possible to ascertain. This should have been considered but perhaps not relied on. In the event of an accident, such as the one that occurred, events happen very quickly and the response from the operators has to be almost automatic; not much time for thought and analysis is possible. This can be achieved only through extensive simulator training on a continuous basis or, in the case of an unusual procedure such as this experiment, by thorough briefing and preparation before attempting the experiment.

To summarise, it is clear that the preparation before the accident was inadequate. The planning, approval process, and training of the operators should have been carried out much more carefully.

DISREGARD FOR THE OPERATING PROCEDURE

On more than one occasion, the operating procedures were disregarded during the lead-up to the test. It should have been stressed that because an experiment was to be carried out, this did not excuse the operators from obeying every procedure or regulation that was laid down. Perhaps this could also be related to the insufficient preparation for the test, but the operating staff should have been aware that it is never "OK" to disregard any procedure. A number of instances when procedures were disregarded are documented.

The ECCS was to be switched off during the course of the test. It was, however, disabled approximately 12 hours before the actual test commenced, and the reactor was operated for approximately 10 hours at at least 50% power with the ECCS disabled. This was a violation of existing procedure, although it did not directly contribute to the accident. It can perhaps be argued that when the ECCS was actually disabled, the intention was there to start the experiment shortly thereafter, and once the staff had a change of plan, they had forgotten to re-enable the ECCS. This is very unlikely, though, because there would have been an indicator in a prominent position informing the operators that the ECCS was disabled.

THE CONTINUED OPERATION OF THE REACTOR AT 200 MW (THERMAL), WELL BELOW THE MINIMUM ALLOWED LEVEL OF 700 MW (THERMAL)

The operation of the reactor for any period below the minimum level was not allowed, because at these levels, the effect of the positive coefficient of reactivity becomes much more pronounced—that is, an increase in power results in a still larger power rise. The operators were aware of this regulation but disregarded it. This was possibly because having stabilised the reactor at 200 MW (thermal) gave them confidence that "things were under control." This action contributed to the instability that eventually caused the accident.

CONTINUED OPERATION WHEN REACTIVITY MARGIN WAS INSUFFICIENT

The decision by the operators to continue the test despite the reactivity margin, provided in the printout at the 11th hour, being below the permissible value, is inexcusable. This condition already placed in doubt the safe shutdown of the reactor. The fact that the test was continued beyond this point played a major part in causing the accident.

FAILURE TO ENTER "HOLD POWER" REQUEST DURING CONTROL TRANSFER

This omission caused the reactor power to drop suddenly to well below the allowed minimum power level. Despite subsequent efforts, the operators did not manage to get the reactor power up to a safe operating level. This caused them to continue the experiment with the reactor power level below the minimum and the associated consequences of instability. The reason the request was not entered is not clear. It could have been forgotten, but it is felt that in such a circumstance, the supervisory control system should have queried the operator's actions and asked him about the necessity of the command. This could be viewed as a design deficiency in the supervisory control system. Alternately, this should have been specified in the test procedure, and the operator failed to follow the procedure correctly.

DISABLING OF AUTOMATIC TRIPS

On several occasions, the operators intentionally disabled the built-in protections within the reactor automatic control system. Their reasoning behind this was that they wanted to be able to repeat the test if it failed. The last automatic trip that was disabled, namely, the one that is triggered when both turbogenerators are shut down, was explicitly against the specifications of the test procedure. If this trip had not been disabled, it is claimed that the reactor would have shut down safely. The fact that the operators disobeyed the instructions demonstrates their self-assured attitude that they could control the reactor under all ensuing circumstances. It is felt that these actions are completely opposed to the philosophy behind any control system.

Should the operators have had the facilities enabling the disabling of these automatic trips available to them? One cannot but also speculate whether this was the first instance that trips were disabled, or is this a routine occurrence at Soviet nuclear power plants?

DESIGN CONSIDERATIONS

Although, as was mentioned before, the Soviets never conceded that there were any shortcomings in their design, some factors that at least partly contributed to the extent of the accident are included here.

The positive void coefficient inherent in the RBMK design had a definite part to play in the accident. It is generally felt that any reactors with this characteristic (e.g. also the Canadian CANDU reactor) have to be equipped with specially designed control systems that can cope with this instability at certain operating levels. Moreover, the potential for misunderstanding these physical characteristics by the operators should be catered for by additional design features. It is felt that this may have been lacking in the case of the RBMK.

LACK OF CONTAINMENT

The second factor that had a definite part to play in the subsequent radiological discharge into the environment was the lack of a containment structure around the reactor. Most Western-designed reactors have this to provide protection during the worst kind of accidents. Again, this is a design deficiency of the RBMK reactor.

It is felt that the accident was caused by the fact that operators attempted to perform the test without any regard for safety procedures. Their blinkered approach, which disregarded any dangers, and their nonchalant attitude towards existing safety features made the accident inevitable.

OTHER CONSEQUENCES OF CHERNOBYL

The most emotional effect of the accident at Chernobyl was no doubt the extensive radiation fallout that was recorded throughout the whole of Europe. Initially, the USSR kept quiet about the accident. Approximately two days after the actual accident, radiation monitors at a Swedish nuclear installation gave the alarm. Once it was established that the problem was not originating from the Swedish plant but that the radiation was carried by atmospheric conditions emanating from the Soviet Union, the Swedish authorities complained to the Soviets. The extent of the problem prevented the Soviets from covering up the problem any further.

The release of information during these initial stages was, however, very slow. This led to vivid speculation in the Western media of a meltdown at Chernobyl. People from all walks of life criticised the Soviet programme for inferior plant design and safety precautions, as well as their attempts at trying to cover up the accident. This led to at least some political tension.

The effect on the environment varied from area to area. In the immediate vicinity of the Chernobyl plant, the contamination of the environment was serious. People entered this area only with protective clothing and then only for short periods. Measures were taken to ensure that the water resources close to the power plant, which also supply Kiev, were free from contamination. All people within a 10-km radius of the plant were evacuated on April 27, approximately 36 hours after the first explosions. Evacuation of a wider area, of approximately a 40-km radius, began only on May 2. People in Kiev were ordered to stay indoors as far as possible, and children were given early holidays and sent to summer camps.

Areas farther afield were affected to a smaller degree. Nevertheless, children in countries such as Poland were given iodine to prevent the absorption of the radioactive iodine isotopes present in the atmosphere. Agricultural products from affected areas were not allowed to be sold, and these products from the Eastern Bloc countries were totally banned in Western Europe. Effects on the nuclear programmes of countries varied. In the Soviet Union itself, it was initially reported that all RBMK reactors had been shut down. Later, however, it was reported that all reactors other than those at Chernobyl were operating normally. There were plans to get the other reactors at Chernobyl, even unit 3, which is in the same building complex as unit 4, back on stream. Soviet experts claimed that after reviewing their design calculations, they had found no fundamental flaws with the reactors. It appears that the Soviet programme was to go ahead, despite objections from countries such as Sweden.

In Yugoslavia, however, the Croatian parliament decided on May 6 to completely scrap plans for a new nuclear plant. They appear to have put their programme on hold. The Dutch government decided on May 7 to postpone any expansion planning of the country's nuclear plant. Chernobyl also seems to have been the last nail in the coffin of the Austrian Zwentendorf project, which was waiting for an approval to go into production after protracted negotiations.

In most countries, however, there was no real change of course. Experts satisfied themselves that the design of Western reactors is very different from that of RBMK, and that a similar accident at their plants was impossible.

There was a positive outcome too. The accident heralded a new era of international cooperation on nuclear safety issues. The general feeling at the post-accident review meeting in Vienna was that the Soviets were very open and frank in their report of the accident. The Soviets were also willing to implement the recommendations of the group in their existing nuclear plants as well as to follow the guidelines of the IAEA in their new projects.

But perhaps the most important result of the disaster was political—Chernobyl contributed directly to the collapse of the Soviet Union. Before Chernobyl, the Soviets had never asked for Western help. Economically teetering on bankruptcy, the Soviet Empire was now seen to be technologically bankrupt as well. Nothing could hold it together any longer.

The fault tree of the Chernobyl incident is shown in Figure 5.6.

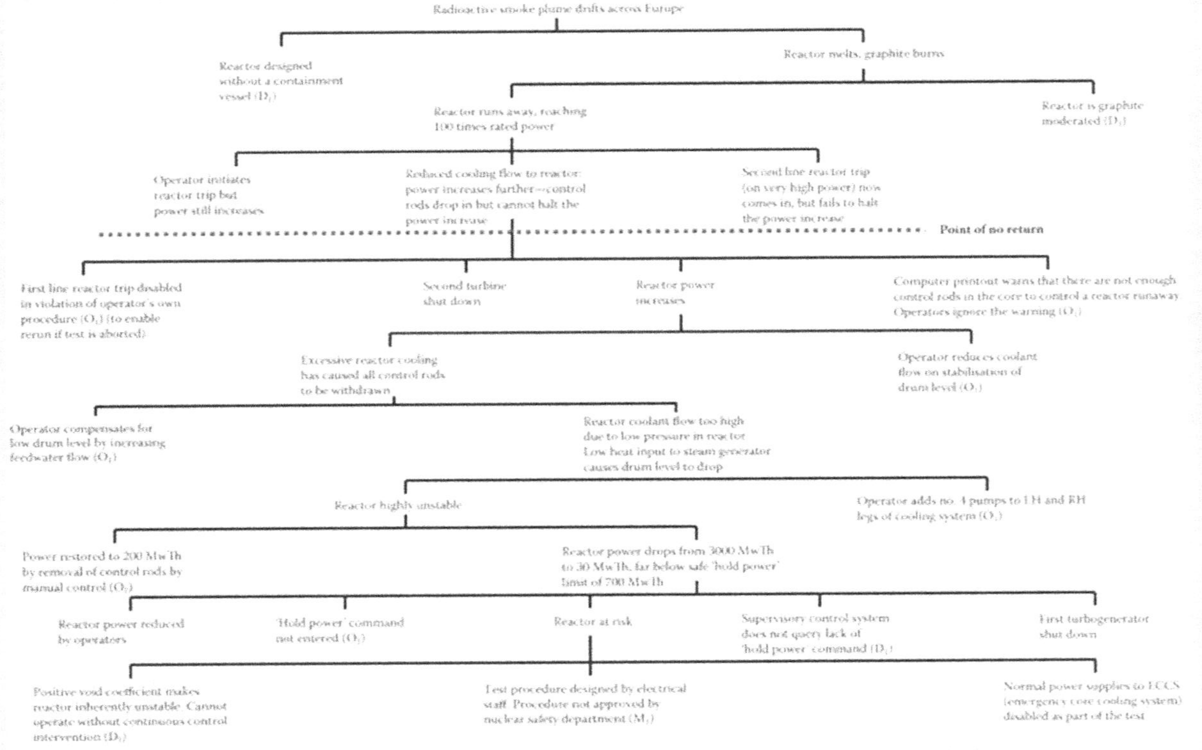

FIGURE 5.6 Fault tree of the Chernobyl accident.

CASE 5.2 PIPER ALPHA

It is unfortunate that society seems to need major events to remind it of its duty and the continuing need for attention to safety.

F. Crawley
Chairman of the Chemical Engineers' Symposium (Jenkins 1991)

To many of the crew of Piper Alpha, as shown in Figure 5.7, the day of July 6, 1988, was just like any other day. None of the crew suspected that by the day's end, the fate of their rig would be sealed in a matter of 22 minutes. This accident, one of the worst in the offshore industry, claimed the lives of 167 men, 2 of whom were on a rescue boat.

On the day of the accident, a maintenance check was scheduled for a safety valve on the backup propane condensate pump. Because it was already quite late, 6:00 p.m., the maintenance crew received permission to halt work for the day and complete the inspection the following day. Because this operation required the removal of the valve, the open end of the pipework was fitted with a blank flange.

FIGURE 5.7 Piper Alpha.

At around 10:00 p.m., the primary pump stopped and could not be restarted. The control crew, unaware that the standby pump line was unavailable, proceeded to start the backup pump. With the pump working against the blank flange, very soon pressure built up sufficiently to cause significant leakage. With the highly flammable propane now in the atmosphere, within seconds it found an ignition source.

The blast ripped through the firewall of module C, setting fire to the oil in the pipes ruptured by the explosion. The fire extinguishing system that should have subdued the fires was not in operation and was not on automatic. In addition, the crew in the command module were incapacitated by the blast and therefore could not switch them on.

At approximately 10:20 p.m., 20 minutes after the first explosion, the fires were burning long enough to weaken and burst the high-pressure gas risers coming from the other nearby platforms. When the gas risers burst, the intensity of the flames dramatically increased.

INTRODUCTION

The Piper Alpha disaster, accepted as the worst offshore tragedy, took the lives of 167 men on July 6, 1988. As with all these tragedies, the disaster and consequent loss of life was the culmination of a series of unfortunate and entirely preventable events. These events can be divided into the categories of machinery failure, irresponsible designs, negligent maintenance procedures, and environmental conditions.

There were two factors that severely hampered the official investigation: Many of the key witnesses were killed in the explosions, and large parts of the remains of the platform were never retrieved because of costs. As such, the official explanation is contested by some of the survivors and a few of the crew of the diving support vessel *Lowland Cavalier*, which was in the vicinity at the time of the accident. These witnesses claim that a smaller explosion occurred 20 minutes before the larger explosion that occurred as a result of the leaking gas from the blank flange.

In November 1988, an investigation was launched, the Cullen Enquiry, to establish the cause of this and two other disasters. A large number (106 in total) of changes to the safety and operational regulations were recommended and accepted by the entire offshore industry (Ross 2008).

PIPER ALPHA BACKGROUND

The Piper and Neighbouring Oil and Gas Fields

In 1973, the Piper oil fields were discovered 176 km (110 miles) northeast of Aberdeen, Scotland. Production started in 1976 with Piper Alpha, the largest and most productive of the rigs in the oil field. It had three nearby neighbours servicing the Tartan and Claymore oil fields as well as the MCP-01

(Manifold Compression Platform) servicing the Frigg gas field. However, only in 1980 was Piper enhanced with the fitment of a gas recovery module that allowed it to join the St. Fergus gas line. The configuration is shown in Figure 5.8.

The fitment of the gas recovery unit allowed both Piper and Tartan to join the St. Fergus line, which was a convenient tie-in point to export gas. Additionally, a gas line linking Piper to Claymore was installed. This was primarily for the purpose of the Claymore gas lift system.

In 1987, the Piper oil field, exploited by Occidental Petroleum, produced approximately 81,000 barrels of oil per day, equating roughly to 20% of the company's daily production of 382,000 barrels. The rig produced oil in the Piper fields from 24 wells, and later gas from an additional 2 wells. The oil rig was connected by an oil pipeline to the Flotta terminal in the Orkney Islands and a gas pipeline to the St. Fergus gas terminal.

Piper Alpha Construction

The production modules, modules A to D at the 84-ft level, were arranged in line so that A is the most southerly and D is the most northerly of the modules. The rig was originally designed so that the most dangerous part, module A, which contained the wellheads, was farthest away from the utilities and control module, module D.

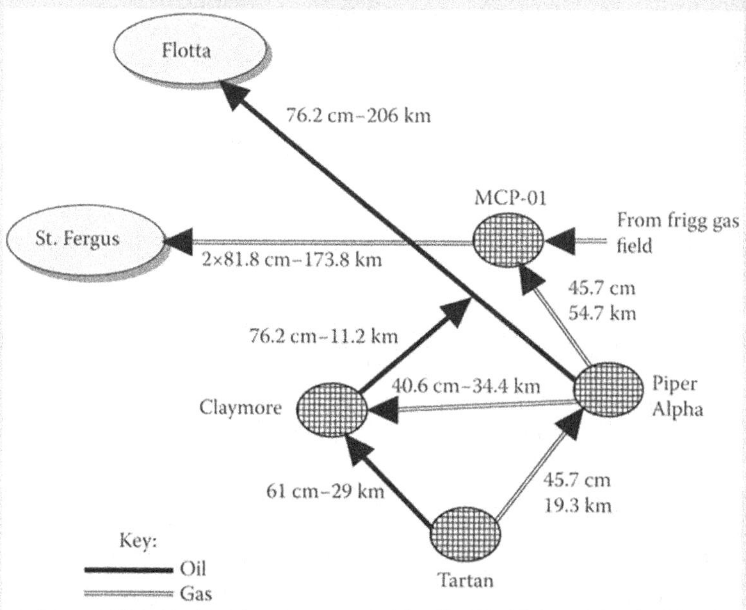

FIGURE 5.8 Rig interconnections showing pipe diameters and distances between rigs.

Module A comprised the wellheads. These were used to control the flow of hydrocarbons and to produce water from the wells. Module B was a production module, where separation of oil from the other fluids occurred, and this was pumped to the plant at Flotta. Module B housed the manifold test separators, the production separators, and the main oil line export pumps. Module C was home to the original gas export and compression equipment. From here, gas from the production separators was compressed for export to the MCP-01 platform and onwards to the shore at St. Fergus. Additionally, a gas conservation module had been installed at the 107-ft level. Module D housed the control room, workshops, and power generation equipment (primary gas turbine generators and emergency diesel generators). The accommodation quarters were located above module D. The modules were separated by firewalls, which were, critically, not blast-proof.

The gas conservation module was out of service at the time of the accident and was under repair. Thus, the gas was being processed only in module C at that time. The part of the process that occurred in module C comprised removing the heavier components of the gas. This was achieved with a Joule-Thomson separator that, by lowering the pressure of the gas, caused the formation of a liquid. The condensate so extracted was collected just below the module in a drum from where it was pumped in the main oil line to Flotta (this was how it was done before the gas conservation unit was installed, and because that unit was out of service at that time, it was again done as such at the present). The condensate was pumped to Flotta using a booster and condensate injection pump. There was an additional backup system; hence, there were two sets of these pump plants on the lower rig.

Backup Propane Pump Safety Valve Maintenance

On July 6, 1988, the backup propane condensate pump, at the 68-ft level (20-m level), just below module C, was to receive a maintenance operation, and this included the inspection of a safety pressure release valve, which required its removal. Thus, the gas flow line running through the backup system would be out of order while the safety valve was removed. Completely separate, and in addition to this, the backup pump was soon to receive a regular and scheduled maintenance operation that would last for two weeks (but had not started yet). The documentation detailing this was filled out, approved (by the maintenance supervisor), and filed that morning.

Near the end of the shift, however, it was becoming apparent that it would not be possible to complete safety valve inspection by shift's end at 6:00 p.m. Therefore, it was decided that the work would be left for the following day, when it could be completed and the hole in the pipeline covered with a blank flange. Critically, this blank flange was not fastened properly. The engineer in charge of the maintenance operation filled out the necessary permit indicating that the backup pump would be out of service because of the missing safety valve. In addition to this, it is speculated that when the engineer brought the documentation to the control room, the supervisor was busy and the engineer decided to sign off the documentation detailing the removal of the safety valve himself, but this was filed in a location different from that of the two-week maintenance permit. The engineer, however, did not inform the supervisor of the missing safety valve and that the backup pump was out of service.

Primary Condensate Pump Failure and Initial Explosion

At approximately 10:00 p.m., the primary condensate pump stopped and could not be restarted. After checking the maintenance work permits, only the two-week permit was found, and the supervisor did not know of the removed safety valve, the permit of which was filed in a different location. Furthermore, the rig workers did not notice the missing safety valve either, because it was situated high up from the platform-level floor and was not easily visible. Thus, seeing that the two-week maintenance had not yet begun, and not knowing about the missing safety valve, the supervisor authorised the use of the backup pump.

Because of the improperly secured blank flange, the high-pressure propane condensate in the line escaped into the atmosphere of module C. Very soon, gas monitors in module C detected the leak, and the operators in the control room were made aware of this by warning lights and alarms. However, before they could react, the stray gas found an ignition source and exploded.

The operators in the control room very quickly shut down the pumps in the system, stopping the flow of oil and gas to the platform. However, there were still oil and gas in the system, totalling nearly 80 metric tons as well as 160 metric tons of diesel located in the tanks above module C.

Explosion in Module B, Firewall Break-Up, and the Fire Suppression System

The explosion that occurred in module C was sufficiently violent to cause break-up of the firewall, which was made up of truss work and panels, between modules B and C. Crucially, the firewall was not designed to withstand an explosion—that is, it was not blast-proof, only fireproof.

Had this firewall been blast-proof, the fire would have been contained in module C and the rig would most likely have been saved from the disaster that followed. The construction of the firewall was thus identified as one of the vulnerable areas of design which greatly contributed to the disaster.

The shrapnel form this blast ruptured oil and gas piping associated with the metering skid in module B and set it ablaze. Crude oil caught fire immediately, and after approximately 10 seconds, a fireball was seen coming from the western face of the B module. This fireball was deemed consistent with a condensate release and was indeed later found to come from a ruptured 4-in (100-mm) pipe.

With the fires now burning, the rig could have been saved from complete disaster had the fire suppression equipment activated.

The fire suppression system on the Piper Alpha, and indeed many oil rigs, is run from diesel-driven pumps. From the control room, it is possible to switch the pumps from fully automatic to manual to completely off.

When divers are working below the sea level, it is standard practice to switch the pumps from automatic to manual when they are working in the vicinity of the spray system intake. This was done purely from a safety point of view so as not to get divers trapped in the entrance to the impellers in case of a fire.

However, the control staff on Piper Alpha had a habit of using the automatic setting regardless of where the divers were working, and indeed, on the day of the accident, the system was switched to manual.

After the initial explosion and consequent oil fires, and with the control crew abandoning the control room, the workers on the platform realised that the fire suppression system had not been activated, and two of them, Bob Vernon and Robbie Carroll, made an attempt to activate it by starting the diesel motors. They were never seen again.

In the vicinity of Piper Alpha, there were two other oil rigs, Claymore and Tartan. These joined the network of oil pumps that fed the onshore line at Flotta Island. The rigs in that network were under instructions that if one of them experienced a severe fire and had to stop pumping, the other would cease production as well. If this was not done, the back pressure in the line would cause a reverse flow of oil on the offline platform, thus further fuelling the fires.

Piper Alpha lost all ability to communicate after the explosions. Because of this, the manager of Claymore believed that the situation on Piper was under control and continued to pump oil into the main oil line. Furthermore, although Tartan had been aware of the explosion, they received orders from ashore to continue production and thus did not cease production either. Therefore, the pressure in the main oil line was sustained and enough to cause back pressure of sufficient strength so as to cause backflow of the oil line from Piper.

In addition, it is highly likely that the emergency shutdown valve on the main oil line failed to close completely, thus allowing a flow of crude oil being pumped by both Tartan and Claymore.

Failure of Gas Risers

With the fires from the oil lines burning on the 68-ft (20-m) level, the close-by gas risers started to weaken as well, and then 20 minutes after the initial explosion, the Tartan gas riser failed catastrophically, releasing an estimated 30 metric tons of high-pressure propane into the inferno and causing an enormous explosion (felt up to a mile away) and fire. In addition, it is speculated that the nearby diesel tanks started to fail and add further fuel to the fire.

At 50 minutes after the initial explosion, a second gas riser failed—from MCP-01, fuelling the fire even more. At this stage, the MSV *Tharos* was nearby and was attempting to rescue the men and assist in fighting the fire. However, the second explosion was so violent—it took the lives of two men on board—that the vessel had to retreat to a safer distance. From here on, it could do nothing but watch the rig burn. The second explosion is regarded as the "point of no return" for Piper Alpha, after which it could not be saved.

Approximately an hour after the initial explosion, the Claymore risers failed as well. Shortly afterwards, the derrick collapsed and the ERQ module toppled into the ocean. This is deemed the end of the Piper Alpha oil rig.

Also located on the 68-ft (20-m) level, along with the condensate pumps and oil lines, were the Tartan, Claymore, and MCP-01 gas risers. These gas risers are the pickup points for the gas lines (propane mostly) that run though Piper and consist mostly (and in this case crucially) of large-bore (18-in or 450-mm) pressurised piping.

The risers are also equipped with emergency shutoff valves (ESVs) that are closed remotely from the control room. Crucially, however, these valves are on the rig and, as seen, are vulnerable themselves and can be weakened in case they are subjected to continuous heat.

Operating companies subsequently identified the need for a subsea isolation valve (SSIV) in order to minimise the risk of a long-lasting fire on board the rig. The reason for this is that the water level above the valve would provide a barrier against the fire and heat.

However, there is a trade-off to be made. The reason that the ESV was located on the platform was to minimise the stored inventory in the pipes. With an SSIV, the inventory trapped in the piping system after the safety valve is much larger, and thus more crude fuel is available to the fire. Since the disaster, many oil companies have fitted these devices in addition to the standard ESV.

Environmental Conditions: Wind Direction

A further factor that led to the major loss of life was the combination of the wind direction and the location of the oil fire.

The oil fire that was burning on the 68-ft (20-m) level under module B caused very thick black smoke. The wind direction was such that the smoke moved towards the helipad, preventing the rescue helicopter from landing and evacuating the men waiting, as per training, in the muster area in the ERQ. In addition to obscuring the helipad, the wind also blew the smoke to the ERQ, quickly filling it with smoke. Upon later retrieval of the ERQ, more than 70 bodies were found of men awaiting evacuation. Their cause of death was determined as asphyxiation by smoke.

The initial investigation concluded that two men, Bob Vernon (the operator that started the backup pump) and Terry Sutton (the artisan that did not tighten the blank flange properly), were to blame for the incident on account of negligent maintenance procedures.

There is some evidence in the form of eyewitness accounts (16 out of the 34 survivors) and alleged radio transmissions that there was a smaller explosion on a different part of the rig that occurred 15 to 20 minutes before the initial explosion, and it is claimed that this fire started the initial large fireball.

One such witness is John Barr, who was supervising a diver on Piper at that time. He said he had ordered the diver to surface 20 minutes before the incident because he had felt an explosion. In addition to him, there was also the crew of the *Tharos*, which was roughly 500 m from Piper Alpha at that time, who claimed to have seen a smaller explosion, and this is, in fact, written up in the vessel's logbook at 9:45 p.m. However, this evidence was later deemed "not necessary for the Cullen Enquiry." Finally, there is Angus Kennedy, an engineer on the supply ship *The Performer*, who said they received a message from Piper Alpha claiming a "small but manageable fire" some 20 minutes before the first explosion.

The source and location of this explosion were never established and were deemed not relevant to the final explanation for the accident. This is seen by these witnesses as a very negligent oversight of the Cullen report.

HOW LOSS OF LIFE COULD HAVE BEEN PREVENTED

Respecting the Work Permit System

The initial explosion, resulting from the condensate leak from the temporary blank flange, was as a direct result of the backup condensate pump being switched on. Had the work permit system been up to standards, the control room operator would have noticed the work permit for the removal of the safety valve and the installation of the blank flange. Instead, he did not even sign it off (the impatient maintenance engineer did), and it was filed in another location. Furthermore, the engineer neglected to inform the supervisor of this, and thus he had no idea that the backup line was out of commission.

A major source of concern and blame was the work permit system on the rig. It had deteriorated to the point of disrespect and was generally seen as merely a formality. As a matter of fact, the work permit documents often were not even filed and were simply communicated verbally.

Blast-Proof Firewalls

It is generally accepted that had the firewalls been blast-proof and not simply fireproof, the fire in module B could have been avoided, as the only source of fire would have been in module C and, on its own, could have been contained, as similar accidents on rigs have proved.

However, the installation of blast-proof firewalls is a very expensive proposition, and because of a negligent and incomplete risk analysis, it was deemed to be too expensive when compared against the risk.

The Fire Suppression System

The fire suppression system, as fitted to the rig, works by sucking up larger quantities of sea water and releasing it on the needed area of the rig in case of a fire. It uses diesel-driven pumps that have settings for *off*, *manual*, and *auto*.

The normal operating procedure is for the system to be switched to the *auto* setting unless divers are in the vicinity of the intakes of the system. However, the crew of Piper Alpha had a habit of switching it to *manual* regardless of where the divers were working, as was the case at the time of the accident.

Had the system been switched to *auto*, the system would have activated and doused the fires raging in the B and C modules, and the rig would have been saved. In addition to this, when some of the men realised that the system was not in operation, two of them volunteered to manually switch on the diesel pumps, and they were never seen again.

Evacuation Procedures and Communication System

The initial explosion and the subsequent fires knocked out the control room, and with it the public address system. This had a direct influence on the evacuation procedures, as the men were told to gather at the muster point in the ERQ near the helipad and await further instructions. The general idea being that the men would be airlifted to safety.

However, had the PA system been in operation, the men could have been informed that the helipad was obscured by thick smoke from the oil fires and the helicopter would not be able to land. Instead, more than 70 of the fatalities were found in the ERQ, having asphyxiated while waiting for the rescue helicopter that never came.

In addition, the second gas riser that ruptured (the MCP-01 riser) exploded with such force that it knocked out the communication link from Claymore to the shore. This severely delayed their ceasing to pump. Instead, they tried raising their management on the shore with a satellite link.

Actions of Neighbouring Rigs

It was protocol that the neighbouring rigs should shut down in case of a major fire. This is to ensure that when the burning rig stops pumping, the back pressure in the line does not cause a reverse flow in the system and fuels the fire.

The other rigs primarily did not stop because:

1. They thought that the ESVs on Piper would have stopped the flow.
2. They thought that the fire suppression system on board the rig was holding up and keeping the blaze under control.
3. The revenue from a rig over a day can amount to a very considerable sum indeed, and once shut down, it would take several days to a week to restart again.

If indeed the other rigs shut down and it turned out in the end that it was not necessary, the control officers were afraid of being blamed for losing the company a lot of money. Compounding the problem was Piper's communication breakdown. Without this, the other rigs could not communicate with them whether they should shut down.

Subsea Isolation Valve

As mentioned, it was afterwards discussed among the petroleum companies that operated oil rigs that an SSIV should be fitted to the oil lines running to the platforms. This would have prevented the fires being fuelled from the failed risers and their ESVs, as the cut-off point to the fires would have been below sea level and thus protected from the flames. The drawback of only fitting an SSIV was that it increased the crude oil stock in the piping system leading to the rig.

The best course of action would thus be to fit both an ESV and an SSIV to the rig and main oil line (MOL). Thus, in the case of the ESV holding up, a minimum of stock within the piping system would ignite, thus limiting the fuel supply of the fire. But in the case of the ESV failing, the rest of the stock in the piping would indeed ignite, but then at least the fuel supply would be cut off by the much more secure SSIV.

If an SSIV had been fitted to the MOL, the burst risers would not have so severely affected the platform, and although it would have been damaged, the platform could have been operational within weeks.

In summary, the main consequences that followed the removal of the safety valve are outlined next in the form of a timeline. As can be seen, the initial events happened in very quick succession, causing a rapid, out-of-control spiral. However, it should be remembered that had the correct procedures been followed, the disaster would have been entirely avoidable.

The sequence of events is portrayed according to the official findings of the Cullen report and does not take consideration of the alternate hypothesis, owing to lack of information.

THE AFTERMATH OF PIPER ALPHA: QUANTITATIVE RISK ASSESSMENT

Definition

Within the UK legal framework, the first definition of *quantitative risk assessment* (QRA) is found in the Offshore Installations (Safety Case) Regulations of 1992. It is defined as "[q]uantitative risk assessment means the identification of hazards and evaluation of the extent of the risk arising therefrom, *incorporating calculations based upon the frequency and magnitude of hazardous events*."

History

After the Piper Alpha disaster in 1988, the North Sea oil and gas industry has been keen on adopting the QRA methods, especially in the North Sea.

However, the QRA methods have been in widespread use before that—as early as 1981—especially by the Norwegian Petroleum Directorate (NPD).

By 1984, the UK Offshore Operators Association Safety Conference devoted an entire session to illustrate the development and application of systematic risk analysis. While many efforts were made in the Norwegian sector to comply with the NPD, within the UK sector, it was more at the discretion of individual operators as there was no formal requirement to submit such risk analyses to UK authorities.

The Industry's Reaction to Piper Alpha

The disaster led to far-reaching changes in the way that health and safety are regulated in the UK offshore oil and gas industry. Following recommendations by the Cullen Enquiry and the formalisation of these into the Offshore Installations (Safety Case) Regulations 1992, the introduction of regulations that require the submission of a safety assessment for any operator present in the UK North SEA brought about the widespread use of QRA. This included the fact that the acceptance standards have been met with regard to the following:

- Integrity of temporary refuge
- Escape routes
- Embarkation points
- Lifeboats

This all places the responsibility on the operator to demonstrate that, in the event of a dangerous situation, the hazards have been identified, risks have been evaluated, and appropriate measures have been taken to reduce the risks to persons to "as low as reasonably practicable."

The QRA risk management flow diagram is illustrated in Figure 5.9. Risk management addresses the risk and attempts to improve it and see whether it is as low as possible. It is also visible from the figure that risk management is an iterative process—if a certain strategy is found lacking, it is iteratively developed until a satisfactory solution is found.

QRA Packages and Techniques

In the immediate aftermath of the Piper Alpha disaster, a number of packages and techniques were developed. The spreadsheet technique was an initial favourite, as it allowed rapid deployment, but it also led to some quality assurance problems. The technique was very complex and routinely required more than a million spreadsheet cells. This made it difficult for the analyst to ensure that the data are managed correctly. This was a cumbersome system as it was extremely difficult to verify, and errors (typing errors, copying errors, etc.) were difficult to find and could have significant consequences.

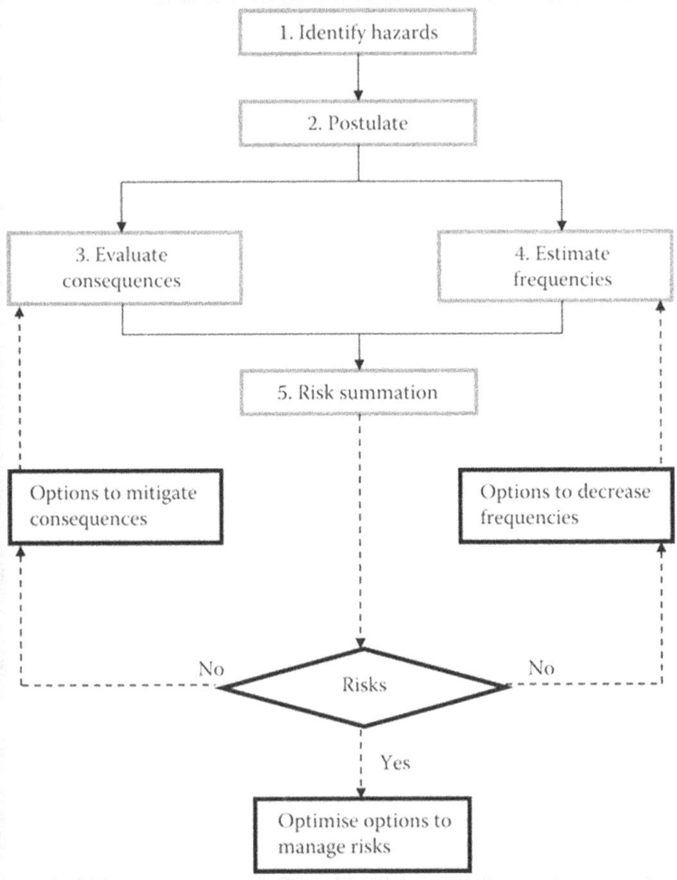

FIGURE 5.9 The QRA strategy flow diagram.

Another favourite technique was the event tree approach. In principle, it was somewhat easier to execute, and the risk of calculation was less than the spreadsheet approach. However, it required greater skill and experience to master and execute properly.

The OHRA Toolkit Project

In 1990, it was recognised by the offshore industry that a need existed for a change in the nature of the computational packages with which QRA was undertaken, especially with regard to transparency and updateability of the analysis. What was envisioned was a package that would allow the user to use the tools of QRA in a more structural and effective manner. This involved developing a completely new platform, using an intuitive "windows"-style user interface for ease of use, and employing the state-of-the-art technology of the time.

Conclusions

July 6, 1988 will forever be remembered as a black day by the petroleum industry in general and the offshore industry in particular, by the friends and families of the 167 men who lost their lives that day, and even by the traumatised victims (two survivor suicides were also recorded).

It was an entirely avoidable disaster that occurred as a result of a sequence of unfortunate events, as is so often the case with these major disasters. Had any one of a number of operating protocols been followed, the rig could have been saved and the loss of life and money could have been minimised. Basic procedures such as respecting of the work permit system, activation of the fire suppression system, and shutdown of the neighbouring rigs would have saved the derrick.

However, key lessons were learned from the accident and in the Cullen Enquiry, and many recommendations were made, all of which were incorporated into the regulations. A significant addition was the requirement, by legislation, for a QRA strategy.

The Piper oil field is still active to this day. After the accident, a replacement rig, Piper Beta, was constructed and is still operational. The rig is located barely a hundred meters from the site of Piper Alpha, where a commemorative buoy is located. Every year, on the morning of July 6, a minute's silence is observed on Tartan, Claymore, MCP-01, and Piper Beta in respect of the deceased, and on the 20th anniversary of the accident (July 6, 2008), the survivors, their families, and the families of the deceased were taken to the site for a memorial service.

A memorial in the Hazlehead Park in Aberdeen was also constructed. The Piper Alpha Families and Survivors Association was created for the survivors and the friends and family of the deceased, and they raised the funds to erect the Piper Alpha memorial in the park.

CLASS ASSIGNMENT

Prepare your own brief analysis of Piper Alpha (±4 pages), highlighting the different types of error that occurred (maintenance, management, operation, design, etc.).

CASE 5.3 FLIXBOROUGH

Author's Note: The Flixborough disaster of the 1970s shows how a lack of understanding of basic engineering principles can have horrific and tragic consequences. More specifically, lack of engineering oversight of a temporary modification led to these consequences. When unsupervised changes occur in the configuration of a dangerous system, watch out!

On Saturday, June 1, 1974, there was an explosion at the Flixborough Works, causing massive destruction. The aftermath of the accident is shown in Figure 5.10; 28 people were killed, 36 people injured, and thousands of houses were damaged within a 5-km radius. The explosion was estimated to be the equivalent of 15 to 45 tons of TNT. It was extremely fortunate that the explosion occurred out of normal operating hours. It is estimated that more than 500 people may have died had the explosion happened during the week.

The plant produced caprolactam, a chemical used in the production of nylon. One of the reactor tanks in the circuit was found to have a severe crack. A hasty decision was taken to remove the tank from the circuit and install a bypass line in order to maintain production. This bypass failed, resulting in a large leak of cyclohexane into the atmosphere, which ignited, causing a vapour cloud explosion.

At the time of the explosion, there was no works engineer at the plant, the previous engineer having resigned a few months before the accident.

The incident led to public comment that there was a lack in industrial safety, in particular in the regulatory and governance aspects. This prompted the Secretary of State for Employment to set up a court of enquiry into the incident to capture immediate learnings from the incident. An Advisory Committee on

FIGURE 5.10 Flixborough plant after the accident: top view.

Major Hazards (ACMH) was also set up to run in parallel. Reports from this committee served to guide UK legislation and regulation and were, at a later stage, adopted by most of the European Union.

Some of the learnings and recommendations include the following:

- The necessity to regulate the design, construction, and operations of plants that are considered to have major hazards. The definition of what constitutes a major hazard is unclear. The ACMH refers to "notifiable installations" and suggest basic criteria.
- Plants must be designed for safe operation and must be carefully assessed for the possibility of rapid catastrophic failures. This was not considered at Flixborough.
- Literature after the catastrophe discussed hazard identification and the various methods of addressing these hazards, noting that not all hazards can be eliminated; however, effective controls can be put into place.
- Finally, there were various technical learnings, such as nitrate stress corrosion and zinc embrittlement of stainless steel, that were taken from the incident.

BACKGROUND

The Works of Flixborough plant were positioned on the river Trent, near the town of Flixborough. The plant was originally owned by Nitrogen Fertilisers Limited, manufacturing ammonium sulphate, and started operation in 1938. Nypro UK then took over the plant. Nypro UK was owned by Dutch State Mines and the British National Coal Board. The plant was modified in the 1960s to produce caprolactam, a chemical used in the production of nylon 6. In 1972, the plant's capacity was increased from 20,000 tons per annum to 70,000 tons per annum. The process of producing caprolactam was also changed. Caprolactam was produced from cyclohexanone, while initially cyclohexanone was produced via the hydrogenation of phenol.

The new process used for the increased output involved the oxidation of cyclohexane. It was a failure in this area of the plant that resulted in the release of a large volume of cyclohexane, which consequently ignited, causing the explosion. It is estimated that the explosion had the energy of 15 to 45 tons of TNT.

ORGANISATIONAL STRUCTURE

The management structure at the Flixborough plant is outlined in the diagram shown in Figure 5.11.

It is important to note that Riggall was the only chartered engineer in the structure. There was no one else in the structure with the capacity and capability to operate on the executive level. Riggall had resigned earlier

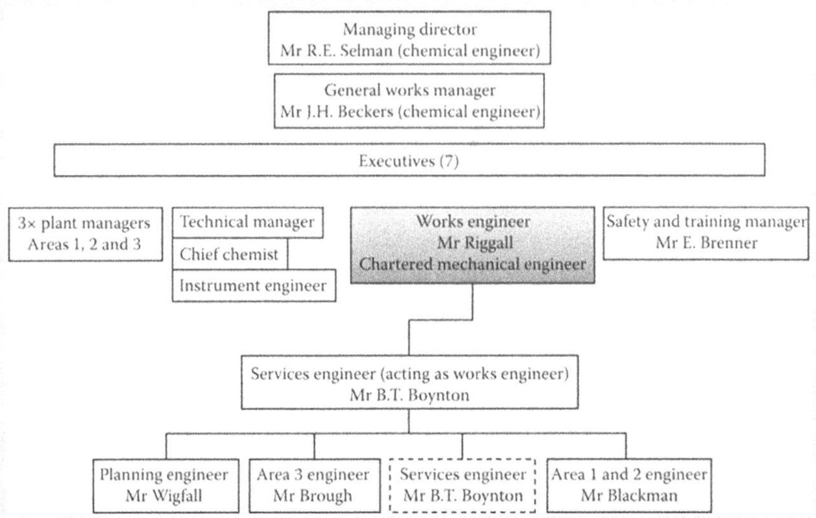

FIGURE 5.11 Flixborough management organogram.

in 1974, and the general works manager was, at the time of the disaster, still looking for a suitable replacement. In the absence of Riggall, B. T. Boynton acted in his capacity, although he did not have the required training, qualifications, or experience. The plant was also in the process of management restructuring, which was due to take place on July 1, 1974. Certain aspects in the engineering department were to be addressed in the restructuring, including the works engineer. In the interim period, the engineering department was told to call on J. F. Hughes, who was assistant chief engineer for one of the National Coal Board's subsidiaries, if any assistance was required.

The general works manager and managing director were both chemical engineers and had limited mechanical knowledge.

PRODUCTION PROCESS (FIGURE 5.12)

The section of the plant where the explosion occurred produced caprolactam. As discussed previously, the process was changed to produce caprolactam through the oxidation of cyclohexane. This involves raising the temperature and pressure of the cyclohexane to around 155°C and 8.8 bar, respectively (to keep the cyclohexane liquid), and passing air through it. This resulted in a partial reaction of cyclohexanone. The Flixborough plant used six reactors in series with around 100 tons of cyclohexane in the system at any time. With this arrangement, the reaction rate was around 6%. Any cyclohexane that was not reacted was recycled and fed back into the first reactor again. Cyclohexane exhibits properties similar to that of petrol.

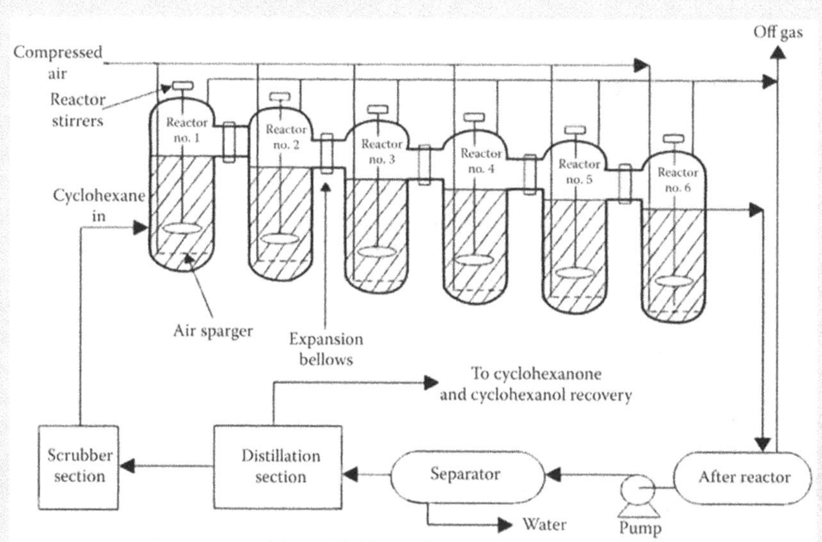

FIGURE 5.12 Cyclohexane reaction process.

Storage of such a large quantity of volatile substance is hazardous, even more so at the temperatures at which the plant operated. It was calculated that with 100 tons of cyclohexane at 155°C and 8.8 bar, 40 tons of cyclohexane would evaporate if exposed to atmospheric conditions. This led to the vapour cloud and explosion.

SIMPLIFIED PROCESS

- Cyclohexane was passed gravitationally from reactor 1 to reactor 6.
- The product from reactor 6 was then passed through a separator, a distillation section, and a scrubber. Cyclohexanone was removed in this process (only approximately 6% of the product).
- The remaining cyclohexanone was then returned back to the start of the cycle again.
- Compressed air was fed into the bottom of each reactor to allow for oxidation.
- Each reactor had an agitator.

PLANT OPERATION

The cyclohexane reaction took place in section 25A of the plant, as is discussed next.

- Cyclohexane was fed into the system from (1) the feed storage facility, (2) recycled cyclohexane from reactor 6, and (3) distilled from the off-gas line.

- Steam was used to heat the charge of cyclohexane in the reactors during start-up and supply additional heat once the reaction process had started. The steam was controlled by an automatic valve that could also be manually controlled. There was an additional parallel block valve bypassing the main control valve. This valve was also opened partially where additional steam was required.
- As discussed previously, compressed air was supplied to the reactors, entering at the bottom of each reactor.
- High-pressure nitrogen was used to elevate the pressure of the system. The nitrogen source was at a pressure of 12 bar.
- The pressure of the entire system was controlled via a pressure control valve that could be either automatically or manually controlled. The valve vented to a flare stack. Two block valves were positioned on either side of this automatic valve. These were generally closed during initial pressurisation at start-up, and after this, they were left open.
- There were numerous safety valves in the circuit. The safety valves were set to open at 11 bar and vented to the flare stack.

Figure 5.13 shows the schematic flow diagram of section 25A, the cyclohexane oxidation circuit, with reactor 5 removed.

PLANT START-UP PROCESS

It is necessary to understand the start-up process, as the leak occurred during start-up. The basic process is discussed in the following steps.

1. The plant is charged with water and pressurised with nitrogen to 4 bar to test for leaks.
2. The pressure increased to 9 bar and tested again for leaks.
3. The system is depressurised, and any leaks are repaired.
4. The system is then charged to the appropriate level with cyclohexane.
5. Nitrogen is used to pressurise the system to 4 bar.
6. Heating commences using the steam.
7. The cyclohexane is cycled at a reduced rate during warm-up.
8. As the charge warms up, the system pressure increases.
9. During this process, the pressure control (block) valves are manually closed until the system has built up pressure.

Overpressurisation can occur if:

- There is initially too much nitrogen in the system, leading to excessive pressure at the required temperature.
- The temperature is raised too high, that is, above an operating temperature of 155°C.
- There is a leak of nitrogen into the system.

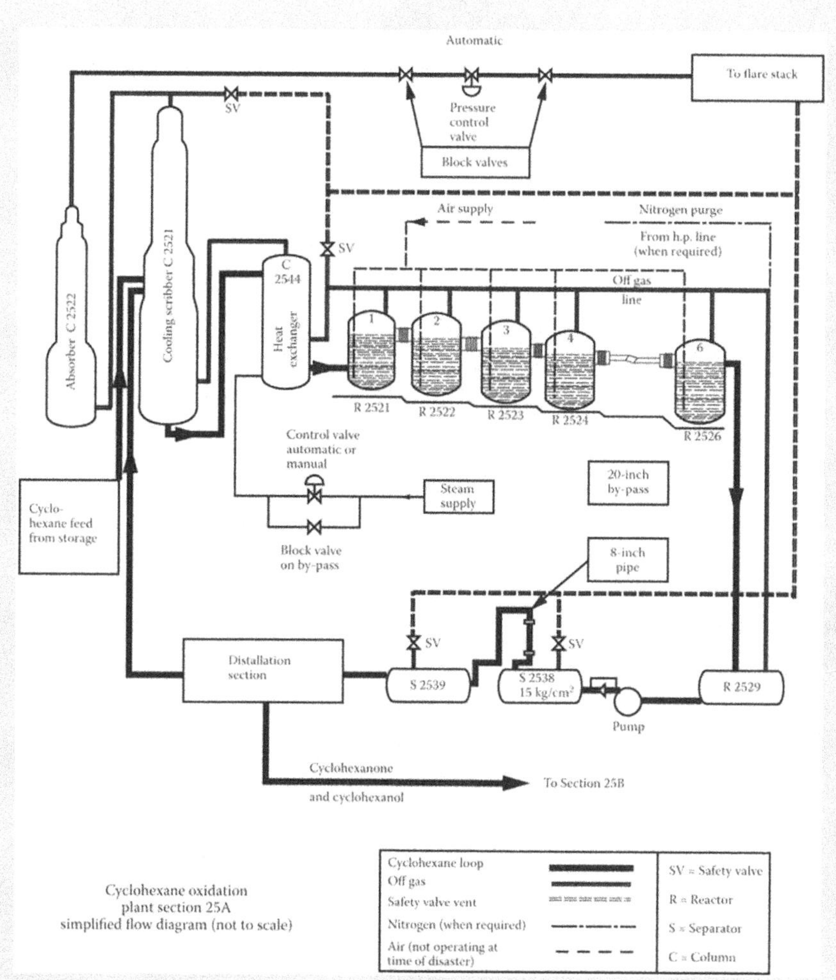

FIGURE 5.13 Reactor train with reactor 5 removed.

DAMAGE TO REACTOR 5

On March 27, 1974, cyclohexane was found to be leaking from reactor 5. Each reactor was 3 m in diameter and 6 m in height, with a 3-mm stainless steel inner shell and a 12.5-mm outer shell. The entire reactor was lagged. The plant manager inspected the reactor shell the following day. He found a 2-meter crack. This crack was identified to be caused by nitrate stress corrosion, a phenomenon still relatively unknown at that stage. It was caused by nitrate-contaminated cooling water that was sprayed on the tank to dilute a small leak. It was then decided to remove reactor 5 from operation.

In order to keep the plant running, a bypass system was proposed . . .

BYPASS CONSTRUCTION (FIGURE 5.14)

On May 28, 1974, the same day that the crack in Reactor 5 was examined, the plant managers and engineering staff decided to construct a bypass pipeline to bridge Reactor 4 and Reactor 5.

There was very little engineering design consideration in the design, construction and installation of this bypass. In fact, the only calculations performed were basic pressure and flow calculations to ensure that the 500-mm pipe, proposed to be used for the bypass as it was the only suitably sized pipe available on site, was capable of handling the flow. The only design work done was to sketch the full-size bypass on the workshop floor. Construction was started immediately, and by the end of May 31, 1974, reactor 5 had been removed. The bypass was installed, and the plant was ready to be started up.

The original connections between the reactors were 700-mm-diameter pipes, with a bellow for expansion. The 500-mm bypass was attached by flanges onto the 700-mm bellows on either side. There was also an elevation difference of 400 mm, requiring the bypass to have a dog-leg. It was appreciated that the bypass must be supported, and a support structure of scaffolding was installed.

FAILURE

At approximately 4:53 p.m. on June 1, 1974, a huge explosion erupted, instantly killing 18 people on site, injuring 36 others, and causing extensive damage in the surrounding region.

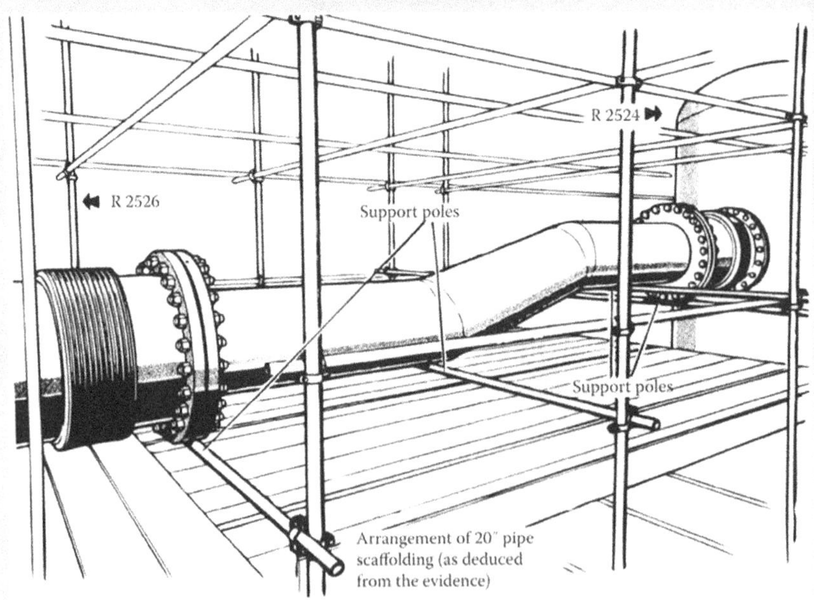

FIGURE 5.14 The bypass.

FIGURE 5.15 View of the scene after the accident.

It was understood that the bypass pipeline had failed, allowing a huge volume of cyclohexane to evaporate. This was then ignited by a source nearby, resulting in the massive detonation. A view of the scene after the accident is shown in Figure 5.15.

The exact cause of the explosion was not known, and the Secretary of State for Employment commissioned an enquiry into the incident to glean any possible learnings from the situation to prevent such an occurrence from ever happening again.

COURT OF ENQUIRY

The court of enquiry was led by R. J. Parker, who investigated the incident based on evidence from the accident scene, film, and footage, as well as interviewing 173 witnesses. The court considered three scenarios that likely caused the failure of the 500-mm bypass line, namely:

1. Rupture of the bypass assembly through internal pressure.
2. Rupture of the assembly in two stages: a small tear in the bellows from the overpressure, leading to an escape and minor explosion, causing final rupture.
3. Rupture of the 200-mm line, leading also to a minor explosion, causing rupture of the bypass assembly.

THE 200-MM HYPOTHESIS

This theory revolves around the possibility of a valve in a 200-mm line leaking. Two bolts were found to be loose on the valve, and it was concluded that these bolts could not have been loosened by the explosion; hence, they were likely to have been loose before the explosion. A leak from this line could perhaps have caused the explosion when ignited.

This theory was proposed by consultants investigating on Nypro's behalf. The proposal was disregarded by the court, stating that it was "a succession of events, most of which are improbable." There was also no conclusive eyewitness evidence that could confirm the theory. It would seem that this hypothesis was put forward by the company to deflect attention away from the temporary bypass. There may have been insurance implications and legal implications—the company was directly responsible for not checking the design of the temporary bypass but not responsible to the same extent for an artisan leaving two bolts loose in a pipe flange.

THE 500-MM HYPOTHESIS

This was the first theory investigated by the court. It proposed that the 500-mm bypass failed in one event because of internal pressure. It was found that under operating pressures, the bellows would be subjected to significant displacement. In the official court of enquiry report, it was indicated by research done by Prof. Newlands that failure of the bellows could occur at pressures only slightly (less than 10%) higher than the operating pressure of the plant. Lifting of the bypass pipe off its support at one end could also occur at these pressures. The pipe had been seen to lift during operation.

This theory is confirmed as a definite possibility by later investigations and calculations. Significant shear force was exerted on the bellows, which were not designed to handle such force. The shear force was attributed to the bending moment that was created by the pressures exerted over the dog leg pipe. This was not taken into consideration in the initial design. The bypass is shown with its dimensions in Figure 5.16.

THE 500-MM HYPOTHESIS IN TWO STAGES

This hypothesis considers the possibility of a tear in the bellows resulting in a leak that caused an explosion, causing the 500-mm bypass line to fail. A tear was found in the bellows. Incredibly, this hypothesis was rejected in the court of enquiry, after tests showed that the bellows would only tear at pressures of above 14.5 bar; this was above the 11-bar safety valve setting, and as such, it was rejected. The enquiry apparently did not consider the lifting of the pipe under pressure, which would have introduced high shear forces into one of the bellows.

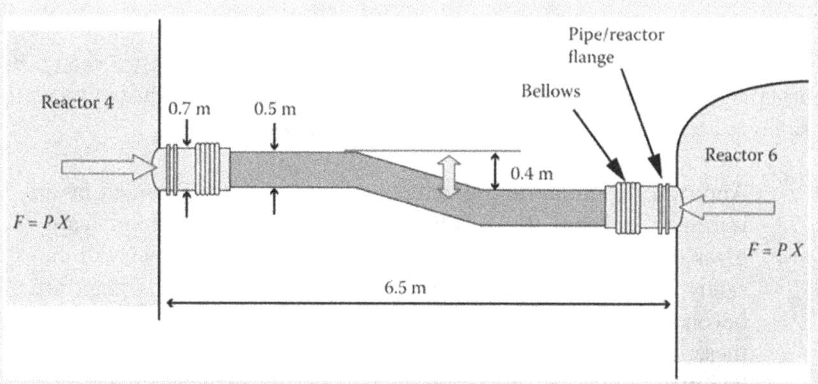

FIGURE 5.16 The bypass with dimensions.

In addition to this, Venart discusses in his paper "Flixborough: The Disaster and Its Aftermath" the possibility of resonant frequencies being achieved in the bypass, owing to unstable flow in the 500-mm line and the relatively low resonant frequency of the heavy system. This would cause increased fatigue in the bellows. Through computational analysis, he proved that failure of the bellows in this manner is possible.

LESSONS LEARNED

With a disaster of this magnitude, it is vital that as much is learnt from the incident as possible; these learnings must then be rolled out through industry for further safety improvement. The Secretary of State for Employment commissioned the court of enquiry as well as set up the ACMH primarily to draw learnings from this incident and others in industry, in order to give guidance for further improvements. There are full reports on the learnings gleaned. Some of the key points will be discussed next.

Hazard Analysis

No complex plant can be made completely free of hazards; however, it is vital that the hazards are assessed. Once they have been identified and assessed, there are different strategies that can be applied, including elimination, control, and containment of the hazard, with vital applications for each strategy.

Design for Safety

It is vital to design and build a plant with safe operation in mind. There is typically a very good return for the initial expense. One of the factors in this design stage is to design in a manner that reduces "critical management decisions (such as shutting down the plant)." This reduces the likelihood of poor or incorrect decisions with potentially disastrous consequences.

Standards and Regulations

Plant and equipment with high potential for accidents or disaster should be strictly governed by standards and regulations to ensure top industry standard safety practice and design. This includes the following:

- Approval for initial design and construction must be granted by an informed authority. In the Flixborough case, the local municipality gives approval for the design and construction of the plant. It can safely be assumed that the design and operation of this plant are beyond the municipality's technical expertise. In situations such as these, the ACMH recommends that the Health and Safety Executive be involved to assess the proposal and determine the design and construction requirements of the plant. The ACMH refers to "notifiable installations" that require specific treatment and approval procedures by the regulatory authority—this is based on specific criteria, such as quantity of hazardous substance held.
- The court of enquiry also recommended that the steam pressure vessel regulations are applied to pressure vessels handling hazardous materials, and where additional precautionary measures are required owing to the nature of the substance, this is complied with as well.
- Additionally, the court of enquiry requested that the British Standards referencing pressure pipework be clarified and that hydraulic pressure testing be made obligatory.

Plant Layout and Construction

In the Flixborough plant, the possibility of such a magnitude of accident was not recognised, and thus there was no strategy to contain it. It is important to identify such possibilities in a plant during the planning, design, and construction phases.

The citing of such plants must also be considered. In the Flixborough situation, it was extremely fortunate that the plant was situated in a relatively remote area.

Management Operations

Two areas were highlighted to be lacking in this incident. First, when a critical post such as that of the works engineer is vacant, caution must be given to decisions that are made that will usually have been made or influenced by that person. External consultation is recommended. Second, it is important that engineers have a broad-based knowledge of the different disciplines, as this will help produce informed decisions.

LESSONS LEARNED—TECHNICAL

Nitrate Stress Corrosion

Mild steel has been proven to show significant accelerated corrosion in the presence of nitrates. This phenomenon was found on reactor 5, attributable to the nitrated cooling water that was used to dilute a leak. Before this, little was known about it, and it was not appreciated.

Zinc Embrittlement of Stainless Steel

Stainless steel that is exposed to zinc at elevated temperatures and stress becomes brittle rapidly and can develop cracks.

Use of Bellows in Piping Systems

Any pipework engineer worth his salt knows to be wary of any design using bellows. They should only be used if essential and then must be constrained to limit their movement. In the original design, with all the reactors in service, they had to be employed to accommodate the increase in the height and diameter of the vessels with temperature, and they were, by the nature of their location, constrained, being unable to move more than a small amount in any direction. In the temporary connection, however, they were not constrained at all. A bellow is only designed to contain pressure and to allow for a limited amount of axial and shear movement. Such restrictions are always given in the bellows manufacturers' catalogues.

CLASS ASSIGNMENT

Construct a free-body diagram of the dog-leg pipe connection and calculate the pressure required to cause it to start to lift. Use the data given in the case. Assume the temporary pipe connection is of the minimum thickness of standard pipe to contain the operating and overpressure conditions and that the pipe is practically empty. Assume that the pipe pivots on the lowest of the three scaffold poles.

CASE 5.4 THE HYATT REGENCY HOTEL, KANSAS CITY

This case is about the collapse of walkways in the atrium of a hotel in Kansas City, Missouri, USA. The case is similar to the Flixborough case in that the failure is so simple and obvious in retrospect. It is amazing to think that such a failure could occur in the most advanced country on earth. All the more reason for persons in developing countries to be extra careful.

In early 1978, plans for the project were prepared, using standard Kansas City, Missouri, Building Codes. Construction on the hotel began in the spring of 1978. Eldridge Construction Company, the general contractor, entered into a subcontract with Havens Steel Company. Havens agreed to fabricate and erect the atrium steel.

The hotel is shown in Figure 5.17.

The design consisted of walkways linking the two sides of the atrium, across the second floor and the fourth floor. The design included long rods

FIGURE 5.17 The Hyatt Regency Hotel.

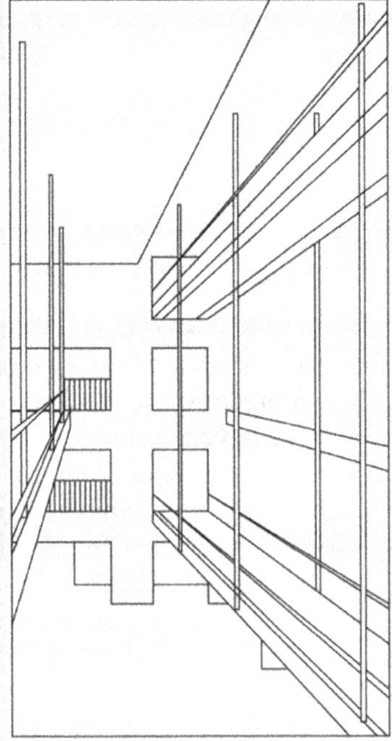

FIGURE 5.18 Layout of walkways in the atrium of the Hyatt Regency Hotel.

anchored in the ceiling. The walkways were suspended from these long rods, as shown in Figure 5.18.

The joint as originally designed is shown in Figure 5.19. This design could only support 91 kN instead of the 151 kN demanded by the building code. The joint would clearly have been a lot stronger if the channels had been placed back-to-back. This would have prevented the buckling of the webs and flanges.

Hence, the original joint as designed was inadequate, but now a second error was introduced. The design as shown in Figure 5.19 is clearly impractical from a construction point of view. Thus, the contractor suggested a change to the design, shown in Figure 5.20.

The joint as modified clearly doubles the load on the bottom nut. This was not picked up by the consulting engineer, who stamped the drawing as approved. The rest, as the saying goes, is history. On July 17, 1981, a dance was being held on the ground floor of the atrium. The second- and fourth-floor walkways across the hotel atrium collapsed, killing 114 and injuring in excess of 200 others.

As a result of the enquiry in November 1984, the engineers Duncan and Gillum, and their company, GCE International Inc., were found guilty of gross negligence, misconduct, and unprofessional conduct in the practice of engineering. Duncan and Gillum lost their licences to practice engineering in Missouri, and GCE had its certificate of authority as an engineering firm revoked.

Duncan and Gillum continued as practising engineers in states other than Missouri.

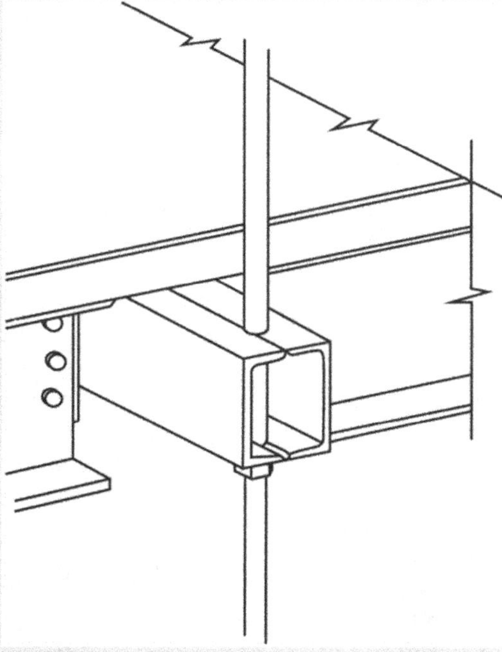

FIGURE 5.19 Joint as originally designed.

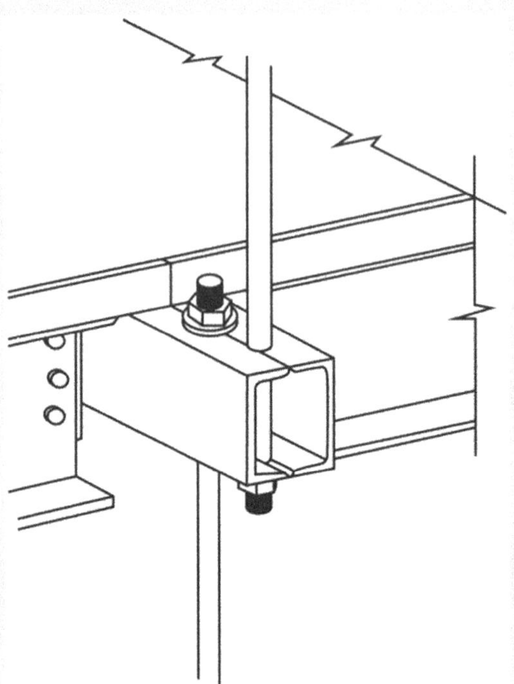

FIGURE 5.20 Joint as modified.

CASE 5.5 THE EMBRAER 120 AIRCRAFT CRASH

Author's Note: The following case is a story of maintenance errors leading to serious consequences. It seems incredible that such unconfigured maintenance practices can exist in the airline industry anywhere, let alone in the United States!

On September 11, 1991, an Embraer 120 (N33701) commercial passenger aircraft suffered a structural break-up in flight and crashed in a cornfield near Eagle Lake, Texas. The three flight crew and 11 passengers were fatally injured. The immediate cause of the disaster was the in-flight loss of a partially secured de-icing boot on the left-hand leading edge of the aircraft's horizontal stabiliser, which led to an immediate and severe nose-down pitchover and the subsequent disintegration of the aircraft. The accident enquiry carried out by the US National Transportation Safety Board revealed that, on the night before the accident, the aircraft had received scheduled maintenance that was to have involved the removal and replacement of both the left and right horizontal stabiliser de-icing boots. Investigators at the crash site discovered that the upper attachment screws of the left stabiliser de-icing boot were missing.

A drawing of the aircraft is shown in Figure 5.21. The de-icing boots can be seen on the leading edges of the wings and stabiliser. A cross section of the horizontal stabiliser is shown in Figure 5.22.

Earlier that year, in August, the airline had carried out a fleet-wide review of the fitness of aircraft de-icing boots for winter operations. During this review, a quality control inspector had noted that both leading edge de-icing boots on the stabiliser of N33701 had dry-rotted pinholes along their entire length. They were scheduled for replacement on September 10, the night before the accident. The work was carried out by two shifts: the second or "evening" shift and

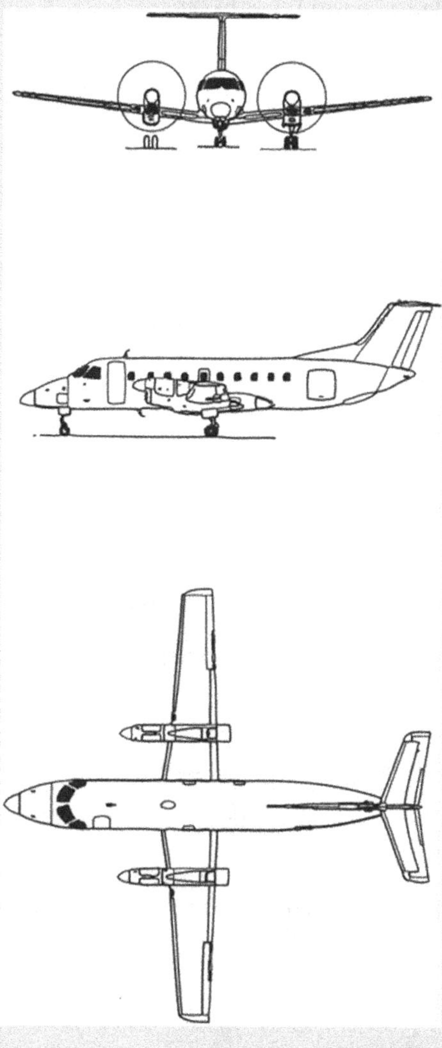

FIGURE 5.21 Embraer 120 (N33701) commercial passenger aircraft.

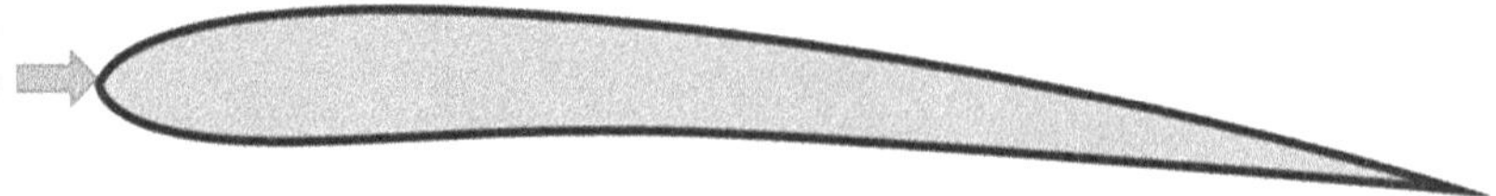

FIGURE 5.22 Typical cross section of a horizontal stabiliser (icing occurs along the leading edge, as shown by the arrow).

the third or "midnight" shift. The aircraft was pulled into the hangar during the second shift at around 9:30 p.m., where work began on replacing the boots.

Statement by the second-shift mechanic:

I and another mechanic, with the help of an inspector, used a hydraulic lift work platform to gain access to the T-tail of the aircraft that was about six metres above the ground. A second-shift supervisor assigned the job. He then took charge of the work on N33701, but did not issue job cards to us. (There were two supervisors on the second shift, one, Harry, dealing with a "C" check on another aircraft and the other, Peter, supervising the work on N33701.) We two mechanics removed most of the screws on the bottom side of the right leading edge and partially removed the attached de-icing assemblies. Meanwhile, the assisting inspector removed the attaching screws on the top of the right side leading edge and then walked across the T-tail and removed the screws from the top of the left side leading edge.

I gave a verbal turnover to the second-shift supervisor (the one working on the "C" check). He told me to give the turnover to a third-shift mechanic. This I did and then left.

Statement by the third-shift mechanic (Billy):

I received the turnover from the second shift mechanic but I was not subsequently assigned to N33701. I do recall seeing a bag of screws on the lift. I then gave a verbal turnover to another mechanic. His name is John.

Statement by third-shift mechanic (John):

I do not remember receiving any turnover; neither did he see any bagged screws.

Statement by third-shift mechanic (Pedro):

The Third Shift Supervisor told me that I would be working on the right hand boot replacement for N33701. He told me to talk to the Second Shift Supervisor to find out what work had been done. So I spoke to Harry and asked him whether any work had been carried out on the left hand assembly. He told me he did not think there would be sufficient time to change the left de-icing boot that night.

I removed the tail leading edge, with the boot from the horizontal stabiliser. I bonded a new de-icing boot to the front of the leading edge at a workbench. But then N33701 was pushed out of the hangar to make room for work on another aircraft. There was no direct light on N33701 as it stood outside the hangar. After the move, I re-installed the right side leading assembly.

Statement by the third-shift inspector:

I went to the top of the T-tail to help with the installation of the de-icing boot and to inspect the de-icing lines on the right side of the stabiliser. I did not spot the missing screws on the top left-hand leading edge assembly. I had no reason to expect them to have been removed and the visibility outside the hangar was poor.

Statement by the third-shift supervisor:

I arrived early for work and saw the second-shift inspector lying on the left stabiliser and observed the two mechanics removing the right de-icing boot. I reviewed the second-shift inspector's turnover form and found no write-up on N33701 because the inspector had not yet made his log entries. I then asked the second-shift supervisor—the one who was dealing with the "C" check on another aircraft—if work had started on the left stabiliser. He looked up at the tail and said "No." I then told this uninvolved second-shift supervisor that we would complete the work on the right boot during my shift, but the work on the left-hand replacement boot would have to wait for another night.

Statement by the second-shift inspector:

At 22h30, I filled out the inspector's turnover form with the entry, "helped the mechanic remove the de-icing boots." I then clocked out and went home. I did not turn the work over to the third-shift inspector. I placed the upper screws removed from leading edges of the stabiliser in a bag and had placed the bag on the hydraulic lift.

The next morning, the aircraft was cleared for flight. The first flight of the morning passed without incident, except that a passenger later recalled that vibrations had rattled his coffee cup. He asked the flight attendant if he could move to another seat. The passenger did not tell anyone about the vibrations, and the other passengers did not notice.

CLASS ASSIGNMENT

1. By means of a sketch, indicate the position of the de-icing boots on the aircraft. Also, by means of a sketch, draw the boot and the detachable leading edge of the stabiliser in cross section, indicating the pneumatic channel and the attachment screws on the upper and lower surfaces of the stabiliser.
2. Study the statements made by the various parties.
3. Prepare a report of the incident. Include the following:
   - A discussion of the responsibilities and errors of the persons involved.
   - Should the management of the company and the workers involved share in the blame for this event? Justify your answer.

- How would you go about ensuring that this type of accident never happens again?
- Draw a fault tree describing the events.

4. There is an important piece of information missing from the investigation. What is it?
5. How did the configuration of this specific aircraft contribute to the accident?

GLOSSARY

Explanation of Terms Used in Case 5.5: The Embraer 120 Aircraft Crash

Atmospheric icing: Atmospheric icing occurs when water droplets in the air freeze on objects they come in contact with. This is very dangerous on aircraft, as the build-up of ice changes the aerodynamics of the flight surfaces and can cause loss of lift, with a subsequent crash.

Not all water freezes at 0°C. Water can exist in the supercooled liquid state. These supercooled droplets are what cause icing problems on air craft. Usually, icing is not a problem in clouds if the temperature in the cloud is –20°C or colder. This is because at temperatures below –20°C, clouds rarely consist of supercooled water droplets, but rather ice particles.

De-icing boot: A *de-icing boot* is a device installed on aircraft surfaces to permit a mechanical de-icing in flight. Such boots are generally installed on the leading edges of wings and control surfaces (e.g. horizontal and vertical stabiliser), as these areas are most likely to accumulate ice and any contamination could severely affect the aircraft's performance. A de-icing boot consists of a thick rubber membrane that is installed over the surface. As atmospheric icing occurs and ice builds up, a pneumatic system inflates the boot with compressed air. This expansion in size cracks any ice that has accumulated, and this ice is then blown away by the airflow. The boots are then deflated to return the wing or surface to its optimal shape. Proper care for de-icing boots is obviously critical. Any holes in the boot will create air leaks that will decrease, if not destroy, any effect that the boots may have. As such, boots must be carefully inspected before each flight, and any holes or cuts must be patched. De-icing boots are most commonly seen on medium-sized airliners and utility aircraft. Larger airliners and military jets tend to use heating systems that are installed underneath the wing's leading edge, keeping it constantly warm and preventing ice from forming.

Pitchover: This is a change in the aircraft's attitude about its lateral axis through its centre of gravity (as opposed to its vertical or longitudinal axes). A nose-up pitchover, if continued, will result in a loop. A nose-down pitchover, if continued long enough, will result in an outside loop or bunt.

Shift turnover: This is the process by which responsibility is passed from the outgoing shift to the incoming one.

Supercooling: *Supercooling* is the process of chilling a liquid below its freezing point, without it becoming solid. This can occur if there are no seed nuclei, such as dust particles, which allow the water to begin to crystallise as ice. The rate of cooling is also important in this process. Droplets of supercooled water often exist in stratiform and cumulus clouds. They form into ice when they are struck by the wings of passing aircraft and abruptly crystallise. (This causes problems with lift, so aircraft that are expected to fly in such conditions are equipped with a de-icing system.)

CASE 5.6 TROUBLE AT THE SMELTER

Author's Note: A complex case that deals, among other things, with the problems of running a multishift operation. And there is a sting in the tail—or should we say tale?

INTRODUCTION

Mark Francis, the reliability and maintenance consultant, met Waylon Davids at a course he was giving on maintenance planning and control. Waylon Davids had worked at West African and Caribbean Aluminium (WACA) smelters for 16 years in various maintenance positions and had recently been appointed manager of the reliability section at the plant. WACA was a company well-known to Mark, who had consulted there several times over the past few years. WACA is the largest smelting and casting company in West Africa, using Jamaica's rich reserves of bauxite as feedstock in the making of aluminium ingots.

Four years before meeting Waylon, Mark had conducted a maintenance audit at the plant and had awarded it a Gold Star for Maintenance Excellence, according to the method Mark had developed. His second project with WACA was to give a course in RCM[1] to a group of artisans, technicians, and engineers at the plant. Subsequent to that, he had trained a group of young engineers in the technology of the plant and how it should be maintained.

WACA was part of an international mining and processing group known for its aggressive and successful management style. Waylon asked Mark to quote for an analysis of failures in the Casthouse, which was accepted. This consulting job required Mark to work with two of Waylon's reliability specialists in doing root cause failure analysis on failures in the Casthouse, with recommendations on how to reduce the failure rates.

Mark's quotation had specified the use of the ASSET methodology to analyse the failures. The ASSET methodology had been developed for use in the nuclear power industry and divided any problem area up into the so-called three Ps of plant, personnel, and procedures. This was explained to WS in

Mark's quote, as was the fact that he could not commence the work immediately because of the pressure of other assignments.

Mark was able to put in one week at the site, where the Casthouse technology was explained to him by the expatriate worker from Europe, Willem Heroult, the reliability specialist attached to the Casthouse. It was during the plant walk-down that Mark first learned of the TPM initiative on castline 1. Willem had nothing but praise for it but said it had been temporarily stopped during the installation of the new lines, numbers 6 and 7. A British consultant had been employed, at considerable cost, to implement TPM in the Casthouse. This had been an initiative of the process department.

CASTHOUSE ORGANISATION

Management of the Casthouse was shared between three groups. There was the process department, which was responsible for optimising the production of ingots. The department was largely staffed by chemical engineers and metallurgists. Then, there was the operations department, responsible for the operators who actually manned the line and produced the product. Finally, there was the maintenance department, responsible for maintaining the Casthouse equipment. The line maintenance staff were all millwrights, who were assisted by the centrally located automation department for maintenance of the various robots on the lines. The automation department staff were, for the most part, electronic technicians. The maintenance department included a planning section, with two planners who had been promoted into the job from being millwrights on the lines.

The operations and process departments reported to the Casthouse manager, Oregon Obaya, while the maintenance department reported to the manager of central maintenance, Cedric Altaba. Just under the maintenance manager was the maintenance superintendent, an experienced former millwright, also an expatriate, Vishnu Batachariah. All persons with the title of manager reported to the expatriate CEO. There were two persons employed in the reliability section under Waylon that would work with Mark directly. Apart from Willem Heroult, the other person was Uriah Johnson.

INITIAL IMPRESSIONS

Mark felt that CS was not the same organisation that he had visited several times before. The excellence that had led to their Gold Star award in previous years now seemed to be lacking. An organisation that had prided itself on its maintenance planning was now foundering in a sea of breakdowns. It was reported to Mark that the two shift artisans in the Casthouse had sometimes to deal with 20 breakdowns in a night.

THE INITIAL INTERVIEWS

When Mark returned to CS after completing his other assignments, he conducted a series of interviews with various persons, to get their impressions of the situation that confronted them. The persons interviewed included craft persons, operators, process engineering staff, reliability engineering staff, the process supervisor, and the maintenance supervisor. He also interviewed the maintenance planner, the SAP coordinator, who was a former Casthouse manager, and representatives of supplier companies like the strapper machine agent.

PLANT WALK-DOWN

Willem Heroult and the Mark performed a plant walk-down during the early phases of the investigation. The most serious non-conformance that was found was a hammer at the dropout section on line 1 that was not hitting the ingot.[2] This has to be taken as indicating a serious lack of control on the part of operations and maintenance, if nothing else.

DIVISION OF RESPONSIBILITIES

The craft people working on the lines were all millwrights, although some were better than others at the electrical side of the work. There was a centralised automation department of four persons, basically instrument mechanics and electronic technicians that were au fait with programming and microprocessors. If a millwright could not solve a problem with a sensor or an item of automatic equipment, he would call for one of these automation specialists.

EXTENT OF THE PROBLEM

At the Monthly Controller Summary Meeting, it was stated that there had been 90 breakdowns per month per line, counting all seven lines.

It was also stated that there are 135 man-hours allocated for a maintenance shutdown per line but that currently shutdowns were being completed in 85 man-hours because of a lack of artisan manpower. There were 12 artisans available, but according to the planning department, 18 were needed.

At this meeting, the maintenance manager, Cedric Altaba, emphasised that six extra artisans would be starting shortly and that they would solve the problem.

MARK'S FIRST PROGRESS REPORT

Mark was asked by Waylon to give feedback to the maintenance manager, Cedric Altaba, on what he had achieved to date. The first thing he brought up, after an introduction laying out the current position, was the fact that there seemed to be significantly more failures on line 1 than on the other lines.

"I have performed an analysis of variance on the lines, and the fact that line 1 experiences more failures seems to be statistically significant. The chances of it being a fluke are less than one in a thousand," Mark said.

This fact was greeted with silence by the audience of Waylon Davids, Cedric Altaba, and the two reliability engineers, Willem and Uriah.

"Do you know of any reason that this might be?" continued Mark.

"You must tell us that," said Waylon.

"The second important thing I think must be said is that the previous TPM initiative on line 1 has been enthusiastically spoken about by everyone that I have approached concerning it—reliability people, the process supervisor, and—"

Mark was interrupted by Cedric Altaba. "I know about this TPM. I used to work for Lever Brothers, and they had TPM. Do you know how expensive it is to implement? And there is only one plant in the whole of the Caribbean that has succeeded with it."

"But you were succeeding with it here—"

Mark continued, before being interrupted again.

"I have been here two years, and I don't know anything about TPM here, and we won't be implementing it," Altaba said.

Mark attempted to continue, "Most of the problems on the cast lines seem to be human-related. . . ."

"Don't tell us that. It is not what you are being paid for." It was Altaba once again.

Mark thought of reminding him that his quotation had stated that he would be using the ASSET methodology as a root cause analysis technique. As mentioned prior, this technique, developed in the nuclear power industry, splits up any problem into three components, known as the three Ps—people, plant, and procedures. It is a useful way of breaking a problem up into smaller components and finding out what went wrong in those three areas.

"Mark," Waylon said, "what we want is a technical analysis of what the main problems are. Perhaps we can look into the people areas later. What are the main *technical* areas where there is trouble?"

Mark replied, "The robot applicator that applies identification stickers to the bundles of ingots, the skimmer that removes dross from the top of the molten ingots, the gantry that loads the ingots onto the trucks, the strapper that straps bundles of ingots together, and the dropout section where ingots are sometimes not released from the moulds."

"At least we have that right," said Altaba. "You know the other team I appointed to analyse failures in the carbon section are doing very well and have already given us valuable information. One serious problem has been solved already. They are young local people—industrial engineering graduates from our local Technikon. Brilliant youngsters—I chose them both personally. When I compare their rate of pay to the rate you are charging, I must say your progress is too slow and you are costing us too much."

"We look forward to a more focused analysis from you at our next meeting, Mark," said Waylon.

After Waylon Davids and Cedric Altaba had left the meeting, Mark turned to Willem and Uriah. "I don't know what these people want!"

Willem replied, "Mark, it is as Waylon said—let's just concentrate on the technical reliability issues, like the other team is doing. For example, they have found that errors and omissions in the work instructions are one of the main causes of problems in the carbon anode production area."

Uriah also contributed, "This is some of the work of the other team that I have been able to get ahold of. Perhaps it will be helpful to us—you see, they have taken hundreds of photos of those sections of the plant they are responsible for. Perhaps we could do the same?"

THE SECOND PROGRESS MEETING

The second progress meeting was attended by the same persons plus the maintenance superintendent, Vishnu Batachariah. Mark explained that he and the two reliability engineers, Willem and Uriah, had completed the analysis of four of the five main problem areas. Mark pointed out the findings, which included the following.

GENERIC PROBLEMS APPLICABLE TO ALL AREAS

1. There is sometimes inadequate time to perform daytime maintenance on "shuts" correctly and completely. This can be caused by a number of reasons, including the following:
   - 1.1. If the furnace does not come back in time, then there is no time for a hot run.
   - 1.2. An inspection reveals work that must be done, but that was not in the tasks for the shut. This leads to work being rushed and inadequately planned.
2. A job that requires minimal time to complete can cause a major production disruption, owing to a number of reasons, such as the need to reheat the line. This can mask the effect of breakdowns if only the time to repair is considered.
3. It is not always possible to recommission equipment successfully when the maintenance person is not allowed to restore power to a section of the line for testing.
4. Unavailability of spares causes problems for two reasons:
   - 4.1. Work cannot be done when required, leading to further deterioration in the plant with possible knock-on effects (e.g. other items are damaged as a consequence).
   - 4.2. "Quick fix" solutions are adopted, which do not last, and the plant breaks down again.
5. Unavailability of operators when required. Because operators are required to be at more than one place on the line, if there is trouble at some point, the operator might not notice it. This is a recurring problem that must be solved. It is suggested that a remote system be incorporated, which warns the operator of problems if he is not at a particular plant. This could be accomplished by radio links from the

various sensors on the line, directed to a beeper kept by the operator. A solution using cell phones might also be possible.

6. Task lists and standard operating procedures (SOPs). Task lists instruct an artisan on how to perform a task. SOPs are the equivalent for the operators. There are problems with these documents in most cases. An initiative is in place to update these documents, but the execution is inadequate at present. For example, task lists have been modified by underqualified and, at this stage, unknown persons. The procedure is for an artisan to write on the task list whatever is necessary to update it. This information is then typed in by another party, whom we shall call the editor. On some of the task lists studied, the following were found:
    6.1. The copying of the material was done slavishly. For example, if the artisan had written, "Please add the following," the person would type in, "Please add the following."
    6.2. Many of the additions were all in upper case, in contradiction to the IOS writing standards.
    6.3. Some of the additions were inadequate, hard to understand, and ambiguous. It is suggested that when a task list is modified, it must be entered into the system by a person who is technically competent. Furthermore, the revisions should occur during an interview between the artisan and the editor so that through feedback and interaction, the task list can be properly generated.
7. The day shift millwrights are not available when "hot runs" are done.
8. One very serious systemic issue is the attempt to keep the 80/20 rule for planned versus breakdown maintenance in place when the plant has fallen into such a state that it is impossible to maintain this rule. The 80% planned work is broken up into 60% scheduled work and 40% loaded work as a result of inspections. Often, there is not enough time to complete the loaded work, which then has to be dealt with by the night shift as a breakdown. The writer contends that the plant should be restored to its original condition, with the suspension of this 80/20 rule in the meantime and the suspension of any associated KPIs.
9. Another important generic issue is the change from the original design intent of the plant, which was to have four lines operative and one as standby. (It is common for such an intent to be abandoned in the interests of production, and WACA is certainly not the only company to have reduced its redundancy like this.) When the new lines 6 and 7 are fully operational, then there should be sufficient redundancy in the system again for complete restoration of lines 1 to 5 to take place.
10. There is no simulation of plant operations. Operators are trained by studying SOPs and undergoing on-the-job training. With the level of computer power available today, it is possible to devise a simulator

that would allow trainee operators to see the effects of maloperation without damaging the real plant. Serious deviations from operating procedure, such as casting at the wrong temperature, inadequate mould filling, and so on, could be highlighted. Such a simulation model might also prove useful to the process personnel.

11. Previous initiatives have produced mixed results. For example, the initiative by a certain consulting company, Havoc Inc., brought more discipline into the maintenance planning process but also resulted in staff reductions that may have had a bad effect on employee morale and definitely resulted in a dilution of the maintenance skill base.

Mark then discussed the four problem areas that had been analysed to that point:

RCFA 1: Gantry

A Pareto analysis of the gantries revealed that the most troublesome areas were:

- Gantry stuck
- Load cell
- Gripper

An analysis of failures per line revealed the following:

- Line 1 100 breakdowns
- Line 2 32 breakdowns
- Line 3 55 breakdowns
- Line 4 36 breakdowns
- Line 5 165 breakdowns

This pattern repeated itself for most of the failure types, with lines 1 and 5 exhibiting more failures than the other lines.

The RCFA on the gantries revealed the following as the main problem areas, with solutions where appropriate:

- Software changes are being continually made to compensate for mechanical wear. This is not getting to the root of the problem. The mechanical components must be restored to original condition instead. For example, continual recalibration can mean that the graphite coupling is loose.
- Unlike other robotic systems in the Casthouse, the gantries have not been upgraded. This is because of cost considerations.
- There has been insufficient training for artisans on the gantries. The subject matter expert (SME)[3] only has six months of experience on gantries, and other artisans even have less experience.

- The fact that there is not enough time for the day shift artisans to complete all the work required in a shut was brought up not only in connection with the gantries but also in all the areas studied. It is recommended that a time-and-motion study approach be made to confirm if this is the case, at least with the gantries, by observing the work actually done during servicing of the gantries on the day shift.
- Internal disconnections in the electrical cables servicing the moving parts have been put forward as a frequent cause of breakdowns. There has not been the time to confirm this assertion.
- Many of the shift team artisans are new—a figure of 50% was quoted for the Casthouse as a whole.
- Because of time pressure for both the day and night shifts, many "quick fixes" are used, which lead to recurrence of the problem after a short time.
- There is not enough time for proper fault finding.
- Carbon build-up in various electrical contacts causes trouble. A maintenance check and refurbishing task must be drawn up for this.
- Task lists are inconsistent or inadequate in some cases. In revising the task lists, it is essential that the artisan and the reviser sit down together and, by a process of back-and-forth communication, establish task lists that will be unambiguous and understandable to a new artisan without the need for further explanation from others.
- Bundle integrity is a problem. If incorrectly sized ingots are not removed from the line but end up being bundled, it is sometimes beyond the capacity of the robot to correctly assemble the bundle, and it "goes on strike," that is, it simply stops working.
- Inadequate spares backup is also a problem, such as robot motors that failed and could not be replaced by the stores. A particular example is the servo motor for the long travel of the gantry.
- A lack of operator vigilance has caused problems in the past. A bundle fell off a trailer because the stacking was taking place too far back on the vehicle.
- Sensors are sometimes "fooled" by passing forklifts. A 500 mm × 500 mm reflective plate should be installed to prevent this.
- Sometimes a trailer is not positioned properly. An additional sensor is required to assist the vehicle driver to position the trailer accurately.
- Wear of components causes other problems. Wear causes the mechanisms involved to judder, and this judder causes load cells to misread and bolts to come loose.
- There are some procedures that need to be written for gantry maintenance; the following are examples:
    - Replacement of a load cell
    - Gripper replacement

    - Coupling replacement
    - Motor replacement
- Consideration should be given to re-establishing a maintenance contract with an outside firm for the upgrade and subsequent maintenance of the gantries, with reliability and maintainability clauses included in the contract. Bonuses should be paid for exceeding the reliability and availability targets. The writer is experienced in the preparation of such contracts and could assist in the preparation and management of such a contract for the gantries. Such contracts are based on the simple principle that what gets measured and rewarded gets done.

At this stage of the meeting, the maintenance superintendent, Vishnu Batachariah, addressed Mark as follows: "Mr Francis, much of the information you are giving us, we already know. What we need is more technical detail. You say, for example, that sensors can be 'fooled' by passing forklifts. What we need to know is what sensors, and where? What is their make and specification number? This is just one example . . ."

Mark thought that this level of detail should rather be provided by the company's own automation people, who would be the ones to correct the situation anyway, but he did not respond.

Waylon now intervened. "Please leave us for a few minutes, Mark. Wait outside and I will call you in later."

After a short discussion, Vishnu and Cedric left the meeting, as did Willem and Uriah. Waylon called Mark back in. "Mark, send us your report and your final invoice."

"You mean you don't want me to complete the final RCA?"

"That's right—we'll see to that ourselves."

CLASS ASSIGNMENT

Read the case and discuss the following specifically:

1. What are the main problems in this case?
2. What do you think about the company's policy of "80/20" and how they were determined to stick to it, come what may?
3. Comment on the relations between the shift crews and the fact that there were so many breakdowns on the night shift.
4. Are there different levels of RCFA?
5. How would you go about solving the problems?
6. How would you apportion responsibility for the failure of this exercise between the consultant and the company?

POSTSCRIPT

Approximately a year after the said visit, Mark was again invited to visit the plant to conduct RCA training. This invitation came from a recently appointed chief engineer, who had been transferred to the smelter from one of the parent company's mining divisions. During this visit, Mark learned that the whole plant had stopped operating for several months as a result of poorly made anodes in the carbon section (the area where [maintenance manager] Cedric Altaba's young technicians had been doing such a good job in sorting out the carbon section's problems). The overseas management were infuriated. Approximately $100 million had to be spent to get the carbon section rebuilt and to get the anode manufacturing process back to producing acceptable-quality anodes. The entire top management was fired. The parent firm promised that if poor maintenance and operation were to stop the plant again, they would close it down. Shortly thereafter, the parent firm disinvested in the plant.

NOTES

1. RCM: reliability-centred maintenance, a maintenance optimisation technique well established in industry.
2. The hammers are required to free the cold ingot from its mould as the mould travels around the head pulley of the conveyor. Most ingots will fall free by themselves, but sometimes the hammer has to jar them loose. If an ingot does not fall but "hangs up" and is transported back to the hot end of the line, the line will have to be stopped.
3. SME: The most experienced millwright on any piece of plant, responsible for the quality of its maintenance and the lead artisan in a repair requiring more persons to assist him.

6 Other Techniques Essential for Modern Reliability Management I

INTRODUCTION

This chapter considers techniques for which the connection with reliability engineering might appear tenuous at first. However, in all cases, the writer has found the techniques important when reliability over the life cycle is considered.

CONFIGURATION MANAGEMENT

Configuration management (CM) is an essential part of modern reliability and maintenance management. It is a difficult subject to accurately define, and various authorities have different definitions as to what is included under CM. To project engineers, it is simply change control. To others, it is simply a cataloguing function. The best description that the author can think of is that it is intelligence, in the military sense, applied to engineering systems. In other words, those in control of the system must at all times know what the configuration of the system is. Has there been a deterioration in plant condition? Has there been a change in the spares holding that might affect the maintenance of the plant? And who knows about this?

The reason for the emphasis on CM in this text is that the writer has discovered in the many incidents he has investigated that incidents are often caused, at least in part, by poor CM or by human unreliability. The two issues are closely related, of course—human errors often cause the configuration to collapse or at least to change.

CM has been defined as a systems engineering process for establishing and maintaining consistency of a product's performance and functional and physical attributes with its requirements, design, and operational information throughout its life. The CM process is widely used by military engineering organisations to manage complex systems, such as weapon systems, vehicles, and information systems. In civilian society, it is being increasingly seen to be an adjunct to planned and predictive maintenance systems.

Codification, which is described elsewhere in this book, is an essential aspect of CM. CM goes further than just codifying items, however. Part of CM is to ensure that all items, as far as is practicable, are physically numbered. This includes all machines and many other items of a plant, including, perhaps, cabling.

The codes applied to plant items such as pumps and motors will not necessarily be the same as the numbers in the parts manuals, as plant coding has to do with position as well as type of the numbered item. Position codes are essential if rotable items are to be controlled, that is, items that may be removed from a geographical position

DOI: 10.1201/9781003326489-6

for overhaul and later returned to stores for subsequent installation in another position. Electric motors that are sent out for rewinding fit into this rotable category, for example.

Various position codes are available for various industries. The KKS code, for example, was developed for German power stations and is now used in power generation plants worldwide.

FLAVOURS OF CM: CMI VERSUS CMII

Classical configuration management, now called by some as CMII, is described in the *MIL-HDBK 61B*. The main phases and therefore interests of CMII are configuration design, configuration identification and confirmation, change control, status accounting, and audits. It lists the following as the advantages of CM:

- Product attributes are defined. Measurable performance parameters are provided. Both buyer and seller have a common basis for acquisition and use of the product.
- Product configuration is documented, and a known basis for making changes is established. Decisions are based on correct, current information. Production repeatability is enhanced.
- Products are labelled and correlated with their associated requirements, design, and product information. The applicable data (such as for procurement, design, or servicing the product) is accessible, avoiding guesswork and trial and error.
- Proposed changes are identified and evaluated for impact prior to making change decisions. Downstream surprises are avoided. Cost and schedule savings are realised.
- Change activity is managed using a defined process. Costly errors of ad hoc, erratic change management are avoided.
- Configuration information, captured during the product definition, change management, product build, distribution, operation, and disposal processes, is organised for retrieval of key information and relationships, as needed. Timely, accurate information avoids costly delays and product downtime, ensures proper replacement and repair, and decreases maintenance costs.
- Actual product configuration is verified against the required attributes. Incorporation of changes to the product is verified and recorded throughout the product life. A high level of confidence in the product information is established.

The handbook goes on to state that in the absence of CM, or where it is ineffectual, there may be:

- Equipment failures due to incorrect part installation or replacement
- Schedule delays and increased cost due to unanticipated changes
- Operational delays due to mismatches with support assets

- Maintenance problems, downtime, and increased maintenance cost due to inconsistencies between equipment and its maintenance instructions
- Numerous other circumstances which decrease operational effectiveness and add cost

The severest consequence is catastrophic loss of expensive equipment and human life. Of course these failures may be attributed to causes other than poor CM. The point is that the intent of CM is to avoid cost and minimise risk. Those who consider the small investment in the CM process a cost-driver may not be considering the compensating benefits of CM and may be ignoring or underestimating the cost, schedule and technical risk of an inadequate or delayed CM process.

As stated preciously, original CM as developed for the US military is now referred to as CMI. CMII is a new process, developed at the University of Arizona in this century. It is a far more radical approach to the subject. It concentrates less on activities such as change control and on catching errors and correcting them, and more on ensuring errors will not occur in the first place. Vincent Guess (2002), in his book, *CMII for Business Process Infrastructure*, makes the following bold claims:

- The product of engineering is documentation.
- The purpose of a prototype is to prove the documentation.
- Requirements must lead; products must conform.
- Business processes must lead; software tools enable.
- Each document must have owners—and the fewer, the better. Two is ideal.
- Continuous corrective action is not continuous improvement.
- A need for corrective action is a symptom that requirements are out of control.
- Intervention resources increase exponentially as data integrity declines.

The implications of this approach are radical and, to some, might even seem unbelievable. The emphasis is on getting the paperwork or its electronic equivalent right and on time. Then, the hardware will follow. This means that items such as maintenance manuals and spare parts lists arrive with or, preferably, before the hardware. Guess justifies his approach by pointing out that in many CMI installations, the company spends 40% of its CM budget on corrective action.

With the powerful software available today, the CMII approach is becoming more viable. For example, with modern, computer-aided design, it is possible to refine a design in the software so that the prototype merely confirms the existing, optimised design that was achieved by iteration in the software.

When CM is specifically applied to maintenance, the following applies.

PREVENTIVE MAINTENANCE

Understanding the "as is" state of an asset and its major components is an essential element in preventive maintenance as used in maintenance, repair, and overhaul and enterprise asset management systems. Complex assets such as aircraft, ships, industrial machinery, and so on depend on many different components being serviceable. This serviceability is often defined in terms of the amount of usage the component has had since it was new, since it was fitted, since it was repaired, the amount of use it has had over its life, and several other limiting factors. Understanding how, near the end of their life, each of these components is has been a major undertaking involving labour-intensive record keeping until recent developments in software.

PREDICTIVE MAINTENANCE

Many types of components use electronic sensors to capture data that provide live condition monitoring. These data are analysed at the site or at a remote location by computer to evaluate its current serviceability and, increasingly, its likely future state, using algorithms that predict potential future failures on the basis of previous examples of failure through field experience and modelling. This is the basis for "predictive maintenance."

Availability of accurate and timely data is essential in order for CM to provide operational value, and a lack of this can often be a limiting factor. Capturing and disseminating the operating data to the various support organisations is becoming an industry in itself.

The consumers of these data have grown more numerous and complex with the growth of programmes offered by original equipment manufacturers (OEMs). These are designed to offer operators guaranteed availability and make the picture more complex with the operator managing the asset but the OEM taking on the liability to ensure its serviceability. In such a situation, individual components within an asset may communicate directly to an analysis centre provided by the OEM or an independent analyst.

INFORMATION

The essence of CM is information. Any change in the configuration must be detected and corrected, or if the change is deliberate, the information about the change must be disseminated to everyone who is affected by the change. For example, if a certain spare part XYZ is no longer obtainable but a part ABC must be used instead, then the drawing must be changed, the maintenance manual must be changed, the store's catalogue must be changed, and the preferred supplier list might have to be changed. This must all happen simultaneously, and the custodians (owners) of these lists must all be simultaneously informed and acknowledge the information. With modern software, this goal is achievable.

THE HUMAN FACTOR

Having agreed with all these, it must still be emphasised that human involvement, interest, and dedication to the CM culture are paramount. A case study will indicate this.

A Case Study of Poor CM

In modern power plants, synchronisation of a generator with the grid is done automatically and electronically. A button is pressed, or an icon on a computer screen is highlighted, and the generator is thereby brought into synchronism with the grid and the breakers close. In older power stations, synchronisation of a generator with the grid is done manually, using a device known as a synchrotach. This is a box on wheels that is wheeled into the control room, and two leads are plugged into two marked sockets on the control desk. On the synchrotach is a dial with two rotating needles, one showing the vector of the grid, and the other showing the vector of the generator. This latter is accelerated or retarded by controlling the steam admission valves to the turbine, until the two needles rotate as one. Then, the breakers are closed, and the generator sends power out into the grid.

It is essential that the generator is in phase with the grid before the breakers are closed, as the generator is much smaller than the grid, say, the difference between 200 and 30,000 MW, for example. If the two systems are not in phase, the grid will damage the generator by forcing it into phase in a fraction of a second.

In the case discussed here, the two ports on the control panel had been tampered with. Someone (it was never established who) was working behind the control panel at some stage before the attempted synchronisation described here. He had pulled the wires out of the sockets and reconnected them incorrectly without telling anyone. He had swapped them over from their correct position.

When the synchronisation was later attempted, when the two needles were rotating in unison, the generator was 180° out of phase with the grid. When the breakers closed, the grid wrenched the generator into synchronisation with itself, breaking the solid steel shaft between the generator and the turbine like it was a stick of chalk.

This accident would not have occurred if the artisan involved behind the control panel had called for help in the way of a circuit diagram and reconnected the wires accordingly. In a CM culture, the person would have realised he had degraded the configuration and have done everything necessary to restore it.

SOFTWARE FOR CM

More important than any software is the human culture that must see that maintaining the configuration of the system, whatever it may be, is of paramount importance. The ERP software should be able to assist in several ways, if properly coordinated. As mentioned before, what is most important is the timeous flow of information concerning changes in configuration to all the stakeholders involved in the change.

A FINAL NOTE

It is beyond the scope of this book to either fully describe or prescribe the setting up or management of a CM system. Various standards are listed in the References that the reader will find helpful. What we do attempt here is to emphasise the importance of CM to both reliability and maintenance.

CODIFICATION AND CODING SYSTEMS

Codification of spares is an essential part of modern maintenance management. It forms a subset of CM, the topic discussed previously in this book.

The History of Coding

Mankind has always had an interest in classification and coding. For example, *taxonomy* is defined as the practice of classifying plants and animals according to their presumed natural relationships. This is the oldest profession on earth, despite what humourists say about prostitution being the oldest. Taxonomy was given a divine imprimatur in Genesis 2, when Adam was instructed to name all the animals in the Garden of Eden.

Coding Systems

Various coding systems have been developed, at various times, in different parts of the world, and for different purposes. A few of the many are discussed here.

Linnaeus

Modern taxonomy began with a Swedish naturalist named Carolus Linnaeus, who, in the 1700s, developed a way to name and organise species that we still use today. His two most important contributions to taxonomy were as follows:

1. The hierarchical classification system
2. The system of binomial nomenclature (in other words, a two-part naming method)

Dewey Decimal Classification System

Several other coding systems have been developed for different purposes in the past. One of the first and most well-known is the Dewey Decimal Classification System for libraries (and today, for information in general). The Dewey Decimal Classification (DDC), which was conceived by Melvil Dewey in 1873 and first published in 1876, is a general knowledge organisation tool that is continuously revised to keep pace with knowledge. The DDC is the most widely used classification system in the world. It is included here because of its ongoing relevance to information retrieval.

The first level of the DDC contains the ten main classes. The second level contains the hundred divisions. The third level contains the thousand sections (as can be seen, a decimal structure is given in the title). The ten main classes are as follows:

- 000 Computer science, information, and general works
- 100 Philosophy and psychology
- 200 Religion

- 300 Social sciences
- 400 Language
- 500 Science
- 600 Technology
- 700 Arts and recreation
- 800 Literature
- 900 History and geography

As with any codification system, consistency is hard to achieve. For example, someone may look for "racing car" books in a library and find some of them under 600, technology, and some under 700, arts and recreation (i.e. sport). This remains a problem in any classification system, even the office filing cabinet or its electronic equivalent. It is essential in any serious coding system that a strict discipline be developed to minimise this type of ambiguity.

UNSPSC

This is the United Nations Standard Products and Services Code. It is a four-level hierarchy coded as an eight-digit number. It has been developed to facilitate e-commerce in products and services and, as such, is a fairly new edition to the family of coding systems. The main motivation for its development was the need for standardisation of purchasing among the many UN agencies. It does not seem to be much used in engineering maintenance circles.

Engineering Codification

Two things are required in most codification systems, namely, a set of numbers and a set of descriptions in English or some other human language. With regard to this second requirement, there are some common-sense facts about codification. What information is required to uniquely identify an item? For some smaller organisations, the following may suffice: name, relevant dimensions, material, manufacturer, and set size. Two examples are given here:

Item Name	Spark Plug	Bonnet Clip
Relevant Dimensions	14 mm, long reach	50 mm
Material	Osmium electrodes	Chrome plated
Manufacturers	NGK, Bosch	—
Number	Four per box	Single

Whatever system is chosen, it must be fully able to describe any item of relevance to that particular organisation.

Neither Linnaeus's nor Dewey's system can be of much use in engineering codification. Several engineering codification systems have been developed, often for spare parts and stores.

For this task, the NATO system and its derivatives are the most popular.

The History of the NATO Coding System

We will present a fairly detailed description of the North Atlantic Treaty Organisation (NATO) system, as it is one of the oldest and best-established systems, there is much open-source material on it, and it has been adapted by many companies in non-military situations.

NATO was formed shortly after World War II to present a united European and American front against communist expansion. The beginnings of the NATO codification system began as long ago as 1953.

The integration of European and North American armed forces was a formidable task. Each country did not even have standardisation of parts supply between the various branches of its own armed forces, let alone with Armies, Air Forces, and Navies of allied countries. Second, even within one country, various suppliers of identical items had their own identification systems. It was thus seen that standardisation of parts supply would save many millions of dollars at three levels, by *eliminating duplication* due to:

- Suppliers' numbering systems
- The fact that, in one country, the various branches of its armed services had their own numbering systems
- The fact that this applied between countries as well

For example, the classification for a certain truck tyre was given by three suppliers, in different countries, as follows:

- Dunlop (UK): 11–00–20SPTGM
- Goodyear Tire CO (U.S.): 11–00–20SRLER
- Goodyear (FRANCE): 11–00–20UNISRL

All these are replaced by the single NATO stock number (NSN): 2610–14–3224604.

It should be noted that NATO stock numbers represent *item of supply* concepts rather than items of production, where an item of supply concept represents a cluster of characteristics related to form, fit, and function. Many items of production may fit a single item of supply concept, as shown previously.

Users of the NCS system include not only the 28 NATO member countries but also the 37 NATO-sponsored countries. It is also grouped into tiers indicating participation and access. The format of a NATO number is as follows:

abcd-ef-ghi-jklm

Each element, *a* through *m*, was originally intended to be a single decimal digit. As inventories grew in complexity, element *g* became alphanumeric, beginning with capital *A* for certain newly added items. By 2000, uppercase *C* was in use.

The first four numbers, *abcd*, form the National Supply Classification Group (NSCG). This relates the item to the NATO Supply Group (*ab*) and NATO Supply Class (*cd*) of similar items that it belongs to.

Some NATO supply groups (*ab*) include the following:

- Group 10: Weapons
- Group 13: Ammunition and Explosives
- Group 22: Railway Equipment
- Group 26: Tyres and Tubes
- Group 31: Bearings
- Group 48: Valves
- Group 76: Publications
- Group 91: Fuels, Oils, and Lubricants

In each of these groups, there will be a Supply Class (*cd*). As an example under Group 13, Ammunition and Explosives, some of the supply classes include the following:

- 1325: Bombs
- 1326: Grenades
- 1370: Pyrotechnics

The nine digits, *ef-ghi-jklm*, comprise the National Item Identification Number (NIIN). This format improves readability but is optional, as NIINs are often listed without hyphens.

The first two digits of the NIIN (the *ef* pair) is used to record which country was the first to codify the item—which one first recognised it as an important item of supply. This is generally the country of origin, meaning, the country of final manufacture. The formal name of the field is CC, for country code. Some country code numbers are as follows:

- 00 and 01: United States
- 14: France
- 18: South Africa
- 44: United Nations
- 99: United Kingdom

Nations Using the Code

- **Tier 1 Nations**. There is a one-way data exchange, and they do not participate in technical NCS management. South Africa is part of this group.
- **Tier 2 Nations**. There is a two-way data exchange and participation in technical NCS management. Australia, for example, is part of this group. Surprisingly, given its antipathy to NATO, Russia is also part of this group.
- **Tier 3 Nations**. These nations are NATO members and have a full membership in the NATO Codification Bureau. The United States, Germany, and the United Kingdom are part of this group.

We can now look at our NATO Stock Number mentioned prior in connection with tyre codes in more detail: 2610-14-3224604.

- 26 is the Supply Group for Tyres and Tubes.
- 10 is the Supply Class.
- 14 is the country code—in this case, France.

The last seven digits are the unique item identification number: 3224604. This specifies the tyre in full.

It is at this stage that we see that apart from the codification, we still need a language description so that human beings can find or classify the item correctly. This is usually done with a Standardised Long Description and a Standardised Short Description. For example, the descriptions that follow could refer to the number 3224604 given previously.

Standardised Long Description

Tyre: Pneumatic, Vehicular. Service Type for Which Designed: Loader. Tyre Rim Nominal Diameter: 25″. Tyre Width: 445 mm. Aspect Ratio: 0.95. Tyre Ply Arrangement: Radial Ply. Rating: 2*. Tyre and Rim Association Number: E3. Tread Material: Standard Tyre. Air Retention Method: Tubeless Tyre. Load Index and Speed Symbol: NA. Tread Pattern: VHB TKPH. Rating: 80.

Standardised Short Description

Tyre Pneumatic: Loader 25″ 445 mm 0.95 2*

Even though this is a fully computerised system, it is still necessary to fully describe every item as in the previous example. Such codification requires a special type of person. Some of the qualities necessary to be a good codifier are as follows:

- Intimate knowledge of the technology being dealt with, with the ability to distinguish between closely related items.
- A real interest in this type of work.
- A significant degree of intelligence.
- A resistance to intellectual fatigue—codification can be a long, hard process.
- Computer skills.
- Technical language skills and, for coding in English, realisation that one item can have more than one name; for example, a *wrench* in the United States is a *spanner* in the United Kingdom, and a *truck* in the United States is a *lorry* in the United Kingdom, although this name is giving way to the US equivalent.

Furthermore, there is more to the job than just sitting behind a computer. If an audit is to be done on a store's system, to eliminate surplus stock, it might first be necessary to personally investigate, by visiting the store, what each coded item actually is. Data washing operations like this are expensive, and consultants charge by the line in the store's catalogue. As of 2015, costs of US$4 per line or more are not uncommon. For a catalogue of 100,000 lines, which is not unusual for a big organisation, the cost for the data washing can therefore be half a million dollars or more. However, such expenditure might still be worthwhile, as shown next.

Technical Dictionaries

To assist the codifiers, technical dictionaries have been developed to assist with the correct naming of items. One such dictionary is the electronic Open Technical Dictionary (eOTD), developed by the Electronic Commerce Code Management Association (ECCMA). Another is the one developed by PiLog, an international data management company headquartered in South Africa.

Advantages of the NATO System

In a study done for the armed forces in the United Kingdom, it was shown that of the 200,000 new items offered to the UK NATO Codification Board for codification each year, approximately 50% are already codified and have been allocated an NSN. Hence, a codification standard has tremendous advantages for reducing duplication of stock.

In another study done in Singapore, it was shown that NSN prevents unnecessary inventory growth and item duplication. Items are managed by NSNs instead of a manufacturer's part numbers. There are also savings in warehousing and item management fees. This is particularly the case when supply bases are supply-managed by contractors, who typically charge $150 per line item per year. Approximately 25% of new items requisitioned were detected with existing NSNs each year. This meant 5,000 items that were not needlessly purchased, translating into a savings of more than US$750,000 per year, just in management costs, not counting the costs of the items themselves.

What the NSN system also facilitates is the identification of the original manufacturers. Suppliers of systems often include parts not made by themselves and add a percentage to the price of such parts, which can be as high as 30%. Purchasing from the original manufacturers can save a considerable amount of money.

A further advantage of the system is that the price of reorders can be easily checked against items previously purchased.

In summary, then, the NSN system can eliminate costs attributed to duplication, reduce store holding costs, and reduce the direct purchase cost of items as well.

Outside of the military, there are tremendous benefits for commercial organisations as well in codification and standardisation, hence the concept of master data.

Master Data

These are data held by an organisation that describe the entities that are both independent and fundamental for an enterprise—data that it needs to reference in order to perform its transactions. Master data include information on customers, suppliers, materials, services, assets, locations, and employees. There is an international standard for master data, ISO 8000, which has the title "Data Quality."

The fundamental principles of ISO Standard 8000 are as follows:

- To set standards and a certification process for master data quality
- To define different areas of data quality, namely:

- Provenance, that is, where the data originated from
- Traceability
- Currency
- Completeness

The benefits of accepting the discipline of ISO 8000 are said to include the following: providing faster access to better-quality characteristic data, faster NSN assignment, more complete records, better search resolution, fewer duplicates, and less need for item reduction studies. Other benefits include the following:

- Higher customer satisfaction
- Savings in design and life cycle costs
- Reduced acquisition lead time
- Increased supportability and safety of systems and equipment

ISO 22745: Standard-Based Exchange of Product Data

The other standard that is applicable to codification is ISO 22745. The ISO 22745 standard provides the framework needed for any organisation to conduct business with internationally recognised data quality. Its most basic purpose is to provide a means to realise the benefits of ISO 8000, which is the ability to specify syntax, semantic encoding, and specification of data requirements for messages containing master data that are exchanged between organisations in the supply chain. Once an organisation begins to standardise the descriptions it uses to describe materials, the organisation can also begin to see cost savings and cost avoidance by implementing business intelligence algorithms to identify conditions such as duplicate items in inventory, purchase price disparities between facilities, vendor reductions, and identification of functional equivalent items.

In summary, ISO 8000 is about master data and ISO 22745 is about technical dictionaries.

Non-Military Applications of Master Data

Various companies around the world have developed their own systems based on the NATO system. An example of such a company is the international resources company Anglo American, which used to have separate store numbering systems for its various subsidiaries, for example, Anglo Base Metals, Anglo Platinum, Anglo Coal, and Kumba Resources. These all now use the same "Anglo Number" for all store items.

There are also consulting companies that will assist organisations to manage master data correctly. One such is the previously mentioned PiLog, a South Africa–based international data management consultancy which has offices in the United States, Russia, the Middle East, and elsewhere.

Coding by Colour or Stamping

Finally, we must mention another type of coding, namely, coding by colour. This can be extremely important to uniquely identify various types of steel, for instance.

TABLE 6.1
Abbreviated List of Classification Codes for Stainless Steels, as Used by SASSDA

Stainless Steel Grade	Enamel Colour	Printer's Ink Colour
202	Broken white	Pantone 7527
3CR12	Black	Pantone black
303	Traffic blue	Pantone 301C blue
304	White	Pantone white
304L	Signal yellow	Pantone 116C yellow
309	Aqua blue	Pantone 322C turquoise
310	Rose	Pantone 203C pink
316	Brilliant green	Pantone 356C bright green
316L	Signal red	Pantone 193C red

The use of the wrong steel in a high-pressure piping system could be disastrous. A mild steel pipe might have the same internal and external diameter as an alloy steel pipe and may be welded into an alloy steel line perfectly, only to explode when under high pressure and temperature. (This did once occur at a company where the author was employed—although he was not involved in any way.) Pipework and bar stock are often colour-coded. Any store audit should include a section on colour-coding. Colour-coded items in the store must be correlated with mill test certificates and other identifying paperwork. As one example, the colour codes for stainless steel, as recommended by Southern African Stainless Steel Development Association (SASSDA), are shown in Table 6.1.

The alternative to colour-coding is stamping, used on stainless steel plates, for example, to correlate them with test certificates in a file.

The colour-coding or stamping information must also be included in the descriptions of the items in the store's catalogue.

Modern Trends in Codification

With SAP becoming ubiquitous in maintenance environments, the SAP numbering system is starting to predominate in maintenance stores. The SAP number, whether it be seven digits or more, uniquely identifies an item. Unfortunately, there is no embedded information. It is just a sequential number. This has serious implications. What is implied is that the two human-readable codes, namely, the Short Description and the Long Description, must be very accurate and complete. Otherwise, the possibility of duplication of items is still present. The information inherent in the previous example of the NATO tyre is a case in point. Study its Long Description again and see the amount of information needed to specify the item unambiguously.

LUBRICATION

The importance of lubrication in reliability and maintenance is often misunderstood or downplayed. The correct way to consider a lubricant is as a component of the machine. Poor lubricant reliability will result, as with any component with poor reliability, in poor reliability of the system of which it is part. This is acknowledged in the MSG system of maintenance optimisation used in the airline industry, where a separate set of questions are asked about lubrication, rather than about the plant itself. However, many reliability and maintenance engineers outside the airline industry leave all lubrication matters in the hands of a preferred supplier, who recommends a set of lubricants based on the machinery in the plant, with some consolidation and second choices so that the number of kinds of lubricant to be held in store is rationalised. However, the ethos of this book, as described earlier, is, "Know your plant and keep it good as new." The author therefore feels that it is essential for reliability and maintenance engineers to have at least some grounding in lubrication.

To understand some of the finer points of lubrication, it is necessary to learn something about how the two main lubricant types, oil and grease, are manufactured.

Manufacture of Lubricants

Lubricants are derived either from crude oil or, in the case of synthetics, from some other base stock. Considering crude oil first, lubricants are derived at the end of a long process beginning as shown in Figure 6.1. In this figure, crude oil, as extracted from the earth, is heated in a crude distillation unit (CDU) to above 350°C. This process is carried out at atmospheric pressure. Hence, the CDU is also known as the atmospheric distillation unit. Crude oil is a complex mixture of hydrocarbons and impurities. The lighter fractions of the crude evaporate and recondense at various elevations in the tower. They are extracted from the sides of

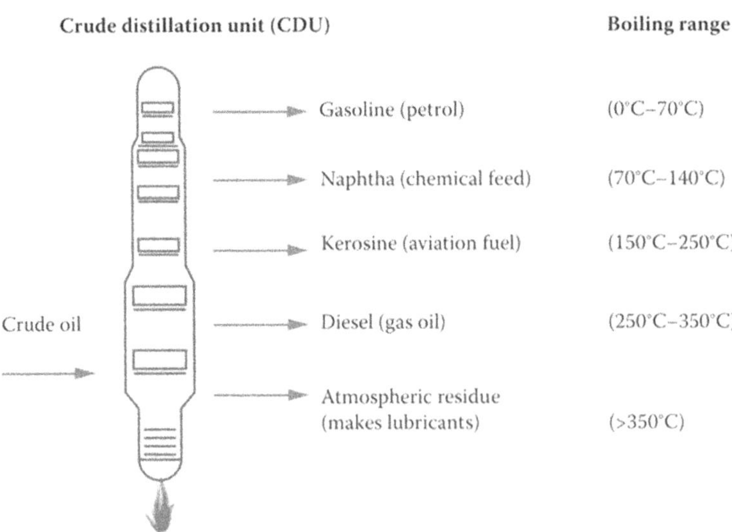

FIGURE 6.1 Crude oil refining.

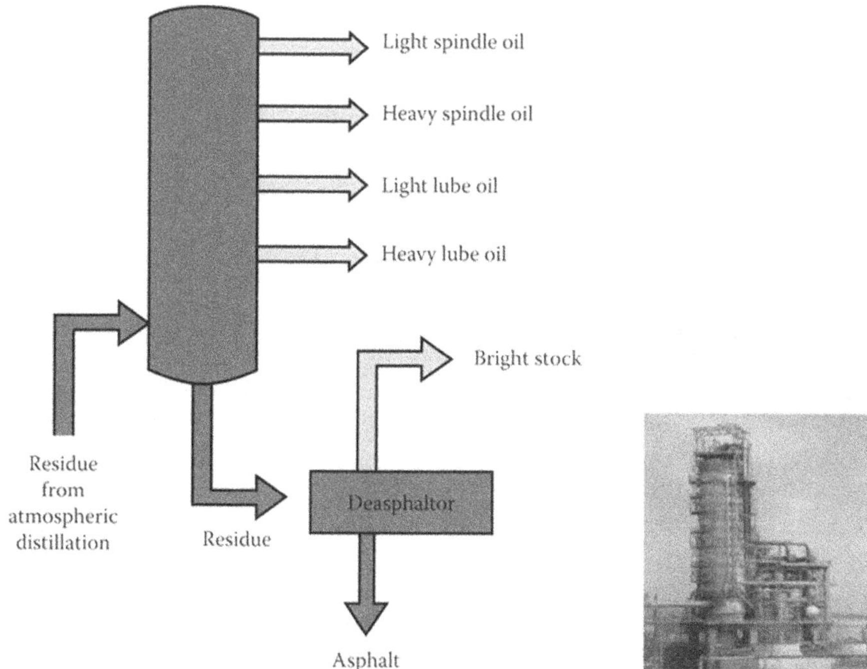

FIGURE 6.2 Oil refining.

the tower and then processed further. It is seen that the lighter elements are either fuels or naphtha, which is the feedstock for various chemical processes. What leaves the bottom of the tower is the material from which lubricants are made and heavier components such as asphalt (Figure 6.2). Gases that exit the top of the tower become liquid petroleum gases or are flared off. Key physical properties of the various fractions are set by distillation (e.g. viscosity, flash, volatility, demulsibility, and colour).

In simplified terms, the product from lower extraction points in the CDU or the atmospheric distillation unit is then further refined under vacuum, as shown in Figure 6.3. The inset photograph shows a typical tower.

In practice, the process is considerably more complex than has been shown. In the vacuum distillation tower, "base oil cuts" with various viscosities are produced. The vacuum residue can also be deasphalted to produce a further base oil cut, bright stock. Asphalt is a by-product and is used for roads and so on. Bright stock can be further refined into lubricating oil.

Mineral-based oils after vacuum distillation still contain some undesirables:

- Large paraffin molecules (wax type) have a high pour point.
- Aromatics, sweet-smelling hydrocarbons such as benzene, which are considered as carcinogenic and form sludge.
- Nitrogen compounds catalyse oxidation.

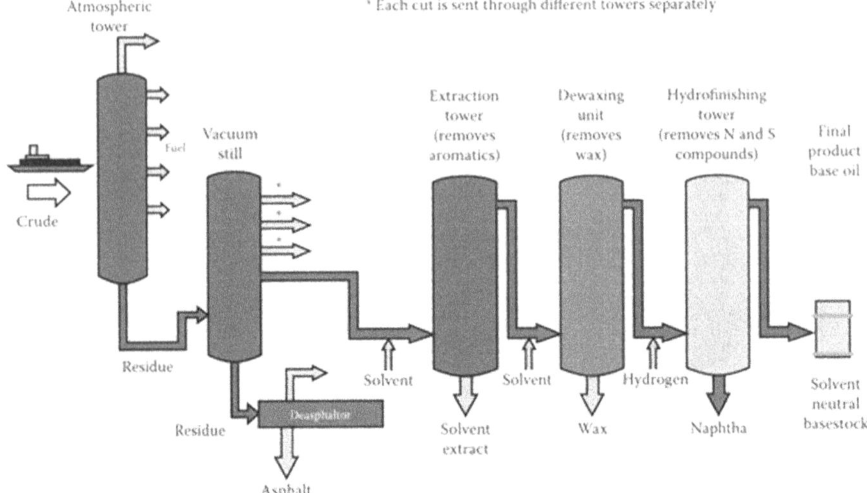

FIGURE 6.3 LUBE OIL REFINING.

However, some components are also desirable:

- Sulphur compounds—some are natural antioxidants.
- Large paraffin molecules—these have a high viscosity index.

The aromatics are extracted with certain solvents, while waxes are removed by solvent dewaxing. Sulphur and nitrogen compounds are reduced by hydrogenation. Further refining is a balance between reducing the undesirable compounds, which will depend in any event on the base stock quality and enhancing the desirable qualities of the oil by additive treatment.

Each of the particular base oil "cuts" is mixed with a special solvent. The solvent is then removed, taking most of the aromatics with it. This improves the oil as follows:

- Viscosity index
- Colour
- Oxidation/thermal stability
- Lowers the carcinogenicity

Process oils are produced as a by-product, which can be used in rubber tyre manufacture.

Waxes are removed as follows: Oil and solvent are mixed and cooled to a set temperature. Large paraffinic molecules precipitate out as wax, which is filtered off. This has the effect of:

- Reducing the pour point
- Reducing the viscosity index
- Improving filterability

The by-product of this process is wax, as used in candles and polishes.

Finally, in the hydrofining process, hydrogen reacts with the base oil in the presence of a catalyst under high temperature and high pressure. This improves colour; reduces sulphur, nitrogen, and oxygen; and reduces acidity. The complete process is as shown in Figure 6.3.

Thus far, we have considered the manufacture of lubricant from crude oil base stock. The alternative, which is becoming more and more important, is synthetic oil, which basically means any oil in which the base stock is not crude oil. Such oils could be called "designer oils," in that the manufacturer is not constrained by the inherent limitations of crude oil but can tailor the oil to more specific requirements. Synthetic oils are generally superior to traditional oils in all respects except one—cost. This limits their applicability.

Tribology

The interaction between lubricant and wearing surface is one of the subjects dealt with in the relatively new field of tribology. *Tribology* is defined as the study that deals with the friction, wear, and lubrication of surfaces in relative motion, such as bearings and gears. Lubrication science is therefore a part of tribology. Where a person might be called a lubrication engineer, he might also be called a tribologist.

In practice, tribology concerns also the maintenance of machines, the lubricant supply and lubrication practices, the reduction of energy, and the care and welfare of environmental issues.

It also serves to reduce and control friction and wear and hence maintain optimum machine performance capability and cost-effectiveness during use. The word was coined by the British physicist David Tabor and lubrication expert Peter Jost in 1964.

The Requisite Qualities of a Lubricant

A lubricant must perform several functions. The most important are as follows:

1. To reduce friction
2. To minimise wear
3. To remove heat
4. To prevent corrosion
5. To remove contaminants

FRICTION

It is generally recognised that there are three types of friction: static, kinetic (or sliding), and rolling. In *static friction*, the surfaces are initially at rest. Force is required to start the surfaces sliding over one another, and this force is equal to or greater than the force of the subsequent sliding friction.

In *kinetic friction*, the force opposing motion is material-dependent and directly proportional to load. It is independent of the relative speed of the surfaces or of the area of the surfaces in contact. The classic formula for friction is

$$F = \mu R,$$

where F is the frictional force generated, μ is the coefficient of friction between the surfaces (a number between 0 and 1), and R is the force normal to the surfaces.

Note that F is independent of area. This puzzles some. But it is logical if one considers pressure. The equation, then, is

$$F = \mu \times \text{pressure between the surfaces} \times \text{the area of the surfaces.}$$

That is, $F = \mu(R/\text{Area}) \times \text{Area}$.

So once again, as the areas cancel out, $F = \mu R$.

Another explanation of this is as follows: There is friction because the surfaces are not completely smooth—peaks on the one surface lock with valleys on the other. As the area increases, the force F becomes greater, because there are more peaks and valleys to interlock. But as the area increases, the pressure holding the faces together under the constant load R decreases in direct proportion.

There are circumstances under which the limitation of μ to unity or less does not apply. It applies to hard surfaces, but not to rubber on tar or concrete. Modern racing cars and motorcycles can corner with forces developed of more than 1g. $F = \mu R$, and R in this case is always the acceleration of gravity g. Hence, in these cases, μ must be greater than unity. This effect is thought to be attributed to the rubber of the tyres forming instantaneous "gear teeth" with irregularities with the road surface.

The third type of friction is *rolling friction*. With rolling friction, deformation of the surfaces occurs, which generates some localised heat. This phenomenon happens even when the surfaces are very hard, as in ball bearings or gear teeth.

WEAR

Friction causes wear. But wear occurs in different ways. It is important to be able to identify the type of wear that is taking place. This will be a factor in the selection of a lubricant and the timing of lubricant change out. The types of wear are as follows:

- **Abrasive:** Scratching and polishing by wear debris or solid contaminants such as sand.
- **Adhesive:** Welding of asperities on one of both of the surfaces, followed by immediate breaking of the weld as the sliding action continues.
- **Corrosive:** Chemical reaction with the surrounding environment.
- **Pitting:** Surface fatigue caused by rolling contact.
- **Cavitation:** Formation and collapse of bubbles on a surface attributed to rapid pressure changes.

LUBRICATION

To overcome or minimise these various types of friction and wear, lubrication is necessary. Lubrication can be by means of a gas, a liquid, or a solid. Gas lubrication does not really concern us here. Petroleum products are good lubricants because of their reasonably high viscosity and their ability to "wet" most surfaces. Water is not a good lubricant because of its low viscosity, which means it squeezes from between two surfaces too readily. Water does have its place in certain low-duty applications, such as the bearings of a swimming pool pump.

Lubricants are mostly oils and greases, the difference being that an oil "runs" while a grease stays in place.

Having categorised friction and wear, we will now categorise lubrication. The basic requirements of a lubricant include the following:

- To separate moving surfaces. For this, the lubricant needs to have a high shear strength.
- To dissipate frictional heat. For this, the lubricant needs to possess good thermal conductivity.
- To control corrosive wear.

Lubrication is described in terms of regimes. The four major regimes, in order of severity, are as follows:

- Hydrodynamic
- Thin-film or mixed
- Elasto-hydrodynamic
- Boundary

Hydrodynamic Lubrication

Hydrodynamic lubrication is where a continuous full fluid film separates the sliding surfaces and is often found in journal bearings. The film thickness may be as little as 5–200 μm, depending on speed, viscosity, and load.

Thin-Film or Mixed Lubrication

Thin-film lubrication is characterised by a combination of both solid and fluid friction and is often found in heavily loaded, slow-moving bearings and gears and particularly under shock loading.

Boundary Lubrication

If conditions for hydrodynamic lubrication are not present, such as higher loads, slower speeds, shock loads, and so on, then metal-to-metal contact can occur, leading to wear. Hence, a method of lubrication must be available and which will still protect the metal surfaces; this is achieved through the use of special additives that react with the metal surface, forming a molecular thin layer and which is able to support the load. This is boundary lubrication. For severe situations, additives with a higher film strength, anti-wear additives, and mild extreme-pressure (EP) agents (e.g. organic phosphates, zinc dialkyldithiophosphate, etc.) may be present in such hydraulic and engine oils. For very severe conditions, very strong additives are required, called EP agents (e.g. sulphur, phosphorus, and chlorine) and may also be used in lubricants for heavy-duty metal forming and removal.

Elasto-Hydrodynamic Lubrication

Elasto-hydrodynamic lubrication (EHL) is an intermediate lubrication mode between hydrodynamic and boundary lubrication. It occurs mainly in rolling contact bearings and also in gears where non-conforming surfaces are subjected to extremely high loads over small contact areas, for example, a ball within a larger race of a bearing (e.g. contact pressure may be up to 500,000 psi or approximately 3400 MPa). EHL is characterised by the fact that surfaces are momentarily elastically deformed under pressure, hence spreading load over a greater area, and also by the fact that lubricant viscosity momentarily increases at high pressure (the lubricant effectively becomes a solid), hence increasing its load-carrying ability. The viscosity increase, and the greater load-carrying area entraps a thin lubricant film between surfaces and, as viscosity increases, enables sufficient hydrodynamic force to generate a full fluid film to achieve separation. Even with good lubrication, however, the repeated elastic deformation can result in metal fatigue and bearing failure.

LUBRICANT SELECTION

Depending on the design of the machine component, the relationship between the moving elements and the lubricant defines the lubrication regime. Each of these regimes will influence the choice of lubricant, the efficiency of the machine component, and the cost of the system. In general, as the lubrication regime becomes more severe in terms of friction, the viscosity of the oil film becomes heavier.

Lubricant Properties

The most important properties of a lubricant are as follows:

- Viscosity
- Thermal stability
- Oxidation stability
- Pour point
- Demulsibility
- Flash point
- Fire point

Of the aforementioned list, the most important property of a lubricant, by far, is its viscosity. Viscosity is measured in various ways.

Viscosity Measurement

Viscosity is a fluid's resistance to flow. There are several measures of viscosity, the most fundamental being absolute viscosity.

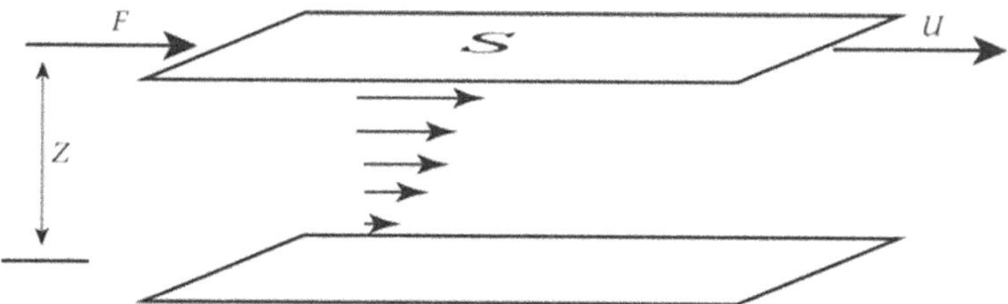

FIGURE 6.4 Viscosity-velocity profile.

Absolute Viscosity

Absolute viscosity (also known as dynamic viscosity) measures the internal resistance in a fluid. It is the tangential force per unit area required to move one horizontal plate with respect to another plate—at unit velocity—when maintaining a unit distance apart in the fluid, as shown in Figure 6.4. Note that this is a theoretical model; it assumes very large plates so that the effects at the edges of the plates may be ignored.

Absolute viscosity can be expressed as follows:

$$\tau = \mu \delta u / \delta z, \qquad 6.1$$

where $\tau = F/S$ is the shear stress (N/m^2), μ is the absolute viscosity (Ns/m^2), δu is the unit velocity (m/s), and δz is the unit distance between layers (m).

Equation 6.1 is known as Newton's law of friction.

In the SI system, the absolute viscosity units are Ns/m^2, that is, Pascal seconds, although this unit is seldom used. Absolute viscosity may also be expressed in the metric CGS (centimetre-gram-second) system as poise (p), where

$$1 \text{ poise} = 1 \text{ dyne s/cm} = 1 \text{ g/(cm s)} = 1/10 \text{ Pa s}.$$

For practical use, the poise is normally too large, and the unit is often divided by 100—into the smaller unit centipoise, where

$$1 \text{ centipoise} = 0.001 \text{ Pascal seconds}.$$

Water at 20.2°C has the absolute viscosity of 1 centipoise. This is purely fortuitous and not part of the design of the system.

Kinematic Viscosity

Another measure of viscosity is kinematic viscosity. This is equal to the absolute viscosity divided by the density of the fluid. It is a practical unit, easily measured by observing the time taken for a fluid to flow through an orifice. Thus, if the density of the fluid is known, the absolute, or we could say the fundamental, viscosity is easily calculated.

TABLE 6.2
Comparison of Viscosity Classification Systems (Approximate)

SUS at 38°C	cSt at 38°C	SAE Motor Oils	SAE Gear Oils	ISO Grades at 40°C	AGMA Grades at 38°C	SUS at 100°C	cSt at 100°C
7,000	1,500	—	—	1,500	—	275	60
4,700	1,000	—	250	1,000	8A	240	50
3,200	700	—	140	680	8	180	38
2,300	480	—	140	460	7	140	29
1,400	320	—	90	320	6	115	24
1,000	230	50	90	220	5	93	18
700	150	40	90	150	4	75	14
460	100	30	80W	100	3	62	11
320	70	20	80W	68	2	54	8
220	48	20W	75W	46	1	48	7

The equation is simply

$$\nu = \mu/\rho,$$

where ν is the kinematic viscosity in m^2/s, μ is the absolute viscosity in Ns/m^2, and ρ is the density in kg/m^3. Note that kilograms here are force units.

Several engineering units are used to express viscosity, but the most common by far are centistoke (cSt) for kinematic viscosity and the centipoise (cP) for dynamic (absolute) viscosity. The kinematic viscosity in centistoke at 40°C is the basis for the ISO 3448 kinematic viscosity grading system, making it the international standard. Another common kinematic viscosity system is Saybolt Universal Seconds (SUS). It is the time that 60 cm^3 of oil takes to flow through a calibrated tube at a controlled temperature, 38°C. The SUS is used for oils with flowing time up to 5,600 seconds, in the range of low to medium viscosity, such as machine oils. Approximate comparison of viscosity classification systems is given in Table 6.2.

THE ROLE OF ADDITIVES

There are many additives in modern lubricating oils. Additives may include the following:

- Viscosity index improvers (VIIs)
- Pour point depressants
- Seal-swell controllers
- Antioxidants
- Metal deactivators
- Anti-foaming agents
- Demulsifiers
- Anti-wear additives

- Extreme-pressure additives
- Corrosion inhibitors
- Detergents
- Dispersants
- Friction modifiers
- Oiliness agents
- Tackiness agents

Unfortunately, additives can also have harmful side effects, so it is important that the oil company blends the correct additives, in the correct amounts, into the base oil. Care must be taken to not interfere with the oil company's formulation by adding extra additives from other, sometimes less-than-reputable companies. The general principle is that the major oil companies, with their large research budgets, probably provide products that cannot be improved upon. Specialist oil companies do have places in niches in the market, but their products should be chosen with care.

Some of the characteristics of some of the additives are now discussed in more detail.

- **Viscosity Index Improvers:** All oils lose viscosity as temperature increases, which is not desirable. The viscosity index is a measure of this loss. The higher the viscosity index, the less the decline in viscosity with temperature. A VII lessens this tendency to lose viscosity with temperature.
- **Pour Point Depressants:** These allow the oil to remain fluid at lower temperatures. They do this by interfering with wax crystal growth in the oil.
- **Seal-Swell Controllers:** These cause very slight swelling of seals, maintaining their life and limiting leakage.
- **Antioxidants:** These protect the oil from oxidation by neutralising oxide-forming compounds. If this is not done, sludge and acids are formed.
- **Metal Deactivators:** These also protect the oil from metal surfaces, which might act as catalysts promoting oxidation. They do this by forming a protective film on the metal.
- **Anti-Foaming Agents:** These alter the surface tension of air bubbles in the oil, causing collapse of the foam. Foaming must be eliminated; otherwise, bearings and other parts will not be lubricated by oil but by air!
- **Demulsifiers:** An emulsion is a liquid dispersed in another liquid, such as water in oil. The water drastically changes the characteristics of the oil, for the worse. Better to have the water, from whatever source, separated from the oil, where it can be syphoned off. Demulsifiers work by changing the surface tension between oil and water.
- **Anti-Wear Additives:** These are surface protection agents that prevent metal-to-metal contact when and if the oil film breaks down, as might happen on highly stressed parts, such as engine camshafts. A typical anti-wear additive is zinc phosphate, which plates out of the oil onto metal surfaces.
- **Extreme-Pressure Additives:** These additives are activated by high pressures and temperatures, where normal anti-wear agents would fail. They react strongly with the metal surface, forming a soft chemical layer that wears

off and prevents welding of the metal surfaces. Phosphorus, chlorine, and sulphur are such additives. A basic way of identifying an EP oil is by its sulphurous smell.

- **Corrosion Inhibitors:** These prevent internal corrosion of metal surfaces by laying down a protective coating.
- **Detergents:** These keep deposits in suspension to prevent the coating of metal surfaces and sludge formation. They do this by developing a non-adhesive film on metal surfaces.
- **Dispersants:** These coat contaminants in order to keep them in suspension.
- *Friction modifiers.* These coat metal surfaces and modify their friction characteristics.
- **Oiliness Agents:** These increase the lubricity, or "slipperiness," of an oil. Oiliness agents are chemically produced fatty esters or vegetable and animal fats (e.g. rapeseed oil, lard oil).
- **Tackiness Agents:** These are used where high speeds result in oil fly-off and mists or lubrication is awkward as a result of run-off (e.g. vertical slideways). Special high-viscosity polymers are used to increase the "stickiness" of oil. They are an especially useful addition to low-viscosity oils.

DIFFERENT LUBRICANTS FOR DIFFERENT APPLICATIONS

Hydraulic Oils

Hydraulic fluids must transmit fluid power, lubricate moving parts, and serve as a heat transfer medium. There are three major classes of hydraulic fluids—petroleum, aqueous, and synthetic. The choice of fluid depends on viscosity grade, operating conditions, and the environment.

Hydraulic fluids with a petroleum base are the most commonly used hydraulic liquids. They usually contain anti-wear, antioxidants, demulsifiers, corrosion inhibitors, and antifoaming agents. These are usually good for a temperature range of −40°C to +100°C.

Where there is a fire hazard, water-based fluid may be used instead, but there are pressure and temperature limitations (7 MPa, i.e. 1000 psi and 65°C).

Synthetic hydraulic fluids are used in the same services as petroleum-based fluids, but they have a better viscosity index, allowing both use over a higher temperature range and use at lower temperatures. However, there may be compatibility problems with seals, pipe joint compounds, and painted surfaces. Before switching to a synthetic fluid, compatibility should be checked.

As the pump is the most important part of a hydraulic system, the fluid used must meet the manufacturer's specifications. The use of an oil with very low viscosity may cause pump slippage, excessive wear, and leakage. Using a fluid with very high viscosity may cause overheating and cavitation. With petroleum-based fluids, overheating will cause oxidation of the fluid, resulting in gums, sludges, and varnishes. Table 6.3 gives recommended ISO viscosity grades for oils for use in hydraulic pumps.

TABLE 6.3
General Viscosity Requirements for Hydraulic Pumps

Pump Type	Max. Pressure (psi)	Max. Pressure (MPa)	Operating Pressure (psi)	Operating Pressure (MPa)	Recommended ISO Viscosity Grades
Vane	3,000	20	1,000	7	ISO VG 16–58
Gear	3,000	20	500	3.5	ISO VG 46–100
Piston	10,000	70	3,000	20	ISO VG 32–220
Pumps designed for use with water-based fluids	—	—	—	—	ISO VG 2–46

TABLE 6.4
General Viscosity Requirements for Turbine Oils

Method of Lubrication	Viscosity at 40°C (cSt)	Viscosity at 38°C (SUS)
Force-feed		
Direct connection	32–46	150–200
Small geared unit	54–77	250–350
Large geared unit	86–100	400–500
Ring-oiled system	32–46	150–250

Turbine Oils

Turbine oils are used in oil circulation systems to lubricate and cool turbine and generator bearings along with thrust bearings. The bearings in such machines are usually plain bearings. The turbine oils are often also used in auxiliary equipment, such as pumps, hydraulic governors, and gear drives associated with the turbine. Such oils are continuously subjected to heat and wear metals and water from various sources. Because of the large volume of oil in use, it is usually tested and regenerated while in use. It therefore has to be of superior quality to an engine oil, for example, that is discarded after a relatively short time. In particular, the level of oxidation stability and rust protection must be very high.

Turbine oils are usually formulated from highly refined mineral oils in the viscosity grades ISO 32 to 100. Large power turbines use ISO 32 or 46 grades. A quality turbine oil will have an ASTM oxidation life value of 2,000 hours or greater and must separate water quickly. More details of viscosity requirements are given in Table 6.4.

Gear Oils

There are five classifications of gear lubricants:

1. Rust and oxidation inhibited (R&O)
2. Extreme pressure

3. Compounded
4. Synthetic
5. Open gear

R&O oils are typically used in spur, helical, and bevel gearboxes operating under light or moderate loads. Lighter viscosity grades (ISO VG 68–150) are used in higher-speed applications, and VG 220–680 is used in lower-speed gearsets. R&O-type oils are particularly suited to applications where bearings and gears are lubricated using the same oil.

EP gear oils have additives that improve film strength and load-carrying capacity. Such oils are used to lubricate hypoid, worm, and other highly loaded gearsets. There are no problems with steel-on-steel gears, but with bronze gears, the choice of EP oils is limited, and they must be chosen with care.

Compounded gear oils are blended with small amounts of fat or tallow, which improves their lubricity and anti-frictional properties. They are used primarily in worm drives, where the sliding action component of the drive requires extra lubricity agents to reduce friction and heat build-up.

Synthetic gear oils show improved stability and life and can operate over a larger temperature range than mineral oils. However, they are less stable in the presence of moisture, and actually, their lubricating properties are poorer. The user must consult the oil manufacturers and the gear manufacturer before deciding to use synthetics.

Open-gear compounds are heavy, tacky residual products that are used primarily in large, low-speed, heavily loaded gearsets, such as on tube mills, swing gears on power shovels, and driving gears on draglines. The lubricant must resist throw-off but is therefore hard to apply. This problem is solved by "cutting back" the lubricant with a solvent to facilitate application. The solvent evaporates, leaving the lubricant in place.

Compressor Oils

Compressors can be divided into two basic types: positive displacement or dynamic. Positive displacement types can be reciprocating, screw type, sliding vane type, or of the roots type. Dynamic compressors are either centrifugal or axial. Lubrication requirements can be onerous, particularly if very high rotational speed is involved or if the medium being compressed can react with the lubricant, for example, oxygen.

In terms of bearing and gear loads, compressors present no serious problems. However, the heat of compression and the presence of moisture can seriously limit the life of any lubricant.

Automotive motor oils are used in small, low-power compressors, both reciprocating and rotary. High-speed rotary compressors may also use automotive oils, because of their suitability for high temperatures. If there is high humidity or low temperature, however, such oils will not be able to separate water well enough. Manufacturers' recommendations should always be respected.

In refrigeration compressors, oil and refrigerant mix continuously. This is required for proper lubrication of the compressor parts. Refrigeration oils are soluble in liquid refrigerant, and at normal room temperatures, they will mix completely. It is essential that the oil separator is maintained and functions properly. Whatever oil is chosen, it must be able to operate correctly over the whole temperature range to which it will be subjected.

For compressing reactive gases such as sulphur dioxide and chlorine, *white oils* are used. These are highly inert mineral oils that have undergone extensive processing to remove reactive compounds.

In any compressor, the proper synthetic can be used in place of a mineral oil in most cases, and this will increase the time between oil changes. This is the major virtue of synthetics. But their extra cost must be balanced against the increased time between oil changes (TBOC).

Wire Rope Lubricants

These lubricants are used to provide lubrication and rust protection for the rope and to reduce friction between the individual strands. Lighter-bodied oils may be used and applied by spray. Pine tar is often included to increase stickiness and to increase the oil's penetrative ability. Such lubrication serves for outdoor conditions, such as cableways and ropes on sheaves in mine headgears.

For less severe service and in cases where accessibility is good, heavier straight mineral oil may be applied by brush.

Biodegradable Lubricants

These have become more important in the twenty-first century with the increasing environmental demands. Biodegradability measured according to the CEC-L33-T82 standard is as follows:

- Mineral oils: 10%—40%
- Vegetable oils: 70%—100%
- Synthetics: 10%—100%, depending on type

Greases

Grease may be defined as a semisolid material consisting of a thickening agent dispersed in a liquid lubricant. Because they do not run in normal applications, they can be used where an oil, if used, would leak out of the system. Also, greases can provide natural sealing, keeping contaminants out.

Greases consist of three components:

1. A fluid lubricant, which usually makes up 70% to 95% by weight of the grease
2. A thickener, which provides the gel-like consistency that holds the liquid lubricant in place—usually 3%–5% by weight of the grease
3. Additives to enhance certain properties of the grease and usually make up 0.5%–10% of the grease by weight

The thickeners are usually metallic soaps of aluminium, calcium, lithium, or sodium. Certain clays are also sometimes used. The principal additives are antioxidants, rust and corrosion inhibitors, EP agents, and compounds that make the grease tackier.

Grease Tests

The single most important property of oil is its viscosity. The single most important property of grease is its consistency, which is a measure of the hardness or softness of the grease. The standard test for consistency is an ASTM test that measures the penetration of a cone of specified dimensions into the grease when released from a specified height, under a controlled, specified temperature. The result is measured in tenths of a millimetre. The grease is then graded with a National Lubricating Grease Institute (NLGI) grade number, as shown in Table 6.5.

Another important property of grease is its *dropping point*. If the temperature is too high, the grease will change from semisolid to liquid. This will happen over a range of temperature. The dropping point measures the start of this process, which is at a different temperature range for different soap thickeners, as shown in Table 6.6.

Greases cannot be used at temperatures above their dropping points. The maximum usable temperature is considered to be the limit where the grease should not be used without frequent relubrication. As can be seen from the table, the maximum usable temperature is well below the dropping point.

TABLE 6.5
NLGI Classification of Greases

NLGI Grade	Penetration in Tenths of a Millimetre
000	445–475
00	400–430
0	355–585
01	310–340
2	265–295
3	220–250
4	175–205
5	130–160
6	85–115

TABLE 6.6
Dropping Point and Maximum Use Temperatures for Greases

Thickener Type	Dropping Point Range (°C)	Maximum Usable Temperature (°C)
Aluminium	105–110	80
Calcium	99–104	80
Lithium	195–200	160
Sodium	170–175	120
Complex soaps	Above 230	Above

Other Tests for Greases

Several other characteristics are necessary to fully categorise grease as suitable for any specific application. Each of these characteristics has its own set of tests. The characteristics are as follows:

- Shear stability
- Oxidation resistance
- Water resistance
- Oil bleed resistance
- Anti-wear resistance
- Corrosion
- Pumpability

Compatibility

An important note to make is that greases are very often incompatible. If it is necessary to change from one grease to another, the equipment should be thoroughly cleaned and flushed to get rid of all traces of the old grease. Only then should the new grease be employed. It is safest to increase the frequency of regreasing for several cycles until it is certain that all the old grease has been removed.

A Case Study in Lubricant Application

The BMC Mini, designed by Alex Issigonis, was a revolutionary advance in vehicle engineering. By setting the four-cylinder engine transversely, driving the front wheels, much space was loosened up for the passengers, as there was no intrusive gearbox or propeller shaft in the passenger space. However, the design employed a unique gearbox, situated in the engine's sump. Both engine and sump were lubricated by the same grade of oil, which was basically engine oil, which could lead to a shortening of gear life. Issigonis presumably was influenced by motorcycles, many of which had and still have common lubrication for the engine and the gearbox. Other manufacturers who copied the east-west layout of the engine, such as Fiat with its 128, positioned the gearbox in line with the engine, allowing it to be lubricated with gear oils. This layout is now universal for transverse-engined, front-wheel-drive cars. It has the disadvantage, however, of necessitating the use of unequal-length drive axles, which, if there is wear in the linkages, can lead to "torque steer," with the longer axle attempting to turn the car off the straight-ahead position. All engineering is a compromise—solve one problem and we may cause another.

Reverse Engineering

Reverse engineering may be said to be a euphemism for "copying." Both Russia and Japan have been past masters of reverse engineering, and now the Chinese are developing their expertise in the art. Figures 6.5 and 6.6 show the Russian reverse-engineered Tupolev Tu 4 bomber, copied from four Boeing B-29s that had landed in Russia during World War II. More than 200 were made and served in the Red Air Force for several years after the war.

FIGURE 6.5 Russian TU 4 Bomber.

FIGURE 6.6 American B-29 Bomber.

The copying process was a considerable feat—Russia used the metric system, and the suppliers of many components could not supply a direct American equivalent; for example, the gauges of aluminium plate were different. This meant that the Tu-4 was heavier than the B-29.

For our present applications, reverse engineering is necessary or advantageous for several reasons:

- Spares may be discontinued for equipment that a company wants to keep in working order.
- A company may wish to replace OEM spares with others because of high spares prices.
- A company may wish to reverse-engineer a complete product, such as a tractor or lawnmower, then modify and improve the design to avoid patent and licence infringements.

All these may be valid reasons for reverse engineering, but the point to remember is that it must be done properly or not at all. The writer knows of a case where a pump manufacturer would no longer supply parts for a line of oil pumps that the company had discontinued. The customer, however, had a large number of these pumps installed and did not wish the capital expense of replacing them all. Reverse engineering was taken by a specialist outside company equipped with the necessary five-axis machine tools and CAD facilities. The pumps were of the progressing cavity principle: a single-helix rotor that revolves eccentrically inside a double-helix stator. The pump impellors were of stainless steel, which the specialist company had chemically analysed. New rotors were made out of available stainless billet, of almost-identical specification to that of the manufacturer.

The pumps operate with very tight clearances, which were accurately reproduced by the specialist company. The oil pumped was at high temperature. This is where the reverse engineering broke down. The coefficient of expansion of the two steels was slightly different, and within a short period of operation, a pump with the new spare rotor seized and caught fire.

This is not to say that reverse engineering should not be attempted. Just do it right. In another example, impellors for some centrifugal compressors are prohibitively expensive when bought from the OEM. Some of the price might be justified as amortisation of R&D costs, but in countries with weak currencies, uses of such compressors are hard-pressed to pay for the spare parts. There are at least two companies in South Africa, for example, that supply such impellors of excellent quality at a fraction of the price.

For a reverse engineering set-up, the following are the most important activities:

- Geometric measurement
- Material investigation—chemical analysis, microstructural analysis, and so on
- Life limitation calculations—fatigue, creep, brittle fracture, corrosion, and so on
- Manufacturing process verification—casting, welding, forging, and so on
- Manufacturer appointment—quality implications
- Legal implications

Reverse engineering should be considered by any organisation involved in maintenance optimisation. Properly conducted, it can even lead to improvements over and above what the OEM can supply. Ideally, reverse engineering should form a separate department in the company.

Further discussion on this subject is beyond the scope of this text. As seen from the previous bullet list, textbooks, on all the aforementioned subjects, at least at the undergraduate level, are required to do the topic justice. The reader is referred to the Bibliography for further reference to the subject.

Regression Analysis

Regression analysis is a technique used to determine the relationship between variables, for example, between surface hardness and life. The simplest form of regression is single linear regression, but one can also have multiple linear regression and various forms of non-linear regression, such as quadratic regression.

To illustrate the method, we will use the following example, relating surface hardness to life in weeks for an abrasion-resisting part in a machine. The data for the linear regression are given in Table 6.7, and the calculations using these data are shown in Table 6.8.

Using the simplest model, namely, linear regression, we have $y = a + bx$.

The required values of a and b can be found by solving the following pair of simultaneous equations:

$$\Sigma y = na + b\Sigma x,$$

where n is the number of readings and

$$\Sigma xy = a\Sigma x + b\Sigma x^2.$$

TABLE 6.7
Data for a Linear Regression

	Sample 1	Sample 2	Sample 3	Sample 4	Sample 5
x = Hardness, Rockwell C scale	50	25	75	125	100
y = Life of part (weeks)	25	23	34	40	35

TABLE 6.8
Linear Regression Calculations

Hardness	Life		
x	y	xy	x^2
50	25	1,250	2,500
25	23	575	625
75	34	2,550	5,625
125	40	5,000	15,625
100	35	3,500	10,000
$\Sigma x = 375$	$\Sigma y = 157$	$\Sigma xy = 12{,}875$	$\Sigma x^2 = 34{,}375$

To fit the best possible line through the data, we will use the method of least squares. Substituting in the previous equations, we get

$$157 = 5a + 375b \quad 6.2$$
$$12{,}875 = 375a + 34{,}375b \quad 6.3$$

Multiplying equation 6.2 by 75:

$$11{,}775 = 375a + 140{,}625b$$
$$12{,}875 = 375a + 34{,}375b \ldots$$

Subtracting, we get:

$$1{,}100 = 6250b.$$

Therefore, $b = 0.176$.
Therefore, from equation 6.3, we get $12{,}875 = 375a + (34{,}375 \times 0.176)$.
Therefore, $a = 18.2$.
Hence, the regression equation is $\mathbf{y = 18.2 + 0.176x}$.

Using the Regression Equation

What if the hardness is 87.5?

Then,

$$\begin{aligned} y &= 18.2 + (0.176 \times 87.5) \\ &= 18.2 + 15.4 \\ &= 33.6 \text{ weeks.} \end{aligned}$$

Note that the regression equation should not be extrapolated.

Correlation

So far, we have not considered how accurate predictions made from a regression equation are likely to be. If the data cluster very closely around the regression line, then the line is a very good fit and quite accurate predictions may be expected. On the other hand, if the data are widely spread, the line is not a very good fit and the predictions made from it may be very inaccurate.

How well the regression line fits the observed data depends on the degree of *correlation* between the two variables. This is measured by the correlation coefficient. This is denoted by the letter r and is calculated as follows:

$$r = C_{xy} / \sqrt{(C_{xx} C_{yy})},$$

where

$$C_{xy} = \Sigma xy - n * x_{ave} * y_{ave}$$
$$C_{xx} = \Sigma x^2 - n * x_{ave}^2$$
$$C_{yy} = \Sigma y^2 - n * y_{ave}^2$$

If there is perfect correlation, the value of r is either +1 or −1, where the + or − sign indicates the slope of the regression line. If y decreases as x increases, the slope of the line is negative. To ensure that this is the case, only the positive root in the previous formula (3) is taken. If there is no correlation between x and y, the r value is zero. For the previous example, the r value can be shown to be 0.971, which indicates a high degree of correlation. However, some of this correlation might just be a fluke, and we have to test for this next. The correlation table (Table 6.9) will enable us to do this.

In this case, we must enter this table at the row with $n - 2$ degrees of freedom, where n is the number of readings, in our case 5. Therefore, we enter the table at row 3. Normally, a 0.05 significance is used. Therefore, from the table, the critical value of r is 0.8783. The calculated value of r (0.971) exceeds this, and so a significant degree of correlation has been established. (But beware of spurious correlation, that is, when the two variables x and y are both influenced by a third variable—take the humorous example of the increase in lecturers' salaries and the sales of liquor. They do seem to increase in lockstep! However, they are spuriously correlated, as both are influenced by an increase in the standard of living in a country.)

Confidence Limits

The larger the correlation coefficient, the closer the association between the two variables. This means that increasing accuracy of prediction is obtained as r gets closer to +1 or −1. However, there is another way of testing the accuracy of predictions made from the regression equation.

This consists of finding the standard deviation of the observed y values around the regression line. This is known as the *residual standard deviation*, since it measures the amount of variation left in the observed y values after allowing for the effect of the x variable.

We can use the residual standard deviation to calculate tentative confidence limits for our predictions. Assuming the errors to be normally distributed, we can expect our predictions to be within one residual standard deviation of the true figure 68% of the time, within two residual standard deviations approximately 95% of the time, and so on. The formula from which the residual standard deviation is calculated may be expressed in the following form:

$$\text{Residual SD} = \sqrt{[C_{yy}(1 - r^2)]/(n - 2)}$$

$(n - 2)$ occurs in the place of $(n - 1)$ in the ordinary standard deviation calculation because another parameter, b, must be estimated from the data. This uses up another degree of freedom. As a general rule, for n observations of a model:

$$\text{Degrees of freedom} = n\text{—number of restrictions,}$$

TABLE 6.9
Significant Value of the Correlation Coefficient

df	$r_{0.1}$	$r_{0.05}$	$r_{0.02}$	$r_{0.01}$	$r_{0.001}$
1	0.98769	0.99692	0.999507	0.99987	0.99999
2	0.90000	0.95000	0.98000	0.99000	0.99900
3	0.8054	0.8783	0.93433	0.95873	0.99116
4	0.7293	0.8114	0.8822	0.91720	0.97406
5	0.6694	0.7545	0.8329	0.8745	0.95074
6	0.6215	0.7067	0.7887	0.8343	0.92493
7	0.5822	0.6664	0.7498	0.7977	0.8982
8	0.5494	0.6319	0.7155	0.7646	0.8721
9	0.5214	0.6021	0.6851	0.7348	0.8471
10	0.4973	0.5760	0.6581	0.7079	0.8233
11	0.4762	0.5529	0.6339	0.6835	0.8010
12	0.4575	0.5324	0.6120	0.6614	0.7800
13	0.4409	0.5139	0.5923	0.6411	0.7603
14	0.4259	0.4973	0.5742	0.6226	0.7420
15	0.4124	0.4821	0.5577	0.6055	0.7246
16	0.4000	0.4683	0.5425	0.5897	0.7084
17	0.3887	0.4555	0.5285	0.5751	0.6932
18	0.3783	0.4438	0.5155	0.5614	0.6787
19	0.3687	0.4329	0.5034	0.5487	0.6652
20	0.3598	0.4227	0.4921	0.5368	0.6524
25	0.3233	0.3809	0.4451	0.4869	0.5974
30	0.2960	0.3494	0.4093	0.4487	0.5541
35	0.2746	0.3246	0.3810	0.4182	0.5189
40	0.2573	0.3044	0.3578	0.3932	0.4896
45	0.2428	0.2875	0.3384	0.3721	0.4648
50	0.2306	0.2732	0.3218	0.3541	0.4433
60	0.2108	0.2500	0.2948	0.3248	0.4078
70	0.1954	0.2319	0.2737	0.3017	0.3799
80	0.1829	0.2172	0.2563	0.2830	0.3568
90	0.1726	0.2050	0.2422	0.2673	0.3375
100	0.1638	0.1946	0.2301	0.2540	0.3211

Note: If the calculated value of r exceeds the tabulated value of r α, a significant correlation has been established at the α significance level.

where a restriction occurs for each parameter of the model that has to be estimated from the observed data.

In our case, the residual SD may be calculated as 1.96, that is, ~2. This refers to the y value. Hence, for two residual SDs, the figure will be 4 weeks. Most of the time, the answer we get from our regression equation will be more accurate than this. We can take the ±4 weeks as a maximum error.

It is easy to perform regression using Excel's Linest and Correl functions.

Analysis of Variance

Analysis of variance (ANOVA) is a powerful statistical tool to ascertain if various relationships are statistically significant. Before the age of useful, inexpensive software, the calculations were tedious. Excel is not good at ANOVA, and other packages must be used. NCSS is recommended for this course. A trial version may be downloaded from the internet. For more information on the mathematics behind the ANOVA technique, the reader is referred to any standard statistical text, website, or the help notes to the NCSS programme.

The example shown here is for four different mines that have made use of three kinds of articulated dump truck (ADT) and have generated MTTF data for each type. What the analysis shows is whether there is any significant difference in either mine or truck type, or both. In this particular case, we have the data shown in Table 6.10.

Analysis of Variance Report

This is given in Table 6.11.

What Table 6.11 shows us is that there is a significant effect that both the site and the vehicle have on the MTTF. This is shown by the *F* ratio for both C1 and C2, that

TABLE 6.10
Analysis of Variance Data

Column in NCSS Spreadsheet	C1	C2	C3
NCSS Nomenclature	Factor 1	Factor 2	Response Variable
Description of Variable	Mine	Vehicle	MTTF (Days)
	Alles Verloren	Cat	16
	Diepgat	Cat	18
	Beginner's Luck	Cat	40
	Kalimba	Cat	14
	Alles Verloren	Vol	19
	Diepgat	Vol	2
	Beginner's Luck	Vol	30
	Kalimba	Vol	30
	Alles Verloren	Bel	45
	Diepgat	Bel	35
	Beginner's Luck	Bel	60
	Kalimba	Bel	40

TABLE 6.11
Analysis of Variance Report

Page/Date/Time	1 2014/02/12 06:32:59 PM
Database	C:\Documents and Settings\Administrator\My Documents\ADT.S0
Response	C3

Analysis of Variance Table

Source Term	df	Sum of Squares	Mean Square	*F* Ratio	Prob. Level	Power ($\alpha = 0.05$)
A: C1	3	976.9167	325.6389	5.78	0.033330[a]	
B: C2	2	1526.167	763.0833	13.55	0.005953[a]	
AB	6	337.8333	56.30556			
S	0	0				
Total (Adjusted)	11	2840.917				
Total	12					

[a]Term significant at α = 0.05.

is, for the mine and the vehicle, respectively. The asterisks indicate that at the 0.05 level, the *F* ratio, which is the ratio of the sums of the squares, is significant.[1] The probability of it being incorrect is given by the very low probabilities of 0.03330 and 0.005953, respectively. By examining the data in Table 6.10, it is seen that the Mine Beginner's Luck and the Bel ADT score are significantly better than the other three mines and two vehicles, respectively. Now, the situation could be investigated further to find out why Beginner's Luck performs better. Perhaps the duty cycle is lower there, or perhaps operations and maintenance are better. One would have to investigate other matters as well—for example, a low MTBF figure is good, but what about the MTTR figure and therefore the availability of the trucks? And what about their first cost and fuel consumption? Are they all in the same tonnage class? And so on.

The NCSS software is very powerful, and in this course, we can only hint at its capabilities. The reader, if interested, should familiarise himself further with the software.

NOTE

1. For more information on the mathematics behind the ANOVA technique, the reader is referred to any standard statistical text, website, or the help notes to the NCSS programme.

7 Other Techniques Essential for Modern Reliability Management II

Good planning and hard work lead to prosperity, but hasty shortcuts lead to poverty.

Proverbs 21:15
New Living Translation

SYSTEMS ENGINEERING

To persons familiar with systems engineering, the placement of this subject in this book might seem odd, as reliability engineering is often seen as a subset of systems engineering, not vice versa. The view taken here is that reliability engineering is a discipline in its own right but that other aspects of systems thinking can assist the reliability engineer in his work.

Systems engineering means different things to different people, and it means different things in different technologies. For example, in power generation plants in Germany and South Africa, a systems engineer is an expert in one or more of the technologies on the plant. Hence, in a coal-fired plant, we find a milling plant systems engineer, a turbine auxiliary systems engineer, and so on.

The other meaning of *systems engineer*, the one we shall use in this text, comes out of the American military. It was being found that with complex weapon systems, it was not enough to concentrate on the weapon itself, for example, a fighter aircraft. The machine itself is only one part of a complete system, which would include all documentation, the logistics and maintenance of the aircraft, training for the crew, and so on.

OPTIMISATION

The main thing to remember about systems engineering is its emphasis on *optimisation* and, therefore, on a consideration of the entire *life cycle* of the product, from conceptual study through specification, design, manufacture, and installation. This is then followed by operation, maintenance, and finally, disinvestment.

To be optimised, the system must perform as per specification, be built on schedule, and not overrun any stage of the allocated budget. Within the specification, attention is paid to parameters such as reliability, maintainability, and operability, parameters usually neglected by traditional hardware designers. *Cost engineering* and *financial management* are also important, as is *project management*.

DOI: 10.1201/9781003326489-7

INTERFACE MANAGEMENT

Another important concept in systems engineering is the emphasis on *coordination of interfaces*. It is found that often problems occur at the interfaces between disciplines. For example, if there is no coordination between electrical and mechanical engineering, one might find cables and piping designed to go through the same place in a wall.

THE SYSTEMS ENGINEERING APPROACH

One approach to the work might be for the systems engineers to first examine the overall system and decide on how best to proceed. They would then draw up specific requirements for each hardware group and specify all the interface conditions between the groups so that no problems will exist there. Then, the hardware groups design their subsystems.

The alternative approach is to have each hardware group design its subsystem, and the systems engineers take these designs and combine them to form a solution.

Both approaches are inadequate. The first approach is inadequate because the systems people cannot be specialists in all the subsystem areas and do not know what can be developed in each. They cannot adequately specify equipment before detailed investigation and development in the specific area. The second approach is also unrealistic because the different disciplines do not at first know what their overall task is and what compromises need to be made with other disciplines.

In practice, therefore, both approaches are used at the same time. The systems engineers develop initial specifications. The hardware engineers describe state-of-the-art estimates of what they think can be achieved. Consistent *communication* between the systems people and the discipline people is essential. Hardware's first estimates of attainable performance allow the systems engineers to revise the overall configurations and specifications. This process continues throughout the project.

As time goes on, the work of each group firms up and becomes less tentative. The optimisation techniques used at the start are continued with modified data. As time goes on, a detailed, reasonably optimum system emerges, which can be expected with confidence to meet the requirements set for it.

SYSTEMS ENGINEERING DESIGN

As a rule, therefore, a systems approach will include a design process consisting of conceptual design, preliminary system design, detail design, development and modification, and a final design freeze. Often, however, we have to work with systems that have not been designed according to this rigorous process.

SYSTEMS ENGINEERS

If we now understand how the systems approach works, what about the systems engineers themselves? They are either engineers trained in systems thinking and the optimising techniques of systems engineering, or they are discipline engineers who have been moved or perhaps even promoted into the systems engineer position because of their aptitude for systems thinking and their mathematical and computer engineering competence.

SYSTEMS ENGINEERING ORGANISATION

Organisation for systems engineering can take many forms but could look like the organisation shown in Figure 7.1. Systems engineering can embrace all the disciplines shown in Figure 7.1, plus others. Some of these disciplines are well covered in other chapters of this book. Furthermore, it is seen that in the systems engineering scheme of things, reliability forms a subset. Other aspects of systems engineering not included elsewhere in this book will be discussed here.

FINANCIAL OPTIMISATION

Financial optimisation is a necessity in systems and reliability work. Financial optimisation often goes under another name, namely, capital budgeting. The implication here is that any organisation only has so much money for new projects, retrofits, and the like. A budget must be drawn up to optimise this financial resource in estimating the probable returns from various projects and apportioning the available budget to those scoring the greatest financial return. Four techniques are generally available to financial managers to enable capital optimisation:

1. Payback period
2. Net present value (NPV)
3. Internal rate of return
4. Profitability index

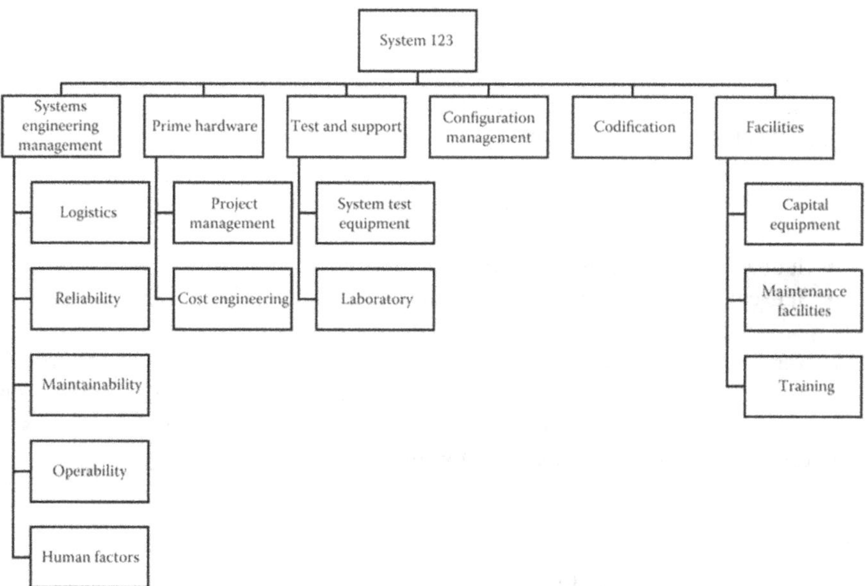

FIGURE 7.1 A possible organisation structure for systems engineering.

Another term used in this field is life cycle costing, or LCC for short, as usually one wants to know the costs and income that any system and project will generate over its life.

Capital budgeting can answer the following questions, among others:

- Should we replace certain equipment?
- Should we expand our facilities by:
  - Renting?
  - Buying new facilities?
  - Extending existing facilities?
- Should we launch a new product?
- Should we go through with the proposed merger?
- Should we refinance our outstanding debt?

Because of the technique's wide applicability, it is well worth becoming an expert in it.

Features of Investment Projects

- Large initial cash outlay
- Expected recurring cash inflows (from either sales or savings, depending on the investment). Depreciation is excluded as it is a non-cash expense.
- Income taxes may play a part in the final decision.

We will now study the three main methods previously mentioned.

Payback

This is the simplest method. If a project costs, say, $1 million, how many years will the profit generated from the project take to pay back the initial investment? If we assume that the profit is $350,000 per year, then the payback period will be just under three years. This may or may not be acceptable to the treasury of the company. The two main problems with payback are as follows:

- It does not consider the time value of money.
- It excludes all cash flows after the payback period.

However, payback is a useful figure to have. For small businesses that have to borrow money for a project, it may be of critical importance for both the lender and the borrower to know if the project can generate sufficient cash to repay the borrowings—after tax, payback is a useful measure for this.

Discounted Cash Flow

This is a more sophisticated form of capital evaluation. In fact, it is, one might say, the correct form, as it is based on the time value of money. This works like interest in reverse. And this interest in reverse is called the *discount rate*. Say, for example, that one invests $100 in the bank now, in what we will call year 0. If the interest rate that the bank is prepared to give us is 5%, then at the end of the year, that $100 will have grown to be $105. Let us now consider the reverse case—if one confidently expects to receive $105 in one year's time, then the value of that $105 is only $100 now. Likewise, if we expect an income of $105 in the second year, it is only worth approximately $95 today in year 0 ($95 × 1.05 × 1.05 = $104.7).

So when we apply discounted cash flow (DCF) to an expenditure and income stream, the earnings in distant years become effectively smaller and smaller. And for any project to recover its initial cost, the income stream must be greater the higher the discount rate. The final measure of the DCF stream is the NPV. This must be positive, or the project is not viable and will be rejected.

As a method of asset valuation, DCF has often been opposed to accounting book value, which is based on the amount paid for the asset. After the stock market crash of 1929, DCF analysis gained popularity as a valuation method for shares. Irving Fisher, in his 1930 book *The Theory of Interest*, and Williams's 1938 text *The Theory of Investment Value* first formally expressed the DCF method in modern economic terms. The economist Dr John Dean published his book *Capital Budgeting* in 1951. Before this, it could be said that there was no theoretically correct way to evaluate investments. One of the first companies to incorporate DCF into their expansion schemes was General Electric, in its time a very progressive company.

Internal Rate of Return

The third method we will discuss is internal rate of return (IRR). In this method, the earnings stream is analysed iteratively at different discount rates until the NPV is zero. If the discount rate is greater than the company's *hurdle rate*, the project is acceptable, as it should earn more money for the company than the average project.

There are, however, problems with IRR. It can, in some circumstances, lead to wrong recommendations. For example, can the very high discount rate of a high IRR be realistically reinvested by the company? Second, if the earnings stream becomes negative at some point, IRR can give the wrong result. As a general rule, a project with a very high IRR should be re-evaluated using NPV instead.

Profitability Index

This is where capital budgeting really comes in. If the company only has so much money for investment, even if there are several attractive projects, the money may not be available to fund them all. The profitability index is defined as:

Profitability index = (present value of cash outflows)/(present value of cash inflows)

On this basis, the projects with the highest profitability indices will be approved until all the funds in the company's treasury are allocated for that year.

Calculations

Generally, capital budgeting calculations are performed using Excel, although printed tables are still available if required. Table 7.1, simulating an Excel spreadsheet, is given, with explanations of what the figures mean.

Class Assignment

Duplicate Table 7.1 in an Excel spreadsheet, using the functions IRR, NPV, and SUM (for payback). Note the following:

- For the IRR calculation in column E, enter lines 1–6 in D. The answer is the 9% in row 7.
- In column F, the values in rows 2–6 are individually calculated to demonstrate the IRR method—how a discount rate of 9% results in an NPV of (approximately) zero. (F7 ≈ D1.)
- In columns G and H, NPVs are calculated directly from the Excel function. It will be seen that at a discount rate of 15%, the project is not viable, but at a discount rate of 5%, it is.
- In column I, payback over five years is calculated simply using the SUM function. If payback over four years is calculated, it will be seen to be negative—hence, the project only pays back in five years.

Determination of the Discount Rate

How is the discount rate determined? If one is a private individual, the discount rate is one's overdraft rate at one's bank. Large companies, however, are usually able to secure finance at much better rates. This is because they are much larger than the average private individual and are also a better risk. Therefore, banks will lend to them at lower interest rates than to private individuals. But companies also

TABLE 7.1
Capital Budgeting Techniques

A	B	C	D	E	F	G	H	I
1	Year 0		–R 70,000					
2	Year 1		R 12,000		R 11,009			
3	Year 2		R 15,000		R 12,625			
4	Year 3		R 18,000		R 13,899			
5	Year 4		R 21,000		R 14,877			
6	Year 5		R 26,000		R 16,898			
7				9%	R 69,309	–R 9,960	R 7,840	R 22,000
8			Project cash flows	IRR calculation	NPV at 9%	NPV at 15%	NPV at 5%	Payback over five years

have access to another form of finance that is generally cheaper than bank finance, namely, the stock market. A company's *cost of capital*, that is, its discount rate, is determined by its cash holdings, bank borrowings, and share capital combined. The cost of capital changes as debt is redeemed, new debt is secured, and so on. The company's financial department will be able to tell its engineers what discount rate to apply at any time. Very often, it is updated every three months or so in a bulletin.

Here follows a representative calculation, for interest's sake, on how the cost of capital might be determined. Assume the following:

- The company has $4 million in short-term loan capital at 15% interest.
- The tax rate is 28%.
- After tax, cost of this capital = 0.72 × 15% = 10.8%.

The company also has:

- $1 million in preference share capital at 10%
- $5 million in ordinary share capital
- Dividends per share = $11
- Earnings per share = $80
- Dividend growth has been 4% per annum

  Therefore,

$$\begin{aligned}\text{cost of ordinary shares} &= (\text{dividends per share})/(\text{net earnings per share})\\ &\quad + \text{growth rate of dividends}\\ &= 11/80 + 0.004 = 0.1415.\end{aligned}$$

The weighted average cost of capital is calculated in Table 7.2.

Therefore, the weighted average cost of capital is 12.4%.

Handling Inflation

Cost of capital = 15%
Inflation = 10%

TABLE 7.2
Cost of Capital Calculation

Type of Finance	Rate (%)	Percentage of Total	After Tax Cost	Weighted Average Cost
Loan	15	40	0.108	0.043
Preference shares	10	10	0.10	0.010
Ordinary shares		50	0.1415	0.071
Cost of capital				0.124

$$\text{Effective discount rate} = 1.15/1.10 = 1.045$$
$$= 4.5\%$$

Thus, inflation *lowers* the effective discount rate.

Handling Tax

Let
S = Sales
E = Cash operating expenses
d = Depreciation
t = Tax rate

Then,

Before-tax cash inflows = S − E
Net income = $S - E - d$
After-tax cash inflows = Before-tax cash inflows − tax
That is, = $(S - E) - (S - E - d)^*t$
That is, = $(S - E) \times (1 - t) + d^*t$
Example:
S = \$12,000
E = \$10,000
d = \$500 per year
t = 28%

$$\begin{aligned}\text{After-tax cash inflow} &= \$(12{,}000 - 10{,}000) \times (1 - 0.28) + \$500 \times 0.28 \\ &= \$(2000 * 0.72) + \$500 * 0.28 \\ &= \$1440 + \$140 \\ &= \$1580\end{aligned}$$

CASE 7.1 THE PULVERISER TENDER

Author's Note: This is a case about total life cycle cost of ownership-capital cost, operating cost, unavailability cost, and maintenance cost. It is about how the choice of technology can have a huge influence on life cycle costs. It is also about how pride, prejudice, and wanting to make one's life easier can have a major financial impact on a company.

THE PERSONALITIES INVOLVED

Andrew Bacon was chief mechanical engineer at Midwest Power when Mark Francis joined the company at the engineer level. (There were four levels in the engineering structure of the company—engineer, senior engineer, principal engineer, and chief engineer.) Andrew appointed Mark as his pulveriser engineer.

Andrew was a graduate of a respected state university, as was Mark. Andrew had begun his career as an apprentice with the Santa Fe railroad company in the Southwest, where he was awarded a scholarship to continue his engineering studies at university. He then joined Midwest Power, one of the largest utilities in the country. Mark was drafted after graduation and served in Vietnam before joining a manufacturing company. After several years in manufacturing and construction, he joined Midwest Power in 1976.

Both persons would have considered themselves both practical and theoretical engineers, able to work with their hands and with their heads, but perhaps at different points along the practical/theoretical spectrum. Andrew was only six years older than Mark, but his drive and ambition had taken him much further. He was chief engineer of this large utility before he was 40, only one step below board level. Andrew was therefore a man who generated considerable respect among his peers, subordinates, and managers at Midwest, as well as among the various suppliers of equipment to the utility.

Mark, on the other hand, had still not risen above the level of engineer. One colleague had described him as "very analytical." Mark was incredulous. "What else is an engineer supposed to be?" he asked.

MIDWEST POWER

Midwest Power was involved in a very large expansion programme, growing at 15% per year in sales and adding capacity accordingly. This rapid expansion had meant that perhaps some of the engineering decisions made by either Midwest or its suppliers were suboptimal.

One particular set of equipment that was proving troublesome at two of the new power plants was the coal pulveriser plant. Coal pulverisers were used on fossil-fired power plants to grind the coal to the consistency of powder before it was blown into the boiler to be fired to generate steam. The steam would then be used to drive a turbine, which, in turn, drove a generator to produce electricity.

PULVERISER TECHNOLOGY

Pulverisers, or mills, as they were sometimes called, come in two basic configurations: vertical spindle and horizontal spindle. There were several variations on each basic type, as shown in Tables 7.3 and 7.4, and in Figures 7.2 through 7.9.

The pulverisers were supplied by the Browne and Whitney Company and had been used successfully for years by Midwest on its older plants, but never in the sizes now employed in the newer plants (Figures 7.2 and 7.3). The scaling up had led to machines that suffered from severe vibration to the extent that they not only damaged the ducting connecting them to the coal plant and

TABLE 7.3
Types of Vertical Spindle Pulverisers

Vertical Spindle Pulverisers					
Specific Name	**Subtype**	**Grinding Medium**	**Conveying Medium**	**Use**	**Manufacturer**
Ball and ring mill	—	Replaceable rings and balls	Air	Mineral processing or coal processing	Browne and Whitney
	Rollers fixed to body of the pulveriser	Replaceable table segments and roller tyres	Air	Mineral processing or coal processing	Konig
	Rollers independent of pulveriser body	Replaceable table segments and roller tyres	Air	Mineral processing or coal processing	DWM (Deutsche Walzenmuhlen)

TABLE 7.4
Horizontal Spindle Pulveriser Designs

Horizontal Spindle Pulverisers				
Specific Name	**Grinding Medium**	**Conveying Medium**	**Use**	**Manufacturer**
Autogenous mill	The ore itself	Water	Mineral processing	Various firms serving the mining industry
Rod mill	Steel rods	Water	Mineral processing	Various firms serving the mining industry
Tube mill or ball mill	Steel balls	Water or air	Mineral processing or coal processing	Fletcher-Wainwright and others

the boilers but, on occasion, would also cause switchgear in the vicinity to malfunction.

Browne and Whitney's main local competition used an alternative type of pulveriser, as made by the Konig Company (Figure 7.4). These pulverisers did not suffer from vibration as much as the Browne and Whitney design did. This design was also more maintainable, as the grinding rollers were attached to the mill body on trunnions and could be swung out of the way for maintenance.

To counter the problems with its larger pulverisers, Browne and Whitney had recently taken out a licence from a European manufacturer, DWM, for an alternative design (Figure 7.5). The DWM suffered from vibration, but not as much as the Browne and Whitney did, but it also did not have the easy maintenance facility of the Konig design.

FIGURE 7.2 The grinding principle of the Browne and Whitney pulveriser.

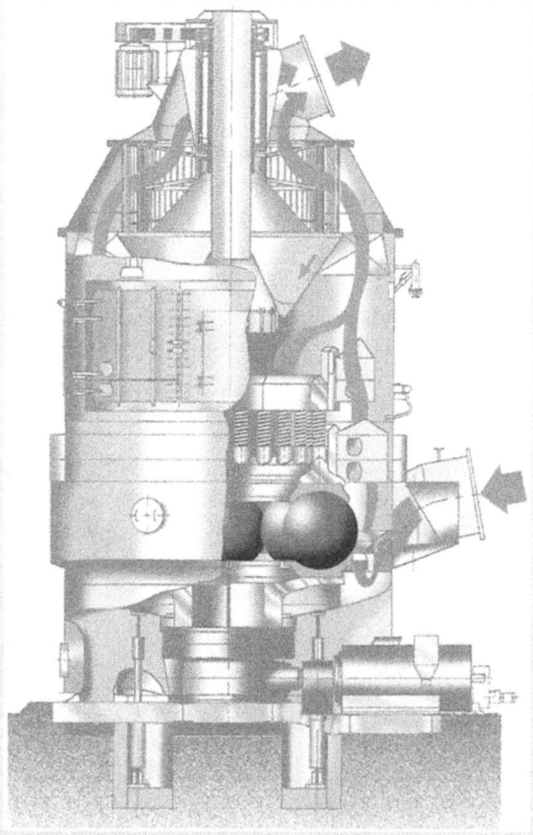

FIGURE 7.3 Cross section of a Browne and Whitney pulveriser. (Prone to vibration because of large unrestrained rotating mass of grinding balls.)

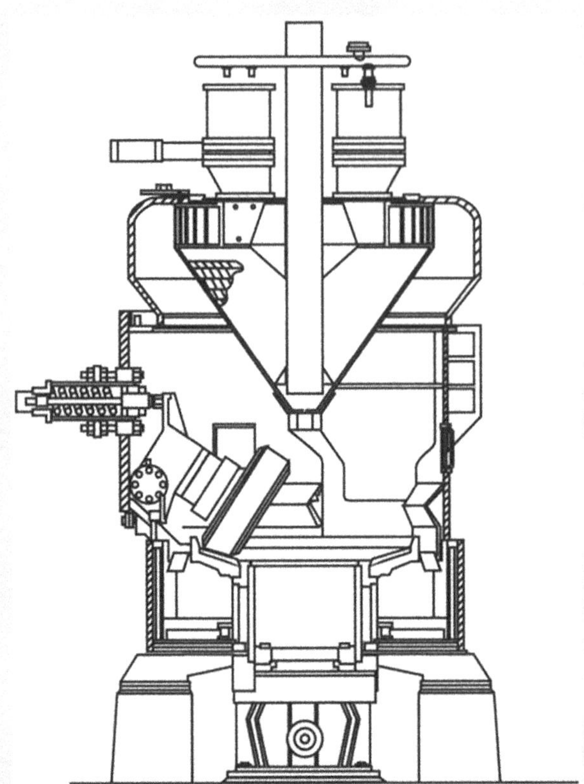

FIGURE 7.4 Cross section of a Konig pulveriser. (Grinding wheels attached to mill body and therefore easily swung out for maintenance; low vibration.)

OPERATION OF A TYPICAL VERTICAL SPINDLE PULVERISER

As shown in Figures 7.3 through 7.5, coal is fed into the machine and falls onto the grinding table, where it moves out by centrifugal force under the grinding wheels (or balls, in the case of the Browne and Wheeler pulveriser). The crushed coal falls off the edge of the table and is taken up by the airstream into the classifier, where cyclonic action separates out the larger particles, which fall back onto the grinding table. The finer particles are blown into the boiler and burnt.

HORIZONTAL SPINDLE MACHINES

There were other small suppliers in the market, offering old-fashioned horizontal spindle pulverisers, otherwise known as tube mills. Horizontal spindle machines used very old technology and were still used in mining and mineral processing applications, with wet feed (Figure 7.6). Some customers still bought air-swept tube mills for pulverised coal application. One supplier of such machines was the Fletcher-Wainwright company (Figure 7.7).

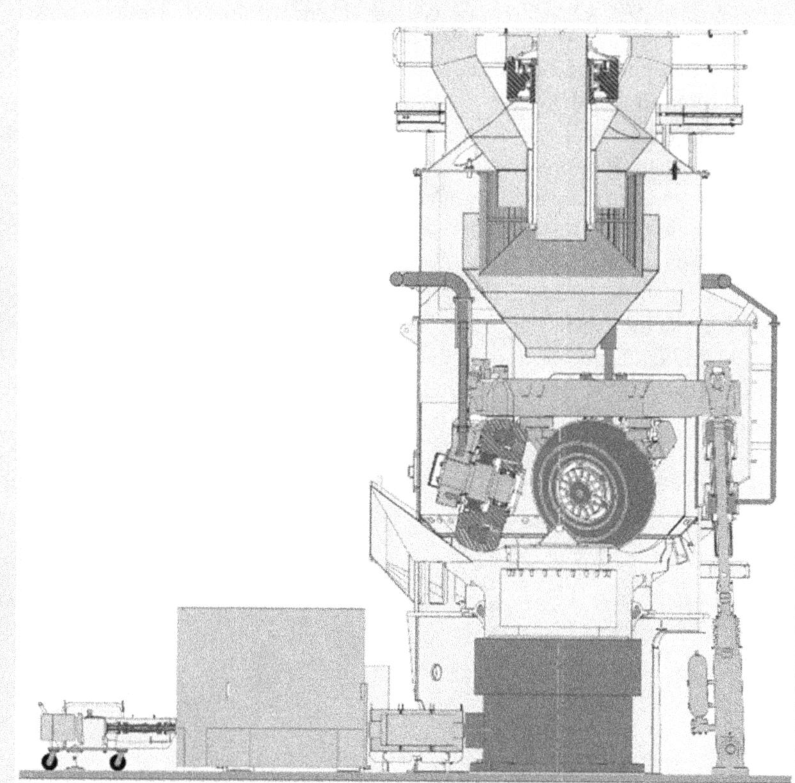

FIGURE 7.5 Cross section of a DWS pulveriser. (Grinding wheels not attached to mill body. Grinding wheel removal is complex. Prone to vibration as for the Browne and Whitney pulveriser.)

Tube mills were known to be old-fashioned but also rugged, reliable, and with the ability to grind anything. A few more conservative power utilities around the world favoured them.

PULVERISER OPERATION: HORIZONTAL SPINDLE TYPE (I.E. TUBE MILL TYPE)

Coal is fed into both ends of the machine. The drum, or tube, is rotated, and the coal and ball mixture is tossed about inside the drum. Coal particles that have been ground fine enough are swept out of the drum by the hot airstream. The pulverised coal is passed through the classifiers, where particles that have not been ground fine enough are separated by centrifugal action and return to the drum for further grinding.

COMPARISON OF THE PULVERISER TYPES: POWER CONSUMPTION

The grinding mechanism in vertical spindle pulverisers is fundamentally different from that in horizontal spindle machines and much more efficient.

In the vertical spindle design, the coal lumps are pulverised by the roller or ball riding over them and crushing them. In the horizontal spindle design, the coal is thrown in the air by the rotation of the drum, along with the grinding balls. Both product and grinding medium then fall by gravity to the bottom of the mill, where they smash into each other. The two modes of grinding are analogous to a continuous mortar-and-pestle action in the case of the vertical spindle machines and, for the horizontal spindle machine, to throwing up some coal and some steel balls into the air, with a random interaction occurring when they reach the ground. Only the force of gravity is available to do the grinding in the tube mill. Far more force is available to grind in the case of the vertical spindle design.

This fundamental difference in grinding action leads to a power consumption for the tube mill of approximately twice that of the vertical spindle designs. On a life cycle basis, this means a very large increase in operating cost over the life of the power plant, which can be in the order of 50 years.

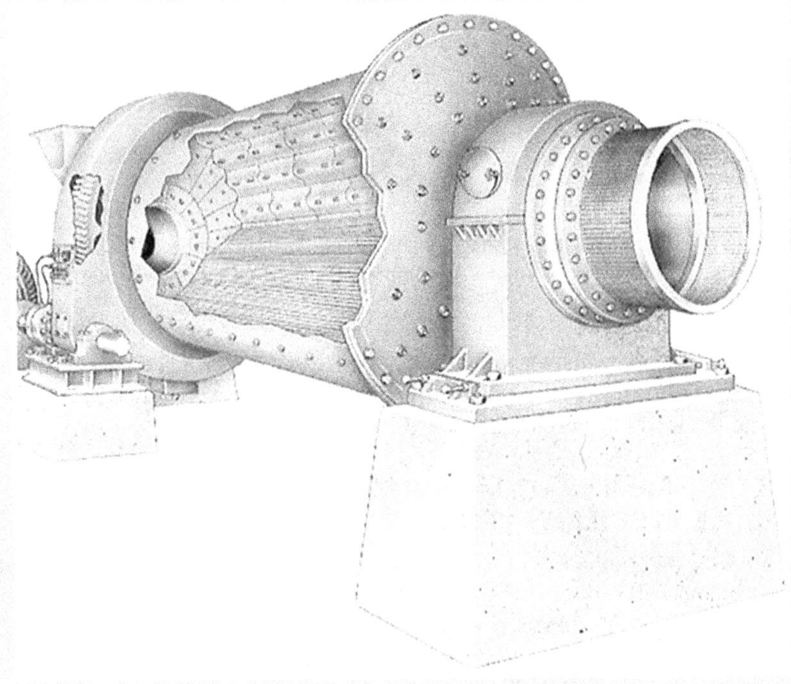

FIGURE 7.6 A typical rod mill design as used in mineral processing. (This is a wet feed mill; water-borne product comes in the far end and is ground by the rod charge before being discharged at the near end. Wet feed is clearly not possible for pulverised coal applications in power plants.)

FIGURE 7.7 Tube mill installation. (Note the very large electric motor.)

COMPARISON OF THE PULVERISER TYPES: MAINTENANCE AND OUTAGE COSTS

The manufacturers of horizontal spindle machines always claimed lower maintenance costs and lower outage costs than vertical spindle machines. As the grinding medium (the steel ball charge) wears down, the balls can be replaced by feeding them in with the coal charge. Hence, there is no need to take the mill off load. Eventually, after several years, the mill liners will have to be replaced. This will require an extensive mill outage and will be expensive.

Another issue to be remembered is that everything placed in the horizontal spindle machine must either remain there or pass through the boiler. Vertical spindle machines have the capacity to reject tramp iron, rocks, and other ungrindable material. Such material falls off the outer rim of the grinding table and into reject boxes, which are emptied periodically by manual, pneumatic, or hydraulic means. Nothing is rejected in a tube mill, which can mean a lot more abrasive material passing through the boiler, with concomitant external wear of the waterwalls, superheater tubes, and reheater tubes in the furnace.

In contrast to tube mills, the vertical spindle pulverisers require frequent changes of grinding elements, both table segments and balls or tyres. The machine must be taken off load to do this. Depending on the abrasiveness of the coal, the grinding element changes might occur every 5,000 hours and take

24 hours to renew, for a Konig pulveriser. Browne and Whitney pulverisers would have longer maintenance-free operating periods (MFOPs) but also have longer MTTRs, so the average availability between the two types of pulverisers would be similar.

Midwest Power allowed for maintenance by equipping their boiler with redundant pulverisers. 4 + 1, 5 + 1, and 4 + 2 were some of the configurations adopted. Another way to extend pulveriser life was to make the pulverisers larger than necessary. At some earlier power plants, the pulverisers were undersized.

SOLUTIONS AT MIDWEST POWER

Andrew Bacon decided that the Browne and Whitney pulverisers were a lost cause and ordered that the last two boiler-turbine units at the latest power plant being built be equipped with Konig vertical spindle pulverisers. He also instructed Mark to prepare a specification for a tube mill installation for one of the boilers at Algonquian power plant, to replace the undersized Konigs at that plant. Three Fletcher-Wainwright machines replaced the five Konigs on unit no. 1 at Algonquian. This installation was to serve as a proving site for the use of tube mills in the future.

Andrew Bacon further stated that all future power plants were to be equipped with tube mills. The next four plants were all equipped with horizontal spindle machines, some 120 machines in all, from Fletcher-Wainwright, and other manufacturers.

"Midwest's policy is to burn low-cost, low-quality, high-abrasive coal—this policy will not change, because of the mines we are tied into with long-term contracts. The coal quality is not going to get any better. Therefore, we will have to use tube mills in the future," Andrew Bacon dogmatically asserted. Few openly questioned this rationale. Mark Francis, who worked for Andrew Bacon at that time, did propose that, in the future, pulverisers should be chosen according to the specific coal at the plant in question and the quality of maintenance. For very poor coals, tube mills could be used; for better coals, vertical spindle mills. He, in fact, drew up an application map, with the areas of use for Browne and Whitney, Konig, and Fletcher-Wainwright pulverisers. In a meeting with the chief executive and Andrew Bacon, one day Mark presented the map. Andrew Bacon's response was, "Ah, the famous performance map again." Nothing came of Mark's proposal.

Mark Francis was never happy with tube-mills-only decision—the tube mills at Algonquian had not yet proved themselves, and no cost analysis was ever done for the four new stations. The pulverisers were simply bought as part of the boiler supplier's contract and were not even priced as a separate item.

Shortly after this decision had been made, Andrew Bacon was promoted to board level as engineering director. After a few years, he had left the company to become a consultant to the power generation industry worldwide, becoming even more successful in that role than he had been at Midwest Power. He was last known to be living in a large mansion on the Pacific coast, among the film community of Southern California.

FAST-FORWARD 25 YEARS

The vibration problems with the Browne and Whitney pulverisers were solved. A specialist company, Berg Inc., was brought in. The pulverisers were mounted on Berg's spring and damper units, and all connections to the rest of the plant were redesigned as flexible. The pulverisers still vibrated, but the damping units kept the vibration to reasonable limits. The machines were also slowed down by approximately 10%, as it was determined that they were approaching a resonant frequency. For a 10% reduction in output, vibration levels were reduced by more than 75%.

On the four new power plants, a generation of engineers grew up knowing only tube mills. Two special interest groups were formed, one for the milling engineers at the older plants equipped with vertical spindle pulverisers, and one for the newer plants equipped with tube mills. These groups collected cost information on their respective pulverisers and had published a report once a year for the past four years on the findings. The data were collected by power plant staff and managed by an outside consultant, Alec "King" Fisher. Alec was an ex-Midwest employee who now made and repaired items in the pulveriser market. He was a tribologist by profession with an intimate working knowledge of coal grinding techniques.

After 25 years, Midwest Power entered a new phase of rapid expansion. The leader of the tube mill special interest group, Martin Brady, was delegated to specify the "ideal tube mill," based on Midwest Power's 25 years of experience with the type. It was confidently expected that the next power plant would be equipped with pulverisers according to his requirements.

However, at Midwest's head office, it was decided that the specification for the new power plant was to be more detailed than any in Midwest's previous history. A separate price was required for pulverisers, and the alternatives of vertical spindle and horizontal spindle had to be offered.

THE PULVERISER COST REPORTS

In the meantime, Mark Francis was still employed at Midwest. He had a number of different postings in the organisation and had at last reached the level of a principal engineer, with a free-ranging portfolio to investigate, optimise, and recommend actions in the maintenance field. He began studying the pulveriser cost reports and was surprised at the information that was forthcoming. He checked with King Fisher as to the accuracy of the information and was assured that for the power plants in question, the costs were correct. At some stations, costs were hidden, it was believed, to show the milling engineer in a better light, but these cases were a minority.

At only one station, Algonquian, was there a direct comparison of costs between vertical and horizontal spindle machines. For the aggregate cost of maintenance, operation, power consumption, and outage, the vertical spindle pulverisers averaged $4.15 per ton of coal processed, as compared to the tube mills at $4.73 per ton of coal. This was for one of the worst coals in the Midwest Power system. Maintenance costs for the two competing technologies were $1.44 per ton of coal for the tube mill and $1.50 for the vertical spindle mills—a 4% difference.

At the station where the Browne and Whitney pulverisers had been replaced by Konigs, the latter did indeed perform better, with total costs of $2.23 per ton of coal versus $2.67 for the Browne and Whitney machines. But compared to other power plants, these Browne and Whitneys were performing well. At another station equipped exclusively with Browne and Whitney pulverisers, the cost was $2.35 per ton of coal. At the newer, tube mill–equipped stations, the costs were $1.95, $2.76, $3.13, and $3.81. Furthermore, these power plants had not yet had many tube mill overhauls. As has been said before, when a tube mill is overhauled, the cost in liner replacement and downtime is very high. The tube mill figures would certainly climb as the power plants got older.

Mark Francis asked if he could present his findings at one of the pulveriser special interest meetings. The thrust of Mark's presentation was that the vertical spindle pulverisers were performing well and should be considered for future power plants. The presentation was met with some incredulity by some members, but not by the majority. King Fisher agreed with the findings—he could hardly do otherwise; the figures used were from his own reports.

Shortly after this, Mark heard that Martin Brady had told King Fisher to no longer compile the pulveriser cost reports, as the costs were not accurate.

When the tenders came in from the boiler suppliers for the new station, they were evaluated by Richard Pepper, a veteran boiler expert who had worked for Midwest some years before and had now returned after working overseas for several years. He reported to Donald Irvine, who occupied the chief mechanical engineer position occupied many years before by Andrew Bacon.

"The figures show that the vertical spindle pulverisers on offer will cost us $400 million less over the life of the new station, on a life cycle–cost basis," said Richard to Mark.

"And this is even allowing for the higher maintenance costs of the vertical spindle pulveriser," he continued.

"Where did you get the maintenance cost estimates?" asked Mark.

"From Martin Brady, the milling expert," replied Richard.

"Well, here are my figures, from Brady's reports. I think they will confirm and reinforce your findings," said Mark.

"Furthermore, the two boiler suppliers both offer both vertical spindle and horizontal spindle pulverisers, but they both recommend the vertical spindle type. They have no hesitation in offering it to us, knowing the kind of coal we will be using," added Richard.

"I am therefore going to recommend to the tender commission that we order the vertical spindle pulverisers for the new power plant," said Richard.

THE ACRIMONIOUS MEETING

Some days thereafter, Donald Irvine and Richard Pepper were called to a meeting with Martin Brady, the tube mill expert, and other power plant representatives. Mark Francis was not invited.

"We hear you are recommending we use vertical spindle pulverisers on the new plant," Martin Brady said.

"Yes," replied Richard.

Martin answered him angrily, "Are you stupid or something? Don't you know how difficult they are to maintain? With a tube mill, all we have to do is put some more balls in it occasionally, while it is on load. The vertical spindle pulveriser has to be taken out of service and its grinding parts replaced every few months!"

"But Mark Francis's figures show that the maintenance costs are not that different—"

Richard Pepper's reply was interrupted by a visibly angry Martin Brady. "Mark Francis's figures are a load of sh**!"

"But they are your figures, gleaned from your reports!" Richard responded.

Martin changed the subject. "Do you mean to tell us that we have been using the wrong kind of pulveriser for the last 25 years?"

"Yes," replied Richard.

A SECOND MEETING

Some hours thereafter, Mark was called to the office of the chief mechanical engineer, Donald Irvine. Richard Pepper was with him, visibly upset. Another veteran of Midwest Power, the combustion specialist, Brian Miner, was with them.

"Mark, we were summoned to a meeting with Martin Brady and other power plant representatives, unhappy with what they had heard concerning our recommendations for pulverisers for the new plant," David Irvine said.

"Unfortunately, you have no credibility with the power plant personnel, so they rejected the arguments put forward in your cost analysis. At the recommendation of the manufacturers, we are visiting their works and their installations at other utilities here and overseas, to try and get further information to resolve this issue."

"What were their main reasons for their insistence on tube mills?" asked Mark.

"One was that with our high ash coals, the abrasion is so high it will wreck any vertical spindle mill," answered David Irvine.

"But ash is not the problem, abrasion is," replied Mark. "I covered that in my report. Some of our higher ash coals have lower abrasion values. Here it is." (Figure 7.8).

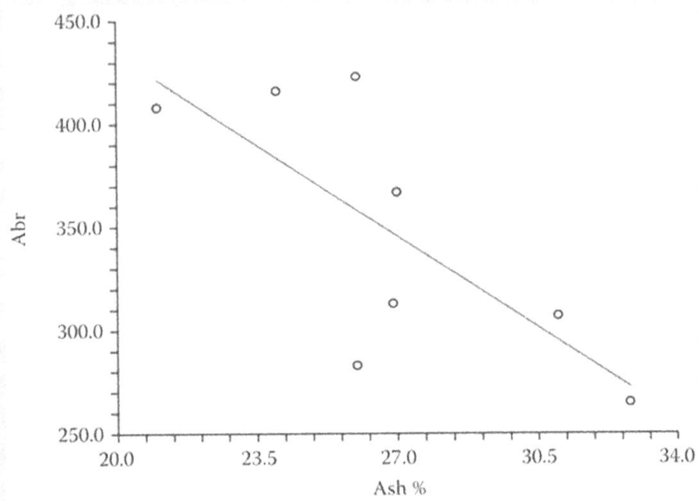

FIGURE 7.8 Abrasiveness index versus ash content for eight coals used by Midwest Power. (The coal for the new station is the dot second from the right.)

"They also say that no manufacturer has ever made a vertical spindle pulveriser in the size range we need," David continued.

"But Konig, for one, has made very large pulverisers for limestone—up to 800 tons/hour. We are looking for a coal pulveriser of just 80 tons/hour," Mark replied.

"Yes, but is limestone as abrasive as high-ash coal?" asked Brian Miner.

QUESTIONS

1. What sort of pulveriser should Midwest choose for their new station? Vertical or horizontal spindle? Give reasons for your answer.
2. Contrast the "engineering style" of Andrew Bacon, Mark Francis, and Martin Brady.
3. In light of Figure 7.8, does Brian Miner's question "Is limestone as abrasive as high-ash coal?" make sense?
4. "Engineering decisions should be based on facts and logic, not emotion" (Mark Francis): Comment on the truth (or otherwise) of this statement.

GLOSSARY

Abrasiveness index: A measure of wear taken as the amount of metal worn down in a miniature mill after grinding a fixed amount of coal. The index is determined by weighing the grinding elements before and after the grinding operation.

Tribology: The branch of engineering that deals with the interaction of surfaces in relative motion (as in bearings or gears): their design and friction and wear and lubrication (*Princeton WordWeb Dictionary*, 2006).

REFERENCES

The figures used in this case are from sales literature from the following pulveriser manufacturers: Claudius Peters, Hitachi Europe, and Loesche GmbH.

SYSTEM MODELLING

Of the specific tools in the systems engineer's toolbox, *system modelling* is one of the most important, as it attempts to optimise a system in a computer before any mistakes can be made in the hardware stage. (Or if optimising is not possible, at least a better understanding of the system is developed.) There are various ways systems can be modelled. These include the following:

- Simulation
- Linear programming
- Dynamic programming
- Decision theory
- Game theory

Some of the techniques are more curious than useful, depending on what is being analysed. But the first two, simulation and linear programming, are almost always applicable and will be discussed further here.

SIMULATION

Because of the ready availability in this century of powerful and cheap computers, *simulation* of systems is readily possible. Simulation allows one to experiment with various designs and design aspects without ever having to cut any metal. Prototypes are now a confirmation of the inherent correctness of the software design, rather than the experimental mules that they were a generation ago.

Simulation software packages are many and varied. Some use high-level simulation languages, which simulate specific technologies or environments, thereby reducing the programming effort. Such packages include programmes to simulate the following:

- The availability of systems and the spare parts levels needed
- Assembly line production of motor vehicles
- Process optimisation of petrochemical plants
- Fatigue analysis of designs with optimisation attributed to iterations of the CAD model, without any prototype testing being necessary

Another type of simulation package uses lower-level language but allows universal simulation. Such packages are known to most ex-students because of undergraduate use. Included in this category are packages such MATLAB® and Mathematica, which include simulation capability. In the commercial environment, Arena, marketed by Rockwell, has almost become the standard in software for generic simulation.

Simulation normally follows one of two processes, Monte Carlo or Markovian. The differences between these two types are described in Chapter 2.

It is always helpful to perform a manual exercise to demonstrate a technique. Although in practice one would always use software, an exercise is given here using manual simulation using past data.

Simulation Example: Stock Control

Suppose a factory has the following simple stock control system:

1. In January, calculate the average weekly demand for each item over the past 12 months.
2. Over the next 12 months, reorder an item whenever the stock level falls below 2 × average weekly demand during the lead time based on the demand calculated in step 1.
3. For each item, the reorder quantity is 10 × average weekly demand calculated in step 1.

With the proposed new system, it is hoped to reduce overall costs:

1. In January, calculate average weekly demand for each item over the last 12 months, as before.
2. Over the next 12 months, reorder an item whenever the stock level falls below

   average lead time demand + 2 × √(average lead time demand).

3. For each item, the reorder quantity is 40 × √(average weekly demand/unit price).

We must now consider what would have happened if the new system had been in operation over the last 12 months.

Considering just one item (call it Item A), we have:
Average weekly demand = 4.5 units
Lead time = 9 weeks
Unit price = $20
With the present stock control system, we calculate that:
Reorder level = 81 units
Reorder quantity = 45 units
With the proposed stock control system, calculate that:
Reorder level = 53 units
Reorder quantity = 19 units

Table 7.5 shows what the actual weekly usage of Item A has been over the last 12 months, together with the stock variation with the present stock control system. Alongside is shown the stock variation that would have occurred if the proposed system had been in operation instead over the same period. The stock level at the start of the period was 93 units.

We can now compare the costs of the two stockholding systems.
Other information required is as follows:
Cost of placing an order: $2
Stockholding cost: 0.15 × average stock value
Cost of a stock-out: 0.5 × value of items demanded during the out-of-stock period
Unit price for Item A: $20
Now, with the present stockholding system, over 52 weeks:
No. of orders placed: 5
No. of items in stock-out: 0
Average stock level: 3,653/52 = 68.52

$$\begin{aligned}\text{Total annual cost:} 5\times 2+68.52\times 20\times 0.15\\ =10+205.56\\ =\$215.56\end{aligned}$$

And with the proposed stockholding system, over 52 weeks:
No. of orders placed: 10
No. of items in stock-out: 0
Average stock level: 1,808/52 = 34.77

$$\begin{aligned}\text{Total annual cost:} 10\times 2+34.77\times 20\times 0.15\\ =20+104.31\\ =\$124.71\end{aligned}$$

Hence, the proposed stockholding system would appear to be better. However, there are several important points to consider.

- The starting value of 93 units was not realistic in the case of the proposed system—we should only compare the two options after the simulation has settled down to a normal pattern of working. If we discard the first 26 weeks of the simulation and redo the calculations, we would find that the total annual costs would change R192.12 and R78.36. Thus, the proposed system remains superior, but this might not always be the case.
- The length of the modified simulation is perhaps too short to be representative of reality, but it is all the information we have. But we can synthesise data on the basis of the existing data to extend the simulation.

TABLE 7.5
Comparison of Two Stockholding Systems

						Present System	Proposed System
Week Number	**Weekly Demand**	**Stock**	**Stock + Orders**	**Orders and Receipts**	**Stock**	**Stock + Orders**	**Orders and Receipts**
0		93	93		93	93	
1	2	91	91		91	91	
2	5	86	86		86	86	
3	9	77	77	Order	77	77	
4	2	75	120		75	75	
5	9	66	111		66	66	
6	4	62	107		62	62	
7	4	58	103		58	58	
8	1	57	102		57	57	
9	0	57	102		57	57	
10	3	54	99		54	54	
11	0	54	99		54	54	
12	6	93	93	Receipt	48	48	Order
13	6	87	87		42	61	
14	8	79	79	Order	34	53	
15	0	79	124		34	72	
16	2	77	122		32	70	
17	4	73	118		28	66	
18	4	69	114		24	62	
19	2	67	112		22	60	
20	4	63	108		18	56	
21	0	63	108		37	56	Receipt
22	9	54	99		28	47	Order
23	3	96	96	Receipt	44	63	Receipt
24	0	96	96		44	63	
25	5	91	91		39	58	
26	6	85	85		33	52	Order
27	8	77	77	Order	25	63	
28	4	73	119		21	59	
29	9	64	110		12	50	Order
30	1	63	119		11	68	
31	6	57	103		24	62	Receipt
32	5	52	98		19	57	
33	6	46	92		13	51	Order
34	6	40	86		7	64	
35	3	37	83		23	61	Receipt
36	5	77	77	Order + Receipt	18	56	
37	7	70	115		11	49	Order

TABLE 7.5 *(Continued)*
Comparison of Two Stockholding Systems

			Present System				Proposed System
Week Number	**Weekly Demand**	**Stock**	**Stock + Orders**	**Orders and Receipts**	**Stock**	**Stock + Orders**	**Orders and Receipts**
38	5	65	110		25	63	Receipt
39	4	61	106		21	59	
40	1	60	105		20	58	
41	6	54	99		14	52	Order
42	1	53	98		32	70	Receipt
43	6	47	92		26	64	
44	7	40	85		19	57	
45	0	85	85	Receipt	19	57	
46	9	76	76	Order	29	48	Order + Receipt
47	5	71	116		24	62	
48	9	62	107		15	53	Order
49	0	62	107		15	72	
50	3	59	104		12	69	
51	4	55	100		8	65	
52	0	55	100		8	65	

- This simulation was only for one store's item. Performing a simulation for every item will be onerous and may yield significantly different results.
- Most important of all, simulation does not optimise. There may be many stockholding policies that are superior to either of those chosen up to now. Hence, simulation, although a very powerful and practical technique, is non-optimising. For optimal solutions, other techniques are applicable.

Class Assignment

Study the data in the preceding example and see if you can estimate a better option than the proposal given. Then using the data given, determine your order and receipt points and thereby prove your option.

Monte Carlo Simulation Using Random Number Generation

If not enough information is available to perform simulation, data can be generated, as in the following example. Consider trucks arriving to deliver goods. These have been timed as shown in Table 7.6.

Random numbers from 0 to 100 are assigned to the probabilities, and then a random number table (Table 7.7) is used to generate a list of random truck arrival times for the following simulation. A random number table can be used to generate a list of random numbers by accessing it in any non-random way. In our example, the number

TABLE 7.6
Timed Truck Arrivals

Truck Inter-Arrival Times	Mean Time (Minutes)	Frequency	Probability = Frequency/50	Cumulative Probability	Random Numbers
0 up to 10 minutes	5	13	0.26	0.26	00–25
10 up to 20 minutes	15	10	0.20	0.46	26–45
20 up to 30 minutes	25	8	0.16	0.62	46–61
30 up to 40 minutes	35	5	0.10	0.72	62–71
40 up to 50 minutes	45	4	0.08	0.80	72–79
50 up to 60 minutes	55	3	0.06	0.86	80–85
60 up to 70 minutes	65	2	0.04	0.90	86–89
70 up to 80 minutes	75	2	0.04	0.94	90–93
80 up to 90 minutes	85	1	0.02	0.96	94–95
90 up to 100 minutes	95	1	0.02	0.98	96–97
100 up to 110 minutes	105	1	0.02	1.00	98–99
Total		50	1.00		

TABLE 7.7
Random Number Table

74	**35**	**59**	**44**	**11**	**26**	**78**	**60**	**91**	**90**	**05**	**43**
22	38	32	08	41	91	14	63	31	15	40	27
84	34	07	60	34	26	80	79	88	64	30	86
21	88	73	30	20	39	04	10	96	49	85	36
04	36	56	62	78	08	98	52	98	09	68	03
12	33	22	47	16	96	20	79	42	48	56	34
35	49	27	91	80	23	58	12	77	71	37	70
57	00	56	82	46	06	88	73	68	13	76	23
54	85	56	76	07	96	83	44	30	90	30	16
48	62	55	32	24	40	58	78	52	35	48	70

41 has been chosen, and all subsequent numbers follow in the column down from there. But we could have used every other number, or we could have chosen numbers by row instead of column, or any other regular pattern we choose.

In practice, computers now generate random numbers for us. Excel, for example, has a random number generating function. For the present, however, using the random number table, we can construct a set of truck arrival times as shown in Table 7.8. This can then be used and extended in our simulation as shown in Table 7.9.

Rules for the simulation are as follows:

1. The working day is from 8:00 a.m. to 5:00 p.m.
2. Trucks arrive only between these times.

TABLE 7.8
Timed Truck Arrivals

Random Number	Truck Inter-Arrival Time	Time from Start of Simulation to Arrival of Truck
41	15	15
34	15	30
20	5	35
78	45	80
16	5	85
80	55	150
46	25	175
07	5	180
24	5	185
26	15	200
91	75	275
26	15	290

3. If any truck is not completely unloaded by 5:00 p.m., work continues until it is unloaded.
4. No trucks await overnight.
5. Lunch is from 12:00 to 1:00 p.m. No work is done during this time.
6. Tea breaks are from 10:00 to 10:15 a.m. and from 3:00 to 3:15 p.m. No work is done during these times.
7. Basic wage is for an eight-hour day. Any work done after 5:00 p.m. is paid overtime.
8. Operating cost per bay = $500 per hour in normal time.
9. Operating cost per bay = $640 per hour in overtime.
10. Cost of truck time = $500 per hour.

Assignment

In Table 7.9, the random numbers have been filled in, to ensure consistency of answers. Complete the table by running the simulation for two days. Then, provide the answers for the following.

- What is the total truck waiting time per day?
- What is the total overtime hours worked?
- What is the cost to the company per day? The company must pay the trucking company for the truck time, plus the wages for its staff.

This will prove to be a time-consuming, somewhat mentally exhausting exercise. But it does demonstrate all the relevant principles of the Monte Carlo simulation, and when completed, it will make the student more aware and more grateful of the power of computers that today do this sort of work for us!

TABLE 7.9
Simulation of Loading Bay Scenario

Random Number	Time between Truck Arrivals (Minutes)	Truck Arrival Time	Bay Number	Unload Start Time	Unload Finish Time	Total Truck Duration (Minutes)
41	15	08:15	1	08:15	09:15	60
34	15	08:30	2	08:30	09:30	60
20	5	08:35	1	09:15	10:30	115
78	45	09:20	2	09:30	10:45	85
16	5	09:25	1			
80	55	10:20	2			
46	25	10:45	1			
07	5	10:50	2			
24						
26						
91						
26						
39						
08						
96						
23						
06						
96						
40						
78						
14						
80						
04						
98						
20						
58						
88						
83						
58						
60						
63						
79						
10						
52						
79						
12						
73						

Simulation techniques have, to some extent, replaced other mathematical, computer-based techniques such as linear programming and dynamic programming. Other simply analytical techniques such as queuing theory are also being replaced by simulation. Nevertheless, these techniques still have their place and are therefore discussed next.

Linear Programming

As powerful as modern simulation is, there are still classes of problems that simulation cannot handle. Take for example the task of optimally fitting 70 applicants into 70 jobs. As we have seen, simulation is non-optimising, and this problem is far too big for the iterations needed to get even an approximately optimal answer. This was in fact one of the problems described by the inventors of the linear programming technique, which can provide the optimum answer for such problems.

History of Linear Programming

Linear programming was developed before World War II by the Russian mathematician A. N. Kolmogorov (perhaps best known for his work in probability theory and the development of the Kolmogorov-Smirnov statistical test). Major progress in the field took place in 1947, when the American George Dantzig developed the simplex algorithm, an efficient way of solving linear programming problems involving limited resources and varying demands.

Two-Dimensional Linear Programming

All linear programming models share the following characteristics:

- One *objective function*, to be maximised or minimised.
- One or more *constraints.*
- *Alternative courses of action* must be possible.
- The objective function and the constraints are linear.

A simple graphical linear programming model will now be discussed. This will demonstrate the principles of the technique as used to solve two-dimensional problems, two variables, one on a vertical axis and one on the horizontal axis. Three-dimensional problems can also be visualised, with a third axis coming out of the paper, but after that, the human mind is at a loss, being unable to visualise problems in four or more dimensions. The mathematics of linear programming, however, know no such limit and can handle multiple variables and multiple constraints.

EXAMPLE 7.1 MANUFACTURING RESOURCES PROBLEM

Consider the following case: A factory consists of a foundry, a machine shop, and a fitting shop. Two products are manufactured, Product A and Product B. Both products require a certain amount of time in each of the three manufacturing areas. The times used in each area per unit of either product are given in Table 7.10. Second, the hours available for manufacture of the two products are given in Table 7.11.

We will assume that there are no market constraints—the company can sell all that it produces. The company wants to know the best amounts to be produced for each product in order to maximise the contribution to overheads (Table 7.12). In this case, it is easy to identify the variables in the problem as

X_A = the number to be produced for Product A
X_B = the number to be produced for Product B

TABLE 7.10
Hours Required per Unit Manufactured

	Product A	Product B
Foundry	1	4
Machine shop	3	4
Fitting shop	4	2

TABLE 7.11
Hours Available for Manufacture

	Hours Available per Week
Foundry	52
Machine shop	60
Fitting shop	60

TABLE 7.12
Contribution to Overheads

	Product A	Product B
Contribution to overheads, Rands per unit produced	480	360

From the preceding tables, we have

$$1 \times X_A + 4 \times X_B \leq 52 \quad 7.1$$

$$3 \times X_A + 4 \times X_B \leq 62 \quad 7.2$$

$$4 \times X_A + 2 \times X_B \leq 60 \quad 7.3$$

The non-negative conditions are:

$$X_A \geq 0$$
$$X_B \geq 0$$

We now have an *objective function* that must be maximised:

$$Z = 480X_A + 360X_B \tag{7.4}$$

As there are only two variables, these relationships can be shown graphically, plotting values for X_A horizontally and values for X_B vertically. Considering the constraint relating to the foundry (equation 7.1), we can draw Figure 7.9.

Similarly, we can draw similar figures for the production regions for the fitting shop and the welding shop. If we combine these figures, we get the feasible region shown in Figure 7.10.

If we study Figure 7.10, we can calculate, using pairs of simultaneous equations, that there are four intersections in the feasible area, which may lead to a maximum:

$$X_A = 0, X_B = 13$$
$$X_A = 4, X_B = 12$$
$$X_A = 12, X_B = 6$$
$$X_A = 15, X_B = 0$$

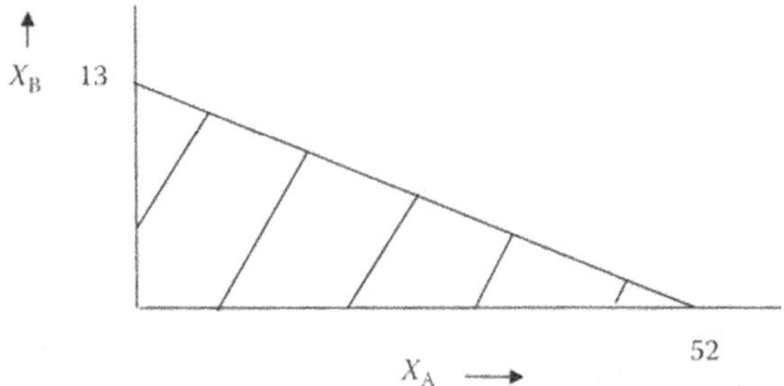

FIGURE 7.9 Feasible production region based on foundry capacity.

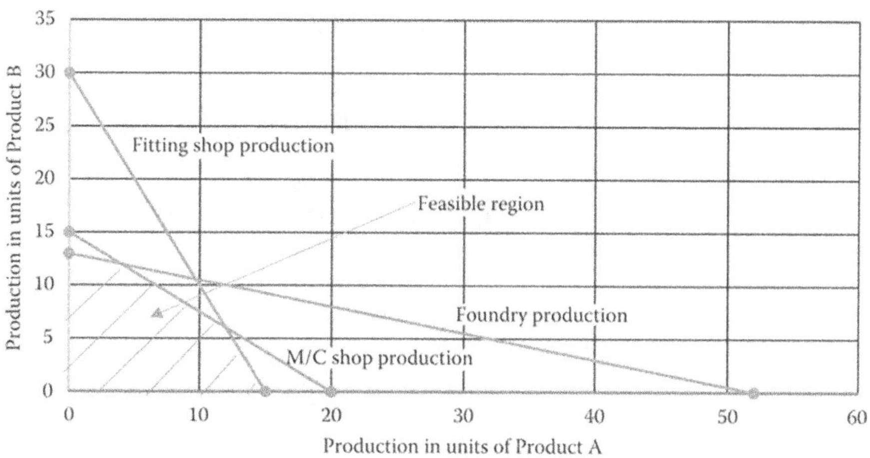

FIGURE 7.10 LP solution showing feasible area of production.

Inserting these values into equation 7.4, we see that of the four possible intersections, the maximum is where X_A is 12 and X_B is 6, giving

$$Z = R\left[(480 \times 12) + (360 \times 6)\right]$$
$$= R7920.$$

Hence, we should produce 12 units of Product A and 6 units of Product B per week.

As was stated previously, this is easy to visualise when there are only two variables in the objective function. Three variables would still be visualisable, with the axis of the third variable coming "out of the paper" as it were, but no more. Linear programming, however, can cater for any number of variables. For problems with more than two variables, an algebraic solution is required. This is usually done using the simplex method or one of its derivatives.

The full description of the simplex method will not be discussed here. It is the algebraic equivalent of the graphical method presented previously and consists of describing the situation in question in terms of a set of equations covering the restraints, the resources, and the feasible outcomes.

The simplex method is iterative. It starts with a trial solution and then tests to see if the solution can be improved. If the solution can be improved, the method obtains the new trial solution. This solution, in turn, is tested to see if it can be improved, and so on. The method is organised in such a way that when no further improvement is available, one can be sure that the solution is then optimal. This process is not carried out in a haphazard fashion. It involves a systematic search for the optimal solution.

It is possible to solve linear programming problems with specialist software, but Excel is also capable of solving large problems. To start with, however, we will solve our previous problem using Excel.

Table 7.13 shows the solution previously solved graphically. The functions in the various columns are as follows:

G7: = C7 * C12 + D7 * D12
G8: = C8 * C12 + D8 * D12
G9: = C9 * C12 + D9 * D12
I12: = C4 * C12 + D4 * D12

The solution can be iterated, but one can also use the Excel Solver.

If we reset the quantities produced, cells C12 and D12, and call up the Solver, we will fill in the four windows on the Solver page as follows:

Set objective: $I$12
By changing variable cells: $C$12:$D$12
Subject to the constraints:
$G$7≤$I$7
$G$8≤$I$8
$G$7≤$I$9
Select a solving method: Simplex LP

TABLE 7.13
Excel Solution for Example 7.1

	B	C	D	E	F	G	H	I
1								
2								
3		Product A	Product B					
4	Contribution	480	360					
5						Resources		Resources
6						Used		Available
7	Foundry	1	4			36	≤	52
8	M/C shop	3	4			60	≤	60
9	Fitting shop	4	2			60	≤	60
10								
11								Total Contribution
12	Solution	12	6					7,920
13								
14								

Also tick the box to make unconstrained variables non-negative.

If we then press the Solve button, we will see that the solution is first shown as 15 of A and 0 of B. We are then asked if we wish to proceed further—the next iteration brings us to 12 of A and 6 of B, and the iteration stops.

ASSIGNMENT

Solve the following problem using linear programming in Excel.

A company produces bicycles, powered bicycles (mopeds), and lawnmowers. The contribution to overheads and profit is as follows:

- Bicycles: $1,000 per unit
- Mopeds: $3,000 per unit
- Lawnmowers: $500 per unit

The use of capital resources is as follows:

- Bicycles: $3,000 per unit
- Mopeds: $12,000 per unit
- Lawnmowers: $1,200 per unit

There are also storage constraints:

- Bicycles: $5 per unit
- Mopeds: $10 per unit
- Lawnmowers: R5 per unit

The resources available are as follows:

- $930,000 for capital
- $1,010 for storage

What combination of bicycles, mopeds, and lawnmowers will maximise the company's contribution to overheads and profit?

EXAMPLE 7.2 OPTIMISATION OF MAINTENANCE IN A LARGE RAILWAY NETWORK

In 2009, the author was part of a team of engineers assessing the operations and maintenance of the Turkish railway network, as shown in Figure 7.11.

Still using Excel, we will now move on to the model developed for the writer by Dr Ian Campbell of the University of the Witwatersrand, Johannesburg, South Africa. This model was to optimise the number and location of diesel repair shops in this rail network. An abbreviated version of the report accompanying this exercise is given here.

The recommendations of the model were startling. Only 3 of 40 repair depots should remain open if costs were to be minimised! An extract of the report is given here.

Linear programming (LP) is used in this project to minimise the cost of operating 40 underutilised diesel locomotive maintenance depots. A second LP model was used to minimise available service time and hence generate an alternative solution. The information given by the Turkish State Railway (TCDD) regarding each depot, its capabilities, number of artisans, costs, and number of operating diesel locomotives allowed for an objective function and constraints to be formulated. Using Excel, optimal solutions were obtained for the two models. The "minimise cost" model yielded a solution that states that only EskiŞehir Depo Müd and Ądana Loko. Bakim Atl. Müd should remain operational at a cost of 1,583,333 TL (Turkish lira), while the other depots should be considered for closure. The "minimise available service time" model showed that Halkali Depo Müd, Samsun Depo Müd, Fevzipasa Depo Şef, Afyon Loko. Bakim Atl. Müd, Tavsanli Depo Müd, and Konya Depo Müd should remain operational, at a service time of 36,603 hours and a cost of 3,250,000 TL. The analysis section of this report discusses the difference between these two solutions, their strengths and weaknesses. From the information used in the Ankara Factory sheet in the Excel worksheet titled "TCDD Optimisation LP," closing the affiliated companies would be acceptable as the Ankara factory would be able to handle the associated increase in its responsibilities.

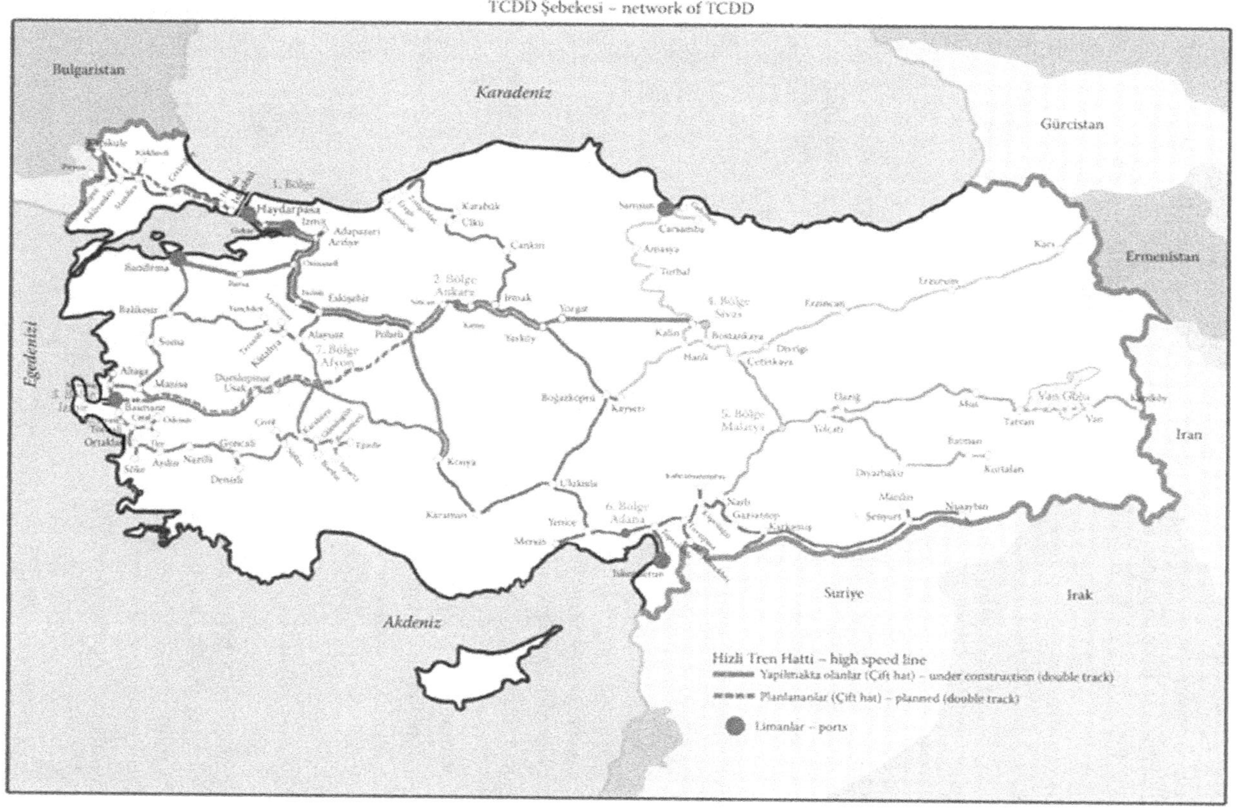

FIGURE 7.11 The Turkish railway network.

MODEL ASSUMPTIONS

- All locomotives have the same type of service requirements. The model can be extended if this is not the case.
- All servicing is done by the same type of personnel. If different personnel (e.g. different skills) are required for different services, the model can be extended to take this into account.
- All locomotives can be serviced at any of the depots. This means that transport costs of moving a locomotive to a service depot are ignored. The locomotives requiring servicing should be fitted into the normal operating schedule such that they can be placed at the appropriate service centre at no cost. If this is not the case owing to operating policies of the railway company, the schedule and operating policies of the railway company would have to be taken into account in the model, which would be an extensive and complex task.
- The model can be solved for any specific region or for the whole country at once. Regional solutions will be required if the regions own their own locomotives and service depots and servicing is not done in other regions.

PARAMETERS REQUIRED

- Locomotives:
  Average mileage per month
  - Average speed
  - Average operating hours/month
  - No. of locomotives in use
- Locomotive services:
  - Technician/mechanic time in hours required for each type of service
- Location/depot/service centre parameters:
  - Fixed costs (salaries, overheads, etc.)
  - Servicing personnel time available for carrying out services
  - Service types that can be done at each location

Some other parameters of interest for the locomotives:

- Average kilometres per month: 6,000
- Average speed: 35 kph
- Average operating time per day: 9.71 hours
- Operating days per month: 30
- Number of locomotives: 592

Manpower:

- Number of artisans: 2,078

Services (for one type of locomotive only—there are five types). These are described in Table 7.14.

TABLE 7.14
Locomotive Service Details

Service Type	Man-Hours	Service Interval (km)
K	2.6	4,250
KB1	31.5	12,740
KB2	63	38,000
KB3	94.5	74,000
GB	157.5	146,000
BGB	315	288,500

ASSIGNMENT

- Comment briefly on any linear programming experience that you have had yourself, if any.
- Also comment on the previous model. What do you think of the recommendation to close 37 of the 40 depots in a cost-minimisation effort?
- The railroad industry in the United States uses the term *feather-bedding.* What might it mean, and does it have any application to this case?
- Do you think Monte Carlo simulation could solve this problem?

Other System Engineering Concerns

Apart from seeing the whole picture, and coordinating it, systems engineers must see that the various activity boxes on a systems engineering organisational chart are in place and that work is proceeding in them. These boxes include:

- Design for reliability/availability
- Design for maintainability
- Design for operability
- Design for supportability

The preceding first two bullet points have been covered in other parts of this text. We will now concern ourselves with the remaining two.

Design for Operability

This part of the process deals with what is known as ergonomics, or on the other side of the Atlantic, human factors engineering. It has to do with designing the system to fit the human being rather than vice versa.

An Example of Poor Operability

As an example of anti-operability, if it may be so called, a Bedford 3-ton Army truck of the 1960s had the gear lever to the left and *behind* the driver. This was a forward-control vehicle, with the engine therefore alongside the driver and the gearbox behind him. The company saved the cost of a linkage from the gearbox to the driver's side. This deficiency could have been eliminated if there had been a human factors specification issued with the tender for the vehicles.

This would not have been likely, however, because in the Army concerned, the driver's manual had not been updated from the previous standard truck that had a "crash" gearbox that required double-declutching when changing gears. The new-for-the-time Bedfords, however, had synchromesh transmissions. But the Army manual said to double-declutch, so all tyro drivers were compelled to do so to get their licences!

Operability Concerns: Human Physical Dimensions

A plant cannot always be designed for the entire population. Submarine and tank crews have in the past been selected by height, with a maximum limit. Miners, too, have been persons of short stature in the past.

People are growing bigger with each generation, and this must be allowed for in designs. Present aircraft seating is a case in point. In the interests of profitability, any person 6 ft (183 cm) in height is forced into discomfort for many hours. Some very tall persons have to travel business class as they simply cannot fit in economy class.

As long ago as 1980, the height of males in the United States at the 95th percentile was 185.6 cm, with a weight of 91.6 kg. Most ergonomists design for the 95th percentile.

Strength Requirements

Strength requirements should also be defined. A maker of hydraulic-powered rock drills had a problem several years ago with the drills being discarded underground because miners could not always open the operating valve. The simple test that the manufacturer introduced to solve the problem was to use one of the office girls to open and close each valve before dispatch. If she could open the valve, burly underground miners could as well.

Job Design

Job design comes out of the work proposed by Hackman and Oldham (1980), stating that work should be designed to have five core job characteristics. These core job dimensions should be the following:

1. *Skill variety.* This refers to the range of skills and activities necessary to complete the job. The more a person is required to use a wide variety of skills, the more satisfying the job is likely to be.
2. *Task identity.* This dimension measures the degree to which the job requires completion of a whole and identifiable piece of work. Employees who are involved in an activity from start to finish are usually more satisfied.
3. *Task significance.* This looks at the impact and influence of a job. Jobs are more satisfying if people believe that they make a difference and are adding real value to colleagues, the organisation, or the larger community.
4. *Autonomy.* This describes the amount of individual choice and discretion involved in a job. More autonomy leads to more satisfaction. For instance, a job is likely to be more satisfying if people are involved in making decisions instead of simply being told what to do.

5. *Feedback*. This dimension measures the amount of information an employee receives about his or her performance and the extent to which he or she can see the impact of the work. The more people are told about their performance, the more interested they will be in doing a good job. So sharing production figures, customer satisfaction scores, and so on can increase the feedback levels.

The five core job dimensions listed prior result in three different psychological states.

1. Experienced meaningfulness of the work
2. Experienced responsibility for the outcomes of work
3. Knowledge of the actual results of the work activity

This concept of job design follows on from the work of Herzberg, discussed in Chapter 8.

Apart from the preceding information, the other aspect of job design is the level of skill, experience, and intelligence required for the job. These three attributes are non-negotiable if plant safety and reliability are to be maintained.

Human Sensory Factors: Vision

Arcs of vision. The operator, without head rotation, should be able to see an unimpeded 70° horizontal arc and a 60° vertical arc.

- *Colour.* Colour is also a factor. Colour perception decreases at the limits of these arcs. Green and red, for example, are only seen in a 60° horizontal arc by most people. Furthermore, the recent experience of a South Africa vehicle manufacturer is enlightening. Cars were coming out of the spray booth in the wrong colours. Management at first thought that this was a deliberate trade union provocation, until they tested the spray booth staff for colour blindness! Some authorities say that 30% of South Africa's African population have some degree of colour blindness. The other area where it is even more important to test for colour blindness is with electricians, where high-voltage cable phases are colour-coded.
- *Lighting level.* Illumination levels vary with the task in hand. Standards are available to determine optimum lighting levels.
- *Hearing.* It can be said that many working environments are too loud. Background noise can interfere with human speech and can therefore be a danger if the environment is a workshop or similar situation. As noise levels increase, persons begin to experience discomfort. When the noise level approaches 120 dB, people begin to "feel" the noise, and at 130 dB, pain begins. Present environmental standards limit the allowable noise levels. Companies take the easy way out by insisting employees wear ear protection. This is easier than quietening the plant. Old-timers who worked in metal pressing, refineries, and power stations normally exhibit deafness, as do old motorcycle racers and heavy metal music aficionados!

At the other extreme, open-plan offices often have deliberately introduced "white noise" to preserve the privacy of conversations. Many persons cannot adapt to it, however, as anyone who has worked in such an environment knows. When the white noise trips out, one hears a collective sigh of relief on the floor.

Noise, therefore, is generally bad and must be eliminated as much as possible even to a further extent than environmental standards dictate.

Other Factors to Be Considered

In workplace design, apart from vision and hearing, as previously mentioned, temperature, vibration levels, and humidity must also be considered.

Design for Supportability: The Concept of Integrated Logistic Support

Integrated logistic support considers all those aspects of the system other than the hardware itself. Items for consideration include the following:

- Maintenance planning.
- Supply support: Spares, change-out modules, consumables, and so on.
- Test and support equipment: This includes condition monitoring equipment, calibration equipment, special tools, lifting equipment, and the like.
- Transportation and handling: This includes transportation of equipment during assembly and erection, as well as the transportation of personnel for erection and for maintenance.
- Personnel and training: This includes the numbers of persons needed for the erection, operation, and maintenance of the facility, plus all their training requirements, including simulators, classroom facilities, computerised training, and so on.
- Facilities: This includes all buildings, air conditioning, heating, and lighting supplies, as well as infrastructure, such as roads, mess facilities, and offices.
- Software and data: This covers operating and maintenance instructions, training manuals, drawings, and specifications. Also included are specific software programs.

All these must be supplied, maintained, and managed over the entire life cycle of the system. This can be achieved by the development of *plans* that are updated throughout the life cycle, such as the *maintenance plan*, the *supply support plan*, and so on.

Some of the items of logistic support have been covered elsewhere in this text, for example, maintenance spares planning and policy.

Queuing Theory

Queuing theory is the mathematical study of waiting lines, or queues. In queuing theory, a model is constructed so that queue lengths and waiting time can be predicted. Queuing theory has its origins in research by A. K. Erlang when he created models to describe the Copenhagen, Denmark, telephone exchange. The ideas have since seen applications including telecommunications, traffic engineering, computing and the design of factories, shops, offices, and hospitals.

Queuing problems are easily solved using simulation, but Erlang's deterministic methods are still useful to anyone with a pocket calculator or an Excel spreadsheet. We will discuss single-line queuing problems with single- and multiple-service bays.

Single-Line Queuing with Single-Service Channel

The following assumptions are made:

- Arrivals are served on a "first in, first out" basis.
- Every arrival eventually gets served—no entity leaves the queue.
- Arrivals are independent of previous arrivals, but the arrival rate is constant.
- Arrivals are described by a Poisson distribution, and from a large population.
- Service times vary, but there is an average service rate.
- Service times are according to the negative exponential distribution.
- The average service rate is greater than the average arrival rate.

The queuing equations are then as follows:

Let λ = mean number of arrivals per hour and μ = mean number of units served per hour.

Then L = average number of units in the system, that is, number in the queue + number being served.

$$L = \lambda / (\mu - \lambda) \qquad 7.5$$

W = the average time that a unit spends in the system.

$$W = 1/(\mu - \lambda) \qquad 7.6$$

L_q = average number of customers in the queue.

$$L_q = \lambda^2 / \mu * (\mu - \lambda) \qquad 7.7$$

W_q = average waiting time in the queue.

$$W_q = \lambda / \mu * (\mu - \lambda) \qquad 7.8$$

The probability that the service is being utilised is ρ, also known as the *utilisation factor.*

$$\text{Hence, } \rho = \lambda / \mu \qquad 7.9$$

P_0 = percent idle time, that is, the probability of no units in the queue.

$$P_0 = 1 - \lambda / \mu \qquad 7.10$$

$P_{n>k}$ = the probability that the number of units in the system is greater than k.

$$P_{n>k} = (\lambda / \mu)^{k+1} \tag{7.11}$$

As an example, let us presume that the repair crew of a factory has to deal with an average of two calls per hour for machine service or repair, and they actually have the capacity to service or repair three machines per hour. Then, we get the following:

$L = \lambda/(\mu - \lambda) = 2/(3 - 2) = 2$ machines in the queue or being served.
$W = 1/(\mu - \lambda) = 1/(3 - 2) = 1$ hour that each machine is in the system.
$L_q = 1.33$ machines waiting in line.
$W_q = 2/3$ hours, that is, 40 minutes wait per machine.
ρ = % of time the shop is busy = 67%.

Therefore, % of time the shop is idle is 33%.

What is the probability that there are more than four machines in system ($P_{n>k}$)? The answer is 13.2%.

Depending on the cost structure of the business, these figures might be acceptable, or they might not be.

Single-Line Queuing with Two Service Bays

The probability that there are zero units in the system is:

$$P_0 = \frac{1}{\left[\sum_{n=0}^{n=m-1} \frac{1}{n!}\left(\frac{\lambda}{\mu}\right)^n\right]} + \frac{1}{m!}\left(\frac{\lambda}{\mu}\right)^m \frac{m\mu}{m\mu - \lambda} \text{ for } m\mu > \lambda \tag{7.12}$$

The average number of units in the system is:

$$L = \frac{\lambda\mu(\lambda/\mu)^m}{(m-1)!(m\mu - \lambda)^2} P_0 + \frac{\lambda}{\mu} \tag{7.13}$$

The average time a unit spends waiting in the system is:

$$W = \frac{\mu(\lambda/\mu)^m}{(m-1)!(m\mu - \lambda)^2} P_0 + \frac{1}{\mu} = L/\lambda \tag{7.14}$$

The average number of units waiting for service is:

$$L_q = L - (\lambda/\mu). \tag{7.15}$$

The average time a unit spends in the queue is:

$$W_q = W - (1/\mu). \tag{7.16}$$

TABLE 7.15
Variation in Service-Level Parameters with One and Two Bays

Symbol	System Parameter	One-Service Channel	Two-Service Channels
P_0	Probability of no units in the system	33%	50%
L	Average number of units in the system	2	0.75
W	Average time a unit spends in the system	60 minutes	22.5 minutes
L_q	Average number of units in the queue	1.33	0.083
W_q	Average time a unit spends in the queue	40 minutes	2.5 minutes

The utilisation rate is:

$$\rho = \lambda / (m\mu). \quad 7.17$$

When the previously mentioned equations are solved for the same parameters as the single-channel example, the differences between one service channel and two are shown in Table 7.15.

It will be seen from Table 7.15 that the most dramatic change in the parameters is the reduction in waiting time W_q from 40 minutes to 2.5 minutes. The decision to go from one to two bays would depend on the costs of the second bay and the cost of waiting time. Queue waiting time is of the utmost importance in service industries, of course, where standing in line is one activity that most human beings tend to hate!

Project Management

Project management can be considered a subset of systems engineering, but it also stands alone.

A *project* is simply defined as a task with a beginning and an end. In its simplest terms, the purpose of project management is to get the project completed on time, within spec and within budget, to the satisfaction of all stakeholders. This is also discussed in Chapter 8 (see the management triangle). The larger the project, the more pressing is the need for good project management.

Almost anything can be defined as a project. A reliability upgrade programme, for example, should make use of project management principles, and hence the inclusion of project management here.

The project management methodology is fairly simple and does not have many special "tools of the trade" to rely on. The main aspects of project management are as follows:

- The work breakdown structure (WBS)
- The critical path network
- Resource allocation
- The spend rate
- Risk analysis

- The matrix organisation

The Work Breakdown Structure

For purposes of illustration, a grossly simplified WBS is shown in Figure 7.12. A real WBS would have possibly hundreds of blocks. The determination of the number of levels in the structure and the number of blocks is determined by the following:

- 100% of the work to complete the project must be covered, no more and no less, with no duplication in the blocks.
- There should be an hour limit placed per block—this really depends on the technology and the size of the project. However, for example, it might be decided that no block should include more than 100 man-hours of work.

The Critical Path Network

The critical path network is derived from the WBS, or for simpler projects, it may be drawn up without a WBS. The network describes all the *activities* necessary to complete the project, with their sequential interrelationships. An example, once again, simplified, is given in Figure 7.12. The network can be drawn with activities shown on the arrows or activities shown on the nodes. It is a matter of personal choice or of the software one is using. It is advantageous to be able to draw the network to scale, and then the critical path and other aspects are clearly shown. However, in large networks, this is not usually possible, as the network's timeline extends off the page or off the computer screen.

Figure 7.13 is a notional critical path network that might have been derived from the WBS in Figure 7.12. It is drawn as an "activity on node" diagram. (The alternative is an "activity on arrow" diagram.) As the key in the figure shows, durations and early and late starts and finishes are shown. By inspection, it will be seen that the critical path is Start–A–G–Finish. There is no float on the A and G blocks where a float is given as Late Start–Early Start or Late Finish–Early Finish. On the other blocks, namely, B, C, D, E and F, there is a float, meaning, that the start dates of these jobs can be delayed or the total time to complete them can be extended.

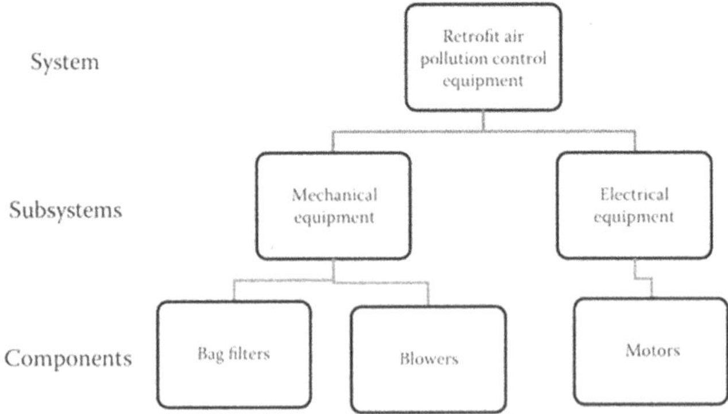

FIGURE 7.12 A simplified work breakdown structure.

In managing a project, attention must be paid to items on the critical path—if their durations are extended for any reason, the end date for the project will be extended. Furthermore, the critical path may change during the progression of the project if one of the previously non-critical activities is delayed beyond its original late finish date.

Activity on Arrow

A critical path network is normally drawn with the activities on the nodes, as in Figure 7.13.

Figure 7.14 is the alternative way of constructing the critical path network, namely, the activity-on-arrow diagram. It is easier to see the float in this diagram.

Project Budgeting and the S Curve

The critical path network can also be of assistance in *project budgeting*, as demonstrated in Figure 7.15. Notice how, even in this simplified example, the budget costs represent an s-shaped curve. This is typical in project work.

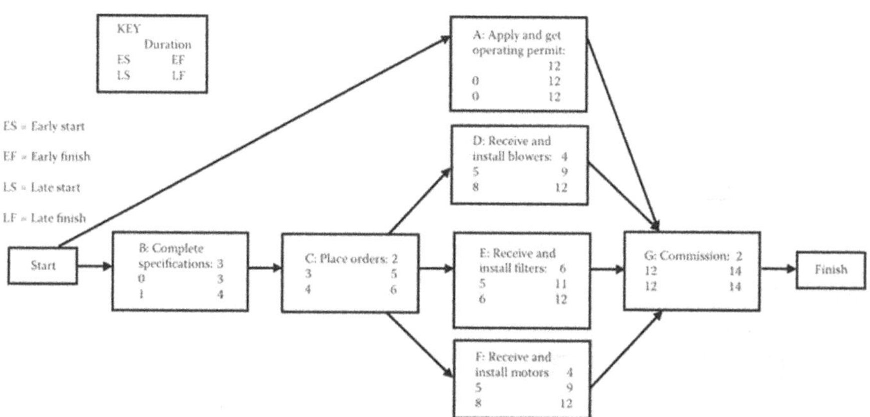

FIGURE 7.13 Critical path network.

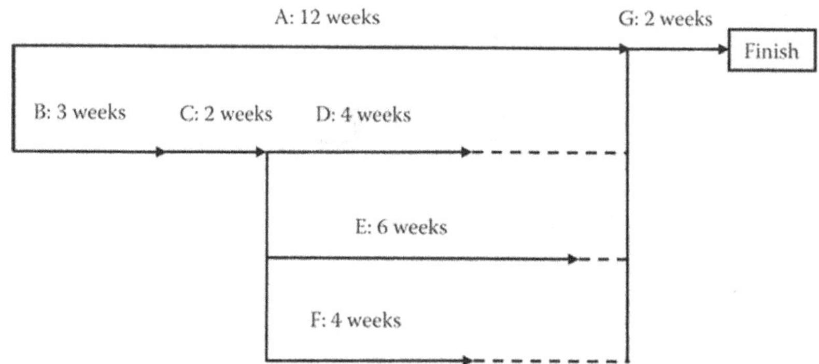

FIGURE 7.14 Activity-on-arrow diagram.

Resource Allocation

The other use for a float is in *resource allocation*. This is best demonstrated using an activity-on-arrow diagram, as shown in Figure 7.16. Considering only activities D, E, and F and assuming that they all require the same resource, represented by the blocks, it is seen if the blocks are moved into the float period in activity D that the resource requirement drops from 3 to 2, except for one week. The float, of course, is now lost, and therefore, D now represents another critical activity.

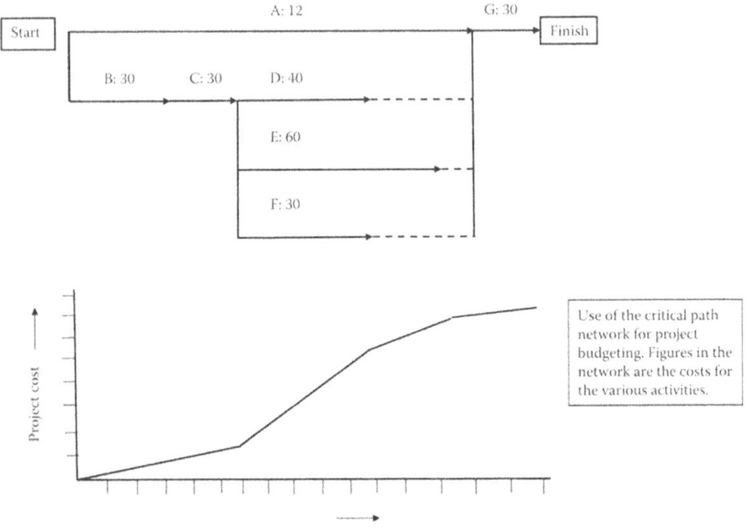

FIGURE 7.15 Project budgeting.

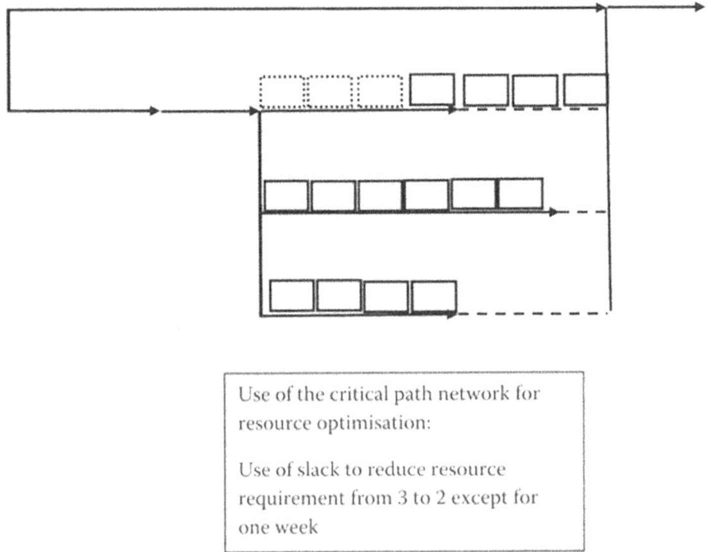

FIGURE 7.16 Resource allocation.

8 Reliability Management

When I look back on my final year engineering class I try, decades later, to think what was the distinguishing feature of the most successful members of that class—those who became CEOs of large companies, or were entrepreneurs who started their own businesses. The answer I come up with is that apart from their technical competence, their greatest attribute was *big heartedness*. This is the ability to not give offence or to take offence—to see through the present human cussedness to the goal that must be achieved. Apart from a love of engineering, they also have a love of people—a combination that is hard to beat.

E. A. Bradley

INTRODUCTION

Reliability engineering has developed in the twenty-first century from being an almost exclusively design function to one encompassing all phases of the system life cycle. For example, many companies now have reliability engineering departments that are responsible for the maintenance budget and the maintenance plan, as well as incident investigations and equipment improvement programmes. In light of such developments, the management abilities of the reliability manager become more and more important. Although management is a human process, by and large, and some persons are inherently better at it than others, it is possible for everyone to learn and practise certain management principles. Some of these are described here.

THE MANAGEMENT PROCESS

The author's view is that management activity can be described in terms of the following simple five-step model:

- Assess the situation.
- Decide on a course of action, leading to a solution.
- Implement the necessary actions.
- Monitor the effects.
- Correct if necessary.

This cycle can be repeated for any management action, whether continuous or unique. This view of management is in line with the teaching of one of the first management theorists, Henri Fayol, who will be referred to again in the following passages.

DOI: 10.1201/9781003326489-8

WHAT IT TAKES TO BE A MANAGER

Many young engineers find the move from doing purely technical work into management somewhat daunting. One should not feel so, however. Much of management is simply common sense. One colleague of the author, with a doctorate in chemical engineering, rose to be the engineering director of a large chemical company. His comment was, "It was my chemical engineering that got me to the top. I learned what I needed to know about finance, marketing, and people management on the way up. It wasn't difficult."

Another colleague, a mechanical engineer, put it differently: "I never considered myself a top flight engineer. But I was always good with people." He was CEO of a large engineering firm before he was 40. The engineering degree gave him entry into the corporate world. His people skills took him to the top. He would have made a success of his life if he had studied law or finance or engineering. A natural people person. That certainly helps.

A HISTORY OF MANAGEMENT THOUGHT

Wherever and whenever there is more than one person involved in working towards a joint goal, management is probably involved. Hence, the building of the pyramids required management, and the organisation of armies has also required management since the earliest times.

WRITERS ON MANAGEMENT

Mary Parker Follett

Mary Parker Follett (1868–1933) was an American management consultant who pioneered the fields of organisational theory and organisational behaviour. In her capacity as a management theorist, Mary Parker Follett pioneered the understanding of lateral processes within hierarchical organisations (which led directly to the formation of matrix-style organisations, the first of which was DuPont, in the 1920s), the importance of informal processes within organisations, and the idea of the "authority of expertise," a factor as real as the formal authority of the organisation chart. She also criticised the overmanaging of employees, a fault today called micromanagement.

In the view of Mary Parker Follett, management is "the art of getting things done through people." Notice that she claimed it as an art, not a science. In fact, modern management is a mixture of both art and science.

Henri Fayol

One of the first management theorists was Henri Fayol (1841–1925), a French mining engineer who was also a successful entrepreneur, establishing a large component of the French steel industry. In 1916, he published his book *Administration Industrielle et Générale*, which was only translated into English in 1949, as *General and Industrial Administration* (Fayol, 1916). Fayol's work was one of the first comprehensive statements of a general theory of management. He proposed that there were five primary functions of management and 14 principles of management.

The functions of management are as follows:

1. To forecast and plan
2. To organise
3. To command or direct
4. To coordinate
5. To control

According to Fayol, the principles of management are as follows:

1. Division of labour. Fayol presented work specialisation as the best way to use the human resources of the organisation.
2. Authority. Managers must be able to give orders. Authority gives them this right. Note that responsibility arises wherever authority is exercised.
3. Discipline. Employees must obey and respect the rules that govern the organisation. Good discipline is the result of effective leadership.
4. Unity of command. Every employee should receive orders from only one superior.
5. Unity of direction. Each group of organisational activities that have the same objective should be directed by one manager using one plan for achievement of one common goal.
6. Subordination. The interests of any one employee or group of employees should not take precedence over the interests of the organisation as a whole.
7. Remuneration. Workers must be paid a fair wage for their services.
8. Centralisation. Centralisation refers to the degree to which subordinates are involved in decision-making.
9. Scalar chain. The line of authority from top management to the lowest ranks represents the scalar chain. Communications should follow this chain.
10. Order. This principle is concerned with systematic arrangement of men, machine, material, and so on; there should be a specific place for every employee in an organisation.
11. Equity. Managers should be kind and fair to their subordinates.
12. Stability of tenure of personnel. High employee turnover is inefficient. Management should provide orderly personnel planning and ensure that replacements are available to fill vacancies.
13. Initiative. Employees who are allowed to originate and carry out plans will exert high levels of effort.
14. Esprit de corps. Promoting team spirit will build harmony and unity within the organisation.

Fayol's work has stood the test of time and has been shown to be relevant and appropriate to contemporary management. Modern management texts just rehash Fayol, at least to some extent.

However, there have been some changes. Consider point 4, that every employee should receive orders from only one superior. In matrix organisations, as used in project management, the employees within the matrix have two persons to report to—their discipline head, for the specification, and their project manager head, for schedule and budget.

Frederick Winslow Taylor

Frederick Taylor (1856–1915) was a mechanical engineer and an American contemporary of Fayol's. He was also president of the American Society of Mechanical Engineers for a time. He is famous (trade unionists might say infamous) for his book *Principles of Scientific Management* (Taylor, 1911). His pioneering work in applying engineering principles to the work done on the factory floor was instrumental in the creation and development of industrial engineering.

After leaving school, Taylor wrote the Harvard entrance exam and passed it. However, instead of attending Harvard, Taylor became an apprentice patternmaker and machinist, gaining shop floor experience. Taylor finished his four-year apprenticeship and, in 1878, became a machine shop labourer at Midvale Steel Works. At Midvale, he was quickly promoted to time clerk, journeyman machinist, gang boss over the lathe hands, machine shop foreman, research director, and finally, chief engineer of the works.

Early on at Midvale, working as a labourer and machinist, Taylor recognised that workmen were not working their machines, or themselves, nearly as hard as they could and that this resulted in high labour costs for the company. When he became a foreman, he expected more output from the workmen. In order to determine how much work should properly be expected, he began to study and analyse the productivity of both the men and the machines (although the word *productivity* was not used at that time and the applied science of productivity had not yet been developed). His focus was on the human component of production Taylor labelled *scientific management*.

Taylor was, in effect, one of the first management consultants. Another famous management consultant, Peter Drucker, writing in 1974, describes Taylor as follows:

> Frederick W. Taylor was the first man in recorded history who deemed work deserving of systematic observation and study. On Taylor's "scientific management" rests, above all, the tremendous surge of affluence in the last seventy-five years which has lifted the working masses in the developed countries well above any level recorded before, even for the well-to-do.

Taylor's scientific management consisted of four principles:

1. Replace rule-of-thumb work methods with methods based on a scientific study of the tasks.
2. Scientifically select, train, and develop each employee rather than passively leaving them to train themselves.
3. Provide "[d]etailed instruction and supervision of each worker in the performance of that worker's discrete task."
4. Divide work nearly equally between managers and workers so that the managers apply scientific management principles to planning the work and the workers actually perform the task.

Taylor thought that by analysing work, the "one best way" to do it would be found. He is most remembered for developing the stopwatch time study, which combined with Frank Gilbreth's motion study methods, later becomes the field of *time and motion study*. He would break a job into its component parts and measure each to the hundredth of a minute. One of his most famous studies involved shovels. He noticed that workers used the same shovel for all materials. He determined that the most effective load was 21½ lb and found or designed shovels that for each material would scoop up that amount. It was largely through the efforts of his disciples (most notably H. L. Gantt) that industry came to implement his ideas.

Taylorism, as it came to be known, was considered by the Left as demeaning in its analysis of workers as machines. Charlie Chaplin, in his film *Modern Times*, mocked Taylorism, both hilariously and effectively, as workplace slavery. This even though Taylor was often able to convince workers that his methods would increase their productivity and, therefore, their pay.

Elton Mayo

Elton Mayo (1880–1949) was a self-taught psychologist who, although Australian-born, rose to be professor of industrial research at Harvard Business School. He is famous for his work at Hawthorn Electric, where he investigated the effect of lighting levels on productivity, finding, to his surprise, that lighting level was irrelevant—that productivity was related to the amount of attention paid to the group of assemblers involved. This led to a more human relations approach to management, as opposed to the earlier works of Taylor. Since then, work has been extended by authorities such as Douglas MacGregor and Frederick Herzberg.

Douglas MacGregor

Douglas McGregor (1906–1964) was a management professor at the Sloan School of Management at the Massachusetts Institute of Technology (MIT). In his book *The Human Side of Enterprise* (McGregor, 1960), he proposed two theories of management, theory X and theory Y. Theory X was that workers were lazy, dishonest, and so on. Theory Y was that workers were self-motivated, careful, quality-conscious, and so on. Whatever the supervisor believed about his workers, that is the way he would manage them. Hence, theory X and theory Y applied to the manager, not the workforce. This aspect of the book has been misunderstood by many.

Frederick Herzberg

Frederick Irving Herzberg (1923–2000) was an American psychologist who became one of the most influential names in business management. He is most famous for introducing job enrichment and the motivator-hygiene theory. His 1968 publication *One More Time, How Do You Motivate Employees?* is the most requested article ever from the *Harvard Business Review*. Unlike some of his peers in business management consulting, his conclusions are backed up by considerable empirical research.

His theory states that there are two scales that apply to every employee:

- A hygiene scale, which can never be at 100%, but the company must score high enough to keep employees happy. Even if this is achieved, however, the employees will still not be motivated to work well. Things on this scale include working conditions, medical aid, and even pay. Money, according to Herzberg, does not motivate.

- A motivation scale, with the things that motivate employees, such as recognition, challenging work, and the possibility of advancement.

Items that appear on the two scales are shown in Table 8.1.

Even though Herzberg's teachings are from the 1960s, the author recognises them as relevant for today, particularly the fact that money is not a motivator. In this respect, the author of this book is definitely "Herzbergian." He has noticed in general that it is not your good workers that complain about their pay, but the poor workers.

Others say that money is a motivator. That can be true, but in the short-term only. An increase in pay, for example, is a proxy for recognition—the recipient works harder for a few weeks or months and then relapses into his or her former state, having become used to the extra money and accepting it as a right. Herzberg himself made this point.

Another point to be made about money is that most of us will never have enough. As our salaries increase, so do our expectations. A yearly holiday at a seaside resort in one's own country is no longer good enough—now an overseas holiday becomes the norm. As our wealth increases, so do our expectations. What, then, are we to do? Wise persons manage within their income, whatever it is.

Where some say the theory that money does not motivate might break down is at the level of chief executive of large corporations. Such people, all around the world, earn obscenely excessive salaries and bonuses. This is a twenty-first-century phenomenon, by and large, and this even applies in government-owned enterprises with protected markets and little stress for the chief executive. The multimillion salaries accruing to these individuals bear no relation to the company performance and are often given even when the company is performing very badly. Has Herzberg's rules broken down in these cases? Most certainly not—it shows that, even when the remuneration is excessive, the chief executive does not necessarily perform. Increasing his out-take from, say, 35 million to 70 million makes no difference. The company still performs badly on his or her watch.

TABLE 8.1
Herzberg's Two Scales

Hygiene Scale	Motivational Scale
Money	Meaningful work
Status	Challenging work
Security	Recognition
Working conditions	Achievement
Fringe benefits	Responsibility
Company policies	Opportunities for growth
Personality issues	The job itself

Another possible exception to Herzberg's work is the example of Henry Ford and the five-dollar-day, where Henry, early in the twentieth century, started paying his workers five dollars a day when other companies were paying much less. Before this, Ford had experienced very high labour turnover in his plant, owing to the boredom and monotony of assembly line work. By changing to a system of high pay, which his employees could not get at any other factory, he effectively created a captive labour market. His workers now put up with the poor conditions because at least one of those poor conditions on their hygiene scale had been increased—they had a lot more money. This did not motivate them to perform better on their job, but it did make them decide to stay.

Yet a further question arises from Herzberg's work: Are some persons simply "unmotivatable"? The author thinks the answer to this is yes—they are those same substandard workers who are always complaining about their remuneration or who job-hop continuously for more money.

Peter Drucker

Peter Drucker was an Austrian-born writer on management who immigrated first to England and then to the United States. He was a leader in the development of management education and invented the concept known as *management by objectives.* He has been described as "the founder of modern management."

His assertions on how to manage are not without controversy. His initial work at General Motors, which led to his book *The Concept of the Corporation* (1993), was not at all well received inside the sponsor company but became a classic outside of it. As far as the author of this book is concerned, management by objectives and its modern equivalents, such as the annual employee performance review, are more trouble than they are worth. On the other hand, another reliability authority, O'Connor (2012), has much admiration for Drucker: "Peter Drucker buried scientific management with the publication of his book *The Practice of Management*, published in 1955." He credits Drucker (1955) with the inspiration for the quality circles concept, where workers are encouraged to solve their own problems rather than the scientific management approach, where workers are encouraged not to think.

J. Edwards Deming

A contemporary of Drucker's and one who shared with him the reconstruction of Japanese industry after the Second World War, Deming is in fact more honoured in Japan than in his homeland, the United States. He is in fact credited by the Japanese themselves for the start of the Japanese Quality Revolution and is honoured in Japan by the naming of the premier Japanese award for quality as the Deming Award. His main contribution is probably the concept of ensuring the quality of production processes rather than inspecting finished products and recycling the out-of-spec items back into production for rework. "Why produce out-of-spec items in the first place? Rather, set up your production so that only in-spec items are produced" has become known as the Japanese Way.

For a reliability engineer, Deming is one of the most important management thinkers to follow because of the close relationship between quality (as built) and reliability (as operated and maintained).

Deming is famous for his essential 14 points on how to run a business. Deming offered these 14 key principles to managers for transforming business effectiveness. The points were first presented in his book *Out of the Crisis* (1986), and they are worth repeating here. Although Deming does not use the term in his book, it is credited with launching the total quality management movement.

The Deming way, if we may call it such, has been around for some time now, and many of his concepts have been adopted in organisations around the world. Nevertheless, it is worthwhile enumerating all his points:

1. Create constancy of purpose towards improvement of product and service, with the aim to become competitive, to stay in business, and to provide jobs.
2. Adopt the new philosophy. We are in a new economic age. (This was written post–World War II.) Western management must awaken to the challenge, learn their responsibilities, and take on leadership for change.
3. Cease dependence on inspection to achieve quality. Eliminate the need for massive inspection by building quality into the product in the first place.
4. End the practice of awarding business on the basis of a price tag. Instead, minimise total cost. Move towards a single supplier for any one item, on a long-term relationship of loyalty and trust.
5. Improve constantly and forever the system of production and service, to improve quality and productivity and thus constantly decrease costs.
6. Institute training on the job.
7. Institute leadership. The aim of supervision should be to help people and machines do a better job. Supervision of management is in need of overhaul, as well as supervision of production workers.
8. Drive out fear so that everyone may work effectively for the company.
9. Break down barriers between departments. People in research, design, sales, and production must work as a team in order to foresee problems of production and usage that may be encountered with the product or service.
10. Eliminate slogans, exhortations, and targets for the workforce asking for zero defects and new levels of productivity. Such exhortations only create adversarial relationships, as the bulk of the causes of low quality and low productivity belong to the system and thus lie beyond the power of the workforce.
11. Eliminate work standards (quotas) on the factory floor. Substitute with leadership.
12. Eliminate management by objective. (In direct contradiction to Drucker.) Eliminate management by numbers and numerical goals. Instead, substitute with leadership.
13. Remove barriers that rob the hourly worker of his right to pride of workmanship. The responsibility of supervisors must be changed from sheer numbers to quality.
14. Remove barriers that rob people in management and in engineering of their right to pride of workmanship. This means, inter alia, abolishment of the annual merit rating.

15. Educate the workforce and allow for self-improvement.
16. The transformation of the company is everybody's job.

We can see from the previous list that Deming and Drucker do not see eye to eye on various points, especially point 12. In this respect, the author is a Deming man, not a Drucker man. It is significant, however, that most companies that have embraced the Deming way use 13 of his 14 points—point 12 is omitted. Here, they stick to Drucker, saying, "What else can we do to measure and assist worker productivity?"

THE CHARACTERISTICS OF A SUCCESSFUL MANAGER

The Management Triangle

We have seen in the preceding passages that the facts about management have been laid down for a long time, since the days of Fayol. The ability to manage, however, is rather like other human attributes—partly inherent, partly trainable. Some are inherently better at it than others. Nevertheless, the ability can be increased with training and experience. What is most fundamental is that all management—in fact, all tasks—have four components, a quality/specification component, a time component, a money component, and a human component. Project managers know this very well: get the project done within specification, on time, and within budget, to the satisfaction of all stakeholders. *Stakeholders* are all the human parties involved—not just the employees, but also the customers, the management, the shareholders in the company, and others. This can be represented by a triangle, as shown in Figure 8.1.

Any job can be regarded as a project. The job may be cyclical, with someone doing the same thing every day, but each day, then, is a project in itself. Hence, a good manager gets his work done, often by others, within specification, time, and budget, to the satisfaction of all.

There is a school of management that says a manager need only concern himself with three of the four factors: budget, schedule, and stakeholders. However, in the case of a technical system, the aphorism given earlier in this text applies: "*Know*

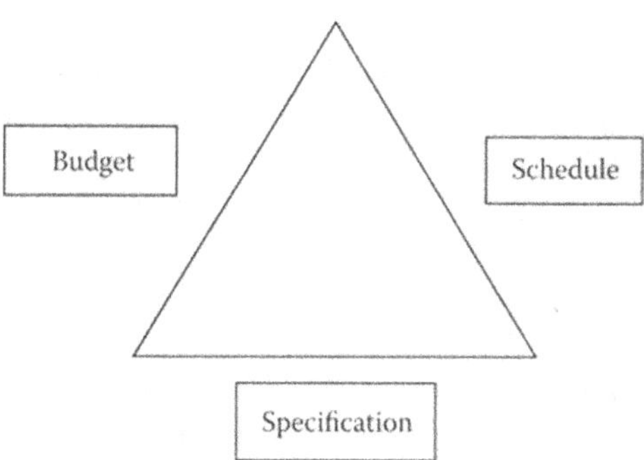

FIGURE 8.1 The management triangle.

your plant." It is impractical and sometimes downright dangerous to allow a complex technical system to be managed by a "one-size-fits-all" manager.

The Management Process

Henri Fayol's five-step management process, as described previously, is correct but has been modified by other authors to include other steps. The author's preferred model is as follows:

1. Investigate what has to be done.
2. Formulate a plan of action.
3. Delegate what needs to be done.
4. Communicate what needs to be done in the form of:

   - What needs to be done by each delegate
   - By when it needs to be done
   - Within what cost constraints

   (It will be seen that this step is the management triangle, as previously described.)
5. Investigate what has been done.
6. Correct if necessary and repeat steps 5 and 6 until completion of the job.

THE CHARACTERISTICS OF A SUCCESSFUL RELIABILITY ENGINEER

Having briefly reviewed the characteristics of a manager, we now turn ourselves to the characteristics of a reliability engineer. They are all important and all necessary:

- Plant knowledge.
- Plant experience.
- A degree in the technological field of the company (e.g. mechanical engineering, electrical engineering, etc.).
- Knowledge of statistics.
- Knowledge of finance subjects, such as discounted cash flow and capital budgeting.
- Knowledge of reliability engineering, preferably with a qualification in such.
- Intelligence. This does not "go without saying" these days. An engineering degree is used to serve as a proxy for this, but with declining university standards, at least in some countries, it is a good idea to have the applicant for a reliability engineer job sit for an IQ exam. Intelligence and problem-solving ability, which are essential for incident investigation studies, are closely correlated.
- Ability to communicate and persuade, verbally and in writing.
- Finally, *interest in the job*. There are far too many persons in the industry today that are square pegs in round holes. If so, please leave and go and find yourself the right hole!

THE CHARACTERISTICS OF A SUCCESSFUL RELIABILITY ENGINEERING MANAGER

It will follow from what has been written prior that a good manager of a reliability section should have all the attributes of a good manager and a good reliability engineer, as far as this is practically possible.

STRESS IN MANAGEMENT

Stress in the workplace is not necessarily a bad thing. Everyone needs some degree of stress to perform well, although the level of stress will vary from person to person. This concept is indicated in Figure 8.2. When we have no stress, our performance is mediocre. As the stress level increases, our performance increases, but if there is too much stress, the performance falls off badly. Reliability managers, of all people, should know this, as it is a human reliability phenomenon.

A good manager is one who knows and understands his men and therefore knows how much to stress them to get the best performance out of them. This stress can be time pressure to finish a job, a challenging task, and so on.

COMMUNICATION

It has been said previously that the ability to communicate well is one of the necessary attributes of a manager, and also of a reliability engineer. It is an unfortunate truth that engineers are, by and large, not the world's best communicators. Other professions such as law and marketing are trained to be good communicators—engineers are not. Those engineers that are good communicators are often the ones who rise to the top of their profession.

As regards written communication, most engineers have been trained to write technical reports. However, to write persuasively and well requires extra training and is outside the scope of this text. It is a truism that to right well, one must read well. The author encourages all who read this to increase their reading habits and increase their range of interests—read magazines, trade journals, biographies, classics, whatever. In so doing, one's word power and ability to express oneself will increase, and

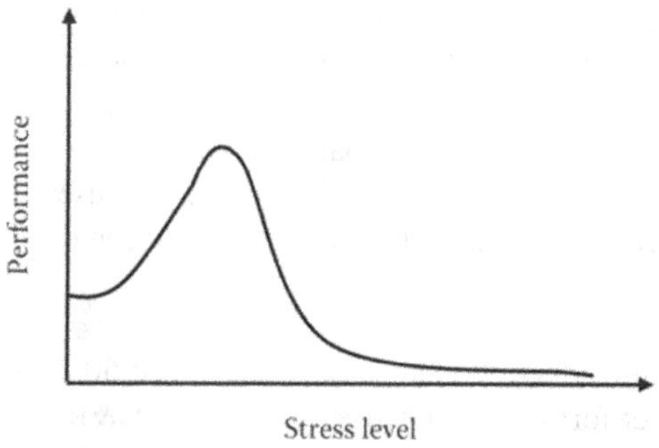

FIGURE 8.2 A generalised stress-versus-performance curve.

with it, one's power of persuasion. The ability to persuade and convince is especially important in the staff positions that reliability persons usually hold. In such cases, one cannot tell others what to do, but one must get them to do it anyway!

READINGS IN RELIABILITY MANAGEMENT

Introduction

These readings are included in this text to show what some others have learned in the field of reliability management. Some of the findings seem to contradict what has been taught in the text. This shows that not all techniques and solutions are universally applicable, or when applied, the technique was not properly understood.

Reading #1: Human Error in Maintenance and Reliability and What to Do about It (Also Known as Rules, Tools, and Schools)

Jack Nicholas Jr

Assignment

This paper is included because of the wisdom it contains, particularly for reliability engineers and managers. The paper is intended for class discussion, if possible. Read the paper that follows. Highlight any items that you believe are particularly important or, conversely, that you disagree with.

This paper describes the cost and consequences of human error in maintenance and reliability (M&R) in a variety of venues, such as utilities, manufacturing, and government. Key elements will focus on the following:

- A long-term strategy and set of tactics to ensure that errors are progressively eliminated, minimised, or mitigated for the duration of an organisation's existence.
- A simple analytical method to detect recurring problems that cause seemingly small delays or reduction in throughput or delivery of a service.
- Whom to turn to first for solutions, especially when safety and equipment reliability are concerned, and how to get their attention.
- What types of policies get the most cooperation from all levels of an organisation when the goal is to minimise human error and maximise profits in an increasingly competitive global economy.
- A systematic root cause analysis technique that focuses first on the human elements rather than on the technical elements of the overall problem.

Introduction

In observations from leadership positions over the past 50 years, I have seen many "new" approaches set forth as the ultimate answer to M&R improvement. Some of the promoters of these new methodologies or approaches have attempted to appeal to all organisations in commerce, academia, and government. These ideas come and go—mostly go—into obscurity. However, a number of fundamental truths and fairly simple concepts seem to me to work best in all venues. This presentation is an attempt to summarise those that I have found to work best and to provide some guidance as to where to find more information on the basics that seem to me to work universally.

These observations are presented in the form of "concepts." The concepts are presented to help illuminate the most common human errors. Errors are committed because of omission or ignorance of the best ways to get consistent results from people who, in the overwhelming majority, in my opinion, want to do the best possible jobs that they are assigned or choose to do. The concepts are interrelated and cannot be encapsulated uniquely. Thus, it is not possible to isolate one and concentrate on it without missing something and not doing so well in attempts to apply each of them.

Concept #1: Exercise and Practice Leadership as Well as Management[1]

Starting at the nominal "top" of any organisation, its "management" group or individual, I find that many problems could be solved easily and most effectively if attention is paid to the functions that those in top positions of organisations should be performing. In my mind, the most important of these functions can be summarised as follows:

- Providing direction, objectives, and goals through effective (two-way) communications
- Obtaining (and keeping) the resources needed for people to do their jobs most effectively and efficiently
- Removing impediments to reaching the ultimate objectives of the organisation
- Adding or removing constraints that are needed to keep the organisation focused
- Thoroughly understanding and constantly refining the processes that the organisation has to execute to serve its customers
- Providing leadership

Unfortunately, for those subject to management with little or no leadership, one finds that a great deal of time is spent on organisation and reorganisation, a waste of time for anything other than communications, in my opinion.

In the course of executing management functions, however, "managers" many times do not (or do not know how to) exercise leadership. The goal of this paper is not to teach leadership but merely to call attention to the absence of it in many instances and where to find information on what it is and how to learn about it.

One of the key ways of exercising leadership is to learn to listen. Listen to those in positions above (who must serve all below), co-equal persons alongside you, and employees in positions below yours (those you serve). Combining the two key elements of leadership and continuous process refinement mentioned prior, the best way to obtain and maintain the processes related to maintenance and reliability (and the overall business processes from which the latter flow) is to engage key employees in technical and first-line supervisory positions.

The initial process diagrams should be complete with the identification of existing impediments, needed (and unnecessary) constraints, and resources needed to make them most effective and efficient (clean and clear) in actual practice. It's the manager's job (and should be his or her "agenda") to make this happen, once the list of impediments, constraints, and resources is prepared. He or she must get busy and act as decisively and rapidly as possible to clear any obstruction to success of the processes needed to serve both internal and external customers. In so doing, the smart manager goes far beyond obtaining "buy-in" but also creates a sense of "ownership" in the processes by all who participate in development and refinement, a far more valuable attribute.

Concept #2: Look First at "Programmatic" Rather Than "Technical" Solutions to Reliability Problems

In the decade after the 1979 Three Mile Island incident involving a nuclear reactor core meltdown, the organisation responsible for regulation of commercial nuclear power plants and related facilities, the US Nuclear Regulatory Commission (USNRC), searched for ways and means to avoid future events of this nature. Many of the solutions involved redesign of power plant control rooms and instrumentation as well as plant safety systems for mitigating and controlling any malfunction at its early stages. This was nothing new to those in the regulatory organisation. It was revisiting reactor and safety system designs that it had concentrated on from the USNRC's beginning even before separation from the Atomic Energy Commission and the Department of Energy. What was new to them was the need to regulate the operation and maintenance of existing plants, something that was in the early stages of development and implementation when the 1979 incident occurred. In fact, the way the applicable law and regulations were written in 1979, USNRC had only limited authority to regulate maintenance practices. It was not until the mid-1990s that, despite intense lobbying and resistance by nuclear utilities and their various industry organisations, the "Maintenance Rule" became law, thus giving the USNRC authority its commissioners felt was needed to assure safety.

Among the many studies performed under USNRC sponsorship was one concerning "programmatic root cause analysis." A model was developed addressing the essential precepts of the causes of human errors in preventive and repair (often called "corrective") maintenance. The model was tested for several years at a nuclear power plant that had new and open-minded management (for reasons that will become obvious as you read ahead). The model provided a set of four "diagnostic query diagrams" titled as follows:

- The Training Query
- The Procedures/Documentation Query
- The Quality Control Query
- The Management Query

Some surprising results came from this study, including but not limited to the following:

- Although technicians performing maintenance were not eliminated as root causes of defective maintenance, their inadequate performance was found to be most likely the effect rather than the root cause of subsequent (infant or premature) equipment failures.
- The queries focused on isolating such root causes as line and upper management performance (including their attitude towards their responsibility for craftsperson performance), procedures and documentation (both the product and the process), training (both delivery and the process), managers of such programme elements, and quality control.
- Quality control was found not to be a primary cause, but a co-cause.
- *Management* was found to be both a primary and a co-cause of inadequate maintenance by technicians.

The landmark finding that management performance was often the root cause of premature equipment failure after maintenance had monumental impact. Often, the continuation or granting of new operating licences for nuclear generating plants and related facilities rested upon the assessment by regulators of attitude of nuclear utility managers at all levels towards this cause. This led in some cases to major shakeups in management teams until those in place could satisfy regulators that they understood their shared responsibility for equipment failures alongside those whose hands were actually on the equipment.

This finding concerning management's direct involvement in equipment failures is seldom, if ever, applied outside the commercial nuclear power industry.

Only now is it being considered for application (along with many other initiatives) to the British Petroleum (BP) refinery at Texas City, Texas, which was involved in a fatal accident on March 23, 2005, that resulted in 15 deaths and a much larger number of serious injuries.

Not only were there costs involved to those on site and BP stakeholders worldwide; the long outage that was needed for disaster recovery (along with maintenance problems at other plants for a variety of reasons) also caused a shortage in refinery products that everyone in the country paid for in terms of a spike in fuel prices for many months after the accident.

Another finding that ultimately came to light from the USNRC-sponsored study was that the majority of root causes of infant or premature equipment failures could be eliminated or at least mitigated and reliability could be improved less expensively and more rapidly by programmatic solutions than by technical (redesign) solutions. That's why the concept is stated as it is.

Those who have developed approaches to root cause analysis of equipment failures all claim that their methods address the queries listed prior, and there is no doubt that this is true in theory. However, in actual practice, management and many other programmatic causes of failure are considered coincidental, if not prohibited, areas of investigation requiring corrective action, that is, unless and until a major fatal accident occurs and an outside, independent panel conducts an investigation, as happened in the BP Texas City case.

Indeed, the programmatic root causes of failures may well go beyond the industries that suffer from them. The rash in 2006 and 2007 of children's toys and costume jewellery items containing lead having to be removed from store shelves certainly has as one of its root causes the lack of oversight of foreign manufacturers by their US partners and the significant reduction in staff and other resources for testing such items at the US Consumer Product Safety Commission (USCPSC). At least one of the major toy distributors in the United States apologised to the American people in testimony before a congressional hearing committee on the subject in mid-2007. On September 11, 2007, the USCPSC and its mainland Chinese government counterparts announced an agreement on work plans concerning safety of toys, fireworks, cigarette lighters, and electrical products.

Findings of contamination in food imports in a large number of products from several countries that got heavy media attention during 2007 may be traced to the lack of US Department of Agriculture or customs inspectors at ports of entry and in US processing plants. Certainly, these may also be traced to lack of management attention or poor attitude, but a co-cause may well be lack of resources resulting from political decisions concerning regulatory agencies charged with inspection and enforcement of the rules. Countries of origin may also share some of this responsibility. In China, a whole new regulatory regime is being created to address this deficiency, and many marginal producers have already been shut down as a result of early action.

Concept #3: Look for Indicators of Small, Seemingly Insignificant, but Repetitious Reliability Problems and Act on the Findings

Another feature of the report on programmatic root causes of equipment failure was the description of an easily implemented analysis method for determining where to apply limited resources to solve many problems of equipment maintenance and reliability. The method was given the title "cluster analysis."

Cluster analysis is explained in just three pages of the report and consists of the following steps:

- Sort the data.
- Identify clusters.
- Determine which clusters are relevant.
- Group the clusters into categories.
- Determine the consequences of relevant clusters.
- Determine technicians involved, when necessary.

Table 8.2 provides an indication of how statistics on relevant clusters looks over time.

This is the sort of analysis that can be carried out at any level of an organisation and should lead to triggering some root cause analysis activity and follow-up. The numbers presented reflect the reduction actually experienced at the plant where this method was tested. As observers become more familiar with the steps of cluster analysis, new clusters will emerge as earlier ones are dealt with.

TABLE 8.2
Cluster Analysis

Problem or Category Cluster	No. in 1988	No. in 1990
Damage	28	15
Tubing/fitting	13	
Packing	2	
Seal/gasket	4	2
Tightening	19	
Wiring	1	
Procedural error	18	10
Diagnostics	2	1
Repair	2	
Weld	2	
Total	98	28

Many of these items in and of themselves may cause small delays in production, but their cumulative effect over time can have quite a substantial effect on the bottom line or mission of the organisation that engages in this relatively easy method of identifying problem needing action.

Concept #4: Don't Be Afraid of Mistakes; Learn from Them

Typically after an incident involving substantial cost to recover, injury, or death, the search is started to find the "guilty" parties so that they can be held accountable. This is the wrong (management) approach in all but those cases where malicious intent is apparent initially or determined to be a cause in the course of investigation of the incident.

Managers sometimes contract for third-party investigators to find the root cause of serious events. This may be OK for the overall look at what happened and to prepare a professional report of findings. However, it has been shown to be the cause for those involved to take a defensive approach that impedes the full story being revealed. This, in turn, inhibits appropriate action being taken to eliminate or mitigate the true root cause, liability claims and related court cases notwithstanding.

Policies and practices that have been proven to avoid repeated problems described prior are discussed next.

Adopt a No-Fault Policy

Adopt a no-fault policy regarding apparent accidents and incidents. The policy should have a corollary provision that emphasises the need to learn and not suffer unnecessarily from undesirable events. Stopping any attempt to "blame" someone will aid in more quickly getting to the truth of what happened and the ultimate solution. Learn from the mistake, correct the problems, and get on with the business of serving customers and providing all stakeholders with the fruits of their investments in the organisation. Those who fail to learn from mistakes and repeat them should be assigned where they can't continue to cause harm to people or equipment or, as a last resort, be let go.

Adopt a Compliance Policy

Implement a compliance policy that applies to the use of all operating and maintenance procedures as written or, if found deficient in some way, as modified by competent personnel following the approved procedures management process. (See concept #5. This assumes that the organisation has adopted a goal of becoming a "procedure-based organisation," which in turn has a formal feedback and follow-up process in place that assures prompt action on all recommendations for changes.)

Practise Peer Review

When a major equipment failure or personnel injury/fatality incident occurs involving one or more personnel and there is an opportunity to interview those involved, institute a practice of "peer review." The purpose of this practice is to fully identify what happened and what should be done to eliminate or mitigate the incident being repeated in the future. The chances of those involved producing an accurate picture of what happened and coming to a conclusion as to what to do to prevent such incidents in the future are greatly increased when they are talking to their peers, without managers or other "outsiders" present. The intent of protecting co-workers and other stakeholders from repeating any errors must be the central goal of such a practice. It can be effective, however, only if the practice is backed by the no-fault policy stated previously. Ultimately, it may be found, as indicated in the concept describing programmatic root cause analysis, that management needs to do something to eliminate or mitigate the problems revealed. This may include providing more training, better documentation, or even changing their attitude concerning provision of other resources needed to ensure no repeat of the incident. The no-fault policy applies to management also, as viewed by those who are subject to its leadership.

Practise Focusing on the Incident at Hand while It Is Being Investigated

Anyone who has been involved in root cause analysis or reliability-centred maintenance (RCM) analysis knows how easy it is to have the process prolonged or even sidetracked by unrelated issues that arise. There is always a strong desire by those assembled to perform such tasks to discuss all the perceived, current problems of the organisation. Those assigned to facilitate such analyses must acquire and liberally apply the skill of diverting such discussions and refocusing attention on the matter at hand. One very effective way of doing this is to start listing, by title only, "Other Items of Interest" for presentation along with the report concerning the incident at hand. Thus, the group discussion can be refocused on the incident, along with assurance that the list of "Other Items of Interest" is prepared for presentation to management along with (or included in) the root cause report.

Concept #5: Become a Procedure-Based Organisation, but Don't Overdo It

In a variety of presentations I have written, co-authored, or contributed to, the emphasis has been on becoming a "PBO—a procedure-based organisation." In this text, an example is provided where such advice was taken too far. A procedure-based organisation produces or receives and complies with detailed written instructions for conducting not only maintenance but also operations and routine checks. This seems so basic that it is overlooked in most organisations, and for all the wrong reasons! It's so much easier than it used to be, given the availability of low-cost word processing and scanning and image-insertion equipment. There is hardly any excuse for not doing it, given the benefits derived in terms of increased reliability and consistent delivery by the operators and maintainers of the maximum possible capacity of a production line. The fundamental approach is depicted in the diagram that follows.

Not only does an organisation have to declare that it is a procedure-based organisation, but it also has to back it up with a working process for procedure and checklist origination, dissemination, feedback, and follow-up. The idea of feedback and follow-up is reinforced in Figure 8.3 by arrows that imply two-way paths for communications. It is not enough just to disseminate an initial set of procedures and checklists. Users must have ongoing evidence that their ideas for improvement are being received, considered, and acted upon promptly. Changes that are concurred in must be seen to be incorporated in revised procedures and checklists coming out of a process that functions as well as is expected of all the maintenance and operations processes it supports. Otherwise, enforcement of a policy requiring compliance will quickly become impossible, because of a perception that management support for the process and related policies is weak or non-existent.

In July 2004, I conducted a one-day seminar in response to a query concerning what it took to become the "world's best maintenance organisation." The activity

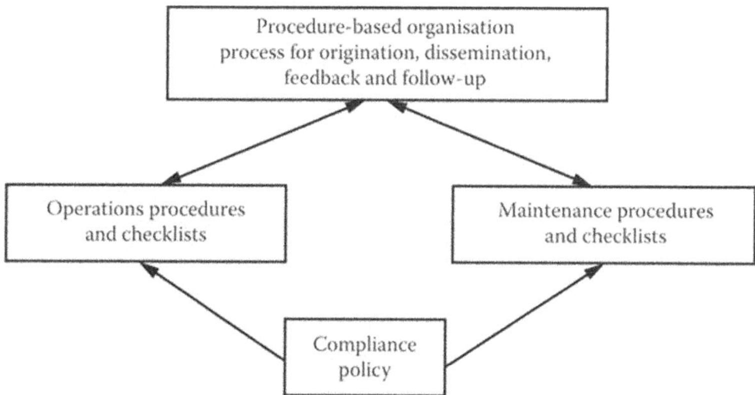

FIGURE 8.3 The procedure-based organisation.

where the seminar was held had been operational for only 18 months after rejuvenating a portion of a steel plant that had a hundred-year history before shutting down and going out of business three years earlier. The new organisation was doing quite well, having returned the equivalent of 80% of its new owner's investment in the short time it had been operating under new management and carefully selected staff. However, all there knew that world steel prices, then inflated due to the "China bubble," could very quickly deflate to where they might not be competitive with foreign suppliers of the products they manufactured. They saw maintenance as an area where their equivalent profit margin (return on investment to their owner) could be improved and their own jobs kept securely in the United States. After attending the seminar, which stressed, among other things, use of detailed procedures and checklists for both operations and maintenance, management decided to apply the principles to start-up of one of their most complex manufacturing processes. The operating and maintenance staff prepared a check-off list for start-up of all systems needed to roll steel bars into coils of wire ready for shipment. Typically, this evolution, which occurred every Monday morning, was fraught with multiple delays while the systems involved were aligned correctly and adjusted to the required level of throughput.

Approximately two weeks after the seminar, I followed up with the company president. He volunteered that they had applied the rolling line start-up check-off list for the first time that week. They decided to run the check-off twice before the first bar of steel was introduced to the line. They found in the first check that they had missed two items. After correcting these items before the second run-through of the checklist, the start-up went without any delay or incident, a first for that plant under the new staff. If ever there was a "Hallelujah moment," for one preaching the benefits of detailed procedures and checklists, that was it for me.

In the summer of 2005, I conducted a procedure and checklist workshop for Gallatin Steel Company in Kentucky, which is owned jointly by Brazilian and Canadian firms. Following the lead of one of its owner companies (Dofasco of Hamilton, Ontario, Canada, which that year had been declared by the *Wall Street Journal* the most profitable steel company in the world), the management decided to embrace a key element of the parent company's success—use of detailed procedures and checklists for maintenance. In the course of the workshop conducted for key technicians and supervisors (with managers present only for the beginning and ending sessions), a detailed process was developed for origination and ongoing support of procedures and checklists. A format and detailed outline was decided upon for the actual documents, and the decision was made to produce all them in-house, using overtime to pay those craftspersons who volunteered to write the procedures.

Two years later, following up with the project manager, I found that the organisation had produced more than 500 detailed preventive and repair maintenance procedures and checklists. In response to a request for an opinion on what the major benefit was from all this effort, he responded by saying that the biggest benefit was the significant increase in confidence that the workforce had gained in performing maintenance. Delays and frustration with not having the correct tools or replacement parts were radically reduced.

The company has been rated by the Kentucky Chamber of Commerce and the State Council of the Kentucky Society for Human Resource Management as one of the best to work for in the state. *Forbes Magazine* ranked Gallatin 16th overall as the best large company to work for in the United States in 2006. Its parent, Dofasco, has consistently received similar recognition in the province of Ontario and in Canada overall as one of the best places to work.

On the downside of this concept, it is possible to demand too much of the craftspersons who are required by a compliance policy to use procedures and checklists. Recently, I was requested to participate in a conference call with the representative of a major corporation that, on any given day, operates approximately 1,100 facilities worldwide. Also on the conference call were representatives of one of their contract maintenance suppliers. The craftspersons of the contractor were resisting the imposition of mandatory check-offs (by initialling) for each step of every maintenance procedure they were required to conduct. In addition, a rigorous audit procedure with punitive provisions for non-compliance by maintenance personnel had been prepared for implementation as part of the customer's compliance policy. The craftspersons who were pushing back had, in my opinion, a good case for doing so. The client had gone way beyond the best practice in use of procedures and imposition of a companion compliance policy.

In organisations engaged in this best practice, several types of procedures (and checklists) are commonly used. These types are summarised in the table that follows. The basic ones are given titles like standard operating procedures (SOPs), special operating procedures (SpOPs), critical operating procedures (COPs), standard maintenance procedures (SMPs), special maintenance procedures (SpMPs), critical maintenance procedures (CMPs), preventive maintenance (PM), or predictive maintenance (PdM) procedures. Standard, PM, and PdM procedures define common, often-repeated, operations, maintenance, or condition monitoring tasks.

All but critical procedures may be written in "two-tier" format. The first tier is an abbreviated version of the second tier that provides a more in-depth explanation and additional steps for use in training of new personnel or occasional review by experienced personnel who may not have performed the standard task for some time.

Operating procedures in organisations following current best practice often contain many routine preventive maintenance tasks that are assigned to operators for completion. Note that individual sign-off on each step is required only for safety or critical task procedures (and checklists). Typical procedure and checklist categories are given in Table 8.3.

During the conference call, I emphasised the need for "trust" in the client–contractor partnership that extended to the conduct of operations and, in this case, maintenance. I recommended the audit requirement be abandoned completely and that the procedures be categorised per the definitions in Table 8.3 with individual steps required to be checked off only for safety and critical maintenance tasks.

TABLE 8.3
Procedures to Match the Task

Procedure or Checklist Type	Where Used	Manner of Use
Safety or critical task	Complex procedures where the safety of personnel and hazards to equipment are the principal concerns	Verbatim compliance. Individual step sign-off.
Standard task	Operating and maintenance procedures for common, often-repeated tasks	Procedures available on file. Used as training documents as well. Can be taken off-site if needed. Captures experience. Utilises the skill of the craft.
Special task	Procedures for major, complex, and infrequent maintenance or operating procedures (e.g. post-maintenance testing)	Procedure is part of a work package. Maintained on file. Used at job site as a reference. Includes post-repair tests.

Concept #6: Eliminate as Much Maintenance as Possible and Increase Emphasis on Reliability

In the past, the traditional view was that the two goals stated in the previous concept statement are contradictory and impossible to achieve. However, this is not the case. More maintenance does not produce more reliability per se. In fact, it can be a root cause of reduced reliability. If the organisation has created the optimum maintenance programme and knows exactly what maintenance to perform (that which is [cost] effective and applicable [i.e. it works]—a result of a proper application of RCM methodology) and exactly how maintenance should be done (a result on proper application of total productive maintenance [TPM] principles), then the stage is set for concentration on the goals stated in the concept above.

A major pillar of TPM and one that is often neglected may be stated as "Manage equipment in order to prevent maintenance." Much can be done at the design stage to eliminate, reduce, or at least minimise the hours spent maintaining equipment through application of maintainability principles and choosing components with generous service factors. However, most of the M&R world is faced with the equipment already in place and in production, acquired on a lowest purchase and installation cost basis. Thus, the challenge is to improve the reliability and maintainability of the equipment we have, not the equipment we'd like to have. Some texts on TPM refer to this as "corrective maintenance," a term that means, in the context of TPM, modifying the equipment in service to improve its design and by extension its capacity to reliably produce a product or service at the lowest possible overall conversion cost.

The cost reduction from increased reliability and decreased maintenance can be significant. It affects the overall conversion cost of a product or service. The reduction in cost directly affects the profit margin and makes it possible for a company to offer cost savings to customers, thus improving competitive position in the marketplace.

Unfortunately, the management error that is often committed is to mandate maintenance cost reduction without compensating by providing a comparable improvement in reliability or maintainability, both of which require labour hours on a never-ending, continuous basis. This is exactly the opposite of what should be done. More often than not, the decision-maker gets away with it in the short term. This is because the "easiest" target for cost reduction is most often maintenance personnel (lay-offs of "excess" personnel). Typically, such action causes a pullback from proactive maintenance and a fallback to reactive maintenance (high priority, if not emergency repairs). This results in a more costly approach as time goes on, especially when lost opportunity costs are considered. The full impact may not be felt for many months and, in some cases, for up to two years. When the percentage of inoperative equipment reaches an intolerable point and maintenance personnel are again augmented so labour hours can again be devoted to proactive measures, it takes approximately two more years to fully recover to the high point of performance where the lay-offs began.

In fact, a study performed at MIT shows that these cyclic events do, in fact, occur. What should be done when excess labour hours become available after proactive maintenance practices take effect? The nature of the jobs experienced maintenance personnel are performing should be changed! Emphasis should be placed on the following:

- Maintenance prevention and elimination
- Reliability improvement and sustainment
- Capacity enhancement

This is done by acquiring and putting in place and using the following:

- **Rules**—related to best practices in maintenance and reliability
- **Tools**—acquisition application and continual updating for maximum productivity
- **Schools**—to teach the new skills needed by modern maintenance organisations

For more on rules, tools, and schools, see the discussion under the next concept for avoiding human error in maintenance and reliability.

A true story involving a 30-year-old aluminium production plant reflects this cyclic effect. In the mid-1990s, the plant changed hands from foreign to US owners. The new owners hired new plant, maintenance, and reliability managers to see what they could do to improve profitability of the aged but still reasonably profitable plant. New management's initial assessment showed that the throughput under the maintenance strategy they inherited was only approximately 50% of the designed-in capacity. Given that the company could sell everything it could produce, the team set out to increase throughput by changing the strategy to a more proactive one.

A vigorous predictive maintenance programme was instituted. Root cause analysis and reliability improvements were made to existing equipment. Within approximately two years, the throughput had been increased to approximately 75% of projected maximum capacity. The owners were making so much money they purchased three more aluminium plants in a distant state. The plant manager was promoted to vice president, and his replacement was recruited from another aluminium producer, with the promise of autonomy in his running the operation.

Within the first week, the newly hired plant manager made his views concerning further improvement well-known. He told the predictive maintenance team that he didn't understand what they were doing and recommended they bid for jobs where they had "real" tools in their hands to "fix" things. Reliability improvement initiatives were put on "hold."

The reliability manager quickly found a new position in the expanding corporate office. His assistant, hired to oversee RCM projects, found employment at a nuclear power plant. The predictive maintenance team leader (in a salaried position) was stuck for a time while he finished a master's degree programme at a local university, but he, too, left for a job managing contract maintenance at a new steel plant.

For months, throughput remained where it was when the new plant manager took over, but in approximately a year, there was a distinct downturn in production. The vice president paid a visit to review performance and found only the original maintenance manager from his "dream team" still in place—but scheduled within two weeks to move to another company where he had accepted an offer of a maintenance manager position.

The vice president found out from the soon-to-depart maintenance manager what had happened, confronted the plant manager, and fired him, assuming the plant manager's duties in addition to his own. The maintenance manager was promoted to be plant manager at one of the newly acquired facilities.

The vice president has been trying to reverse the downward trend in production ever since by building a new team and restoring confidence in the union staff members who remain in or returned to their hard-won, higher-paying predictive maintenance positions.

Concept #7: Don't Forget the Roots of Your M&R Programme Initiatives for Improvement

It is not uncommon, with so many new initiatives being offered in the field of M&R, to see earlier, even highly successful principles and methodologies abandoned or forgotten, with the promotion, retirement, and transfer of those who implemented them.

One of the earliest adopters of RCM methodology developed for commercial aircraft was the US Navy. In the 1970s and 1980s, vigorous efforts were undertaken to change maintenance from a more costly, shipyard-based strategy to one anchored in RCM and operating base support.

By the 1990s, most of those engaged in implementing the "new" RCM-based approach had retired or moved on to other jobs, attributed in part to the post–Cold War drawdown in naval forces and related support facility manning.

By the late 1990s, the Navy found that its maintenance programmes were in need of overhaul and revitalisation in order to ensure reliability in the face of apparent return of intrusive maintenance requirements that were superimposed on the RCM-based strategies (which differed from class to class of ship and submarine). In addition, while the specifications for building ships still contained the Department of Defense–mandated requirement to provide an RCM-based maintenance programme, new methods of contracting for ships often resulted in these efforts being underfunded and inadequately implemented. The shipbuilders often simply implemented original equipment manufacturer (OEM) recommendations, which had been determined in the studies done decades earlier to be heavily tilted towards regular "overhaul," requiring heavy life cycle replacement parts costs. The OEMs did what the shipbuilders asked and were benefiting handsomely as a result.

Luckily for the Navy and US taxpayers, some "old-timers" still remained in civil service who had, by this time, achieved positions with sufficient clout to rectify this problem. They devised a revitalisation initiative to avoid inapplicable and ineffective maintenance and reduce maintenance costs without sacrificing reliability. The initiative was based on three parallel efforts:

- **Rules**—Improving maintenance requirements and plans (including reliability improvements)
- **Tools**—Using computer and diagnostic technology (i.e., condition-based maintenance)
- **Schools**—Educating all levels of maintenance decision-makers in reliability and condition-based maintenance principles

Commercial organisations suffer from the same problems that those in the Navy did in the 1990s. Consumers and promoters of keeping as many core industries in our country as possible pay the price for this error by humans engaged in maintenance and reliability. The error is that they forgot (or never learned about) the past.

Conclusions

The paper has expounded on seven concepts that are essential to minimising error by those engaged in maintenance and reliability. These are summarised next:

- Exercise and practise leadership as well as management.
- Look first at "programmatic" rather than "technical" solutions to reliability problems.
- Look for indicators of small, seemingly insignificant, but repetitious reliability problems and act on the findings.
- Don't be afraid of mistakes; learn from them.
- Become a procedure-based organisation, but don't overdo it.
- Eliminate as much maintenance as possible and increase emphasis on reliability.
- Don't forget the roots of your M&R programme initiatives for improvement.

Having heard about those listed prior, I'm certain those who read this text or hear it presented can come up with many more ideas on reducing the occurrence and impact of human error in maintenance and reliability. However, concentrating on these will make a big difference in achieving the goals and objective of your organisations.

The mantra for modern maintenance and reliability programmes everywhere could well be, **"Rules, tools, and schools!"**[2]

READING #2: WHY MANAGERS DON'T ENDORSE RELIABILITY INITIATIVES

Winston Ledet

Assignment: Read this article and compare the recommendations with the previous article and with the case by Wardhaugh in Chapter 3. Are there any conflicts? In this article, pay particular attention to the *cause of randomness.*

Don Kuenzli, as a plant manager, successfully transformed two oil refineries to highly reliable and sustainable performance by creating a culture of continuous improvement through defect elimination. He had a vision to create "a world-class facility with pacesetter performance" for these refineries. When Don shared his vision with his boss at one of these sites, he was told that they could not afford to create this level of performance. Managers often make the assumption that the best performance is gained by paying a high price. There are many benchmark studies that demonstrate quite the opposite.

In the DuPont benchmark study, the most reliable performance in the world was TPM. It was also the least expensive to achieve and sustain. The assumption that high reliability is expensive comes from the fact that the vast majority of efforts to gain high reliability are misguided. The conventional wisdom is that maintenance best practices are planning, scheduling, preventative maintenance, optimised procurement, and predictive maintenance. While these practices in fact make maintenance much more efficient and effective, they do not address the most important aspect of reliability. Why is the equipment failing in the first place? Other best practices, such as RCM, are a step in the right direction but still do not address the largest root cause. In the ABCs of failure (TMG News, April 2008), we concluded that approximately 84% of the defects that lead to failures are in fact created randomly by careless work practices throughout the entire organisation.

For those who have not seen our earlier article on the ABCs of failure, we concluded that 4% of the defects are attributed to aging of equipment, 12% of the defects are caused by basic wear and tear, which leaves 84% being attributed to careless work processes. If one starts an initiative to improve reliability on the basis of conventional wisdom, he might expect to improve maintenance practices by implementing more preventive maintenance. However, by scheduling more frequent preventive tasks on equipment that we already do some amount of preventive maintenance on, or by expanding preventive maintenance to equipment that was not included in the preventive maintenance programme before, we can only succeed in removing the defects that are created based on the passage of time. That only includes the aging and basic wear-and-tear defects, and they represent only 16% of the defects. People get very frustrated when they go beyond that 16% because it becomes apparent that the probability of adding a defect while overdoing the preventive maintenance is higher than the probability of removing a defect. This also becomes very expensive and wasteful because work is being done to change parts that are in fact not defective. During 27 years of experience at DuPont, we went through approximately seven cycles of increasing preventive maintenance to the point of frustration and then abandoned most of the preventive work. Although important, preventive maintenance can't solve all our problems, and we are wrong to expect it to be more than it is.

The other best practice that everyone recognises as having merit is predictive maintenance. In this case, it is recognised that failures are not predictable with time alone but depend on how long it takes for a defect to propagate to a failure event. It also depends on how good the technology is in detecting that defect before it becomes a failure. Reliability programmes for predictive maintenance concentrate on getting the requisite variety of detection technologies to find defects soon enough to allow for orderly planning and scheduling. The computer model of the DuPont benchmark facilities found that the number of inspections required to ensure that > 90% of the defects would be detected before the failure event occurred was so high that 97% of the time an inspection did not detect a defect at all. The difficulty with this approach is that it is demoralising to sustain that kind of diligence over long periods except where the consequences of failure are catastrophic. Nuclear power plants are a great example of a facility that warrants this level of diligence and a good place to see how predictive maintenance can be very effective.

The problem in other facilities is that the cost of this kind of diligence is not competitive because the processes and equipment are much more complex than the simple process of boiling water to generate electricity. The experience with predictive maintenance at DuPont was similar to the experience with preventive maintenance. We started many predictive maintenance initiatives and succeeded until a routine operation was going, but then someone looked at the results of the inspections and decided that inspections were being overdone because we only found defects 3% of the time. This led to abandoning many predictive maintenance technologies. I once admired the fact that a mechanic in DuPont knew ten different technologies for predictive maintenance. I asked him how he learned so many technologies. He said that he had been doing predictive maintenance for 15 years. I replied, "But this initiative is only one year old." He said, "Yes, but this is my ninth initiative." A few years later, we declared victory, dissolved the corporate maintenance leadership team that was leading the predictive maintenance initiative, and completed the cycle once again.

So why did these initiatives consistently fail? The problem is not that they were pursuing the wrong best practices; it was that they failed to attack the larger problem of the randomness in the failure rates. When 84% of the failures are caused by random lack of discipline to operate, maintain, design, procure, and improve equipment, there is no efficient way to deal with the defects that get generated in these careless work habits. To cope with these defects, many companies try to solve the problem by adding spare equipment. This just adds to the amount of equipment that has to be maintained and therefore increases the expense to procure and maintain this extra equipment.

One of the worst ways to do this is to keep a piece of equipment that has been replaced by a new one. We have seen many sites where the old piece of equipment is kept as a spare. In this case, the maintenance cost is very high compared to maintaining the new piece of equipment, and it is simply there to use when the new piece is out for repairs. In DuPont, the plant that had the best pump life had zero spares. This decision caused them to treat the pumps like the precious assets they were.

As many of you have seen before, we use the stable domains to depict how reliability is generated by the behaviour of the people. Figure 8.4 is a diagram in a simple form to illustrate another dimension to the picture.

People generally agree with this way of looking at how operations and maintenance of a facility can be classified in one of these domains. Over the last 15 years, we have endeavoured to show why the successful sites have skipped going to the planned domain. That domain is inherently unstable because of the randomness of the defects that exist and the other factors mentioned prio. Although the planned domain makes the work more efficient, it does nothing to reduce the amount of work that must be done. In order to see more clearly why this happens, it is better to look at these domains from a different perspective. This other dimension is the amount of activity and therefore cost that is required to attain and sustain each domain. Figure 8.5 shows that view.

This diagram is representative of the improvement realised at the Lima refinery. The number of work orders was reduced by 67% over an eight-year period. This transformation, however, did not go through the planned domain to get to the precision domain. As the diagram shows, the extra work required to do this in the planned domain would make it much more probable that they would have returned to the reactive domain than progressed to the precision domain. If they had undertaken the extra work to get to the planned domain, it would be logical to assume that another increase

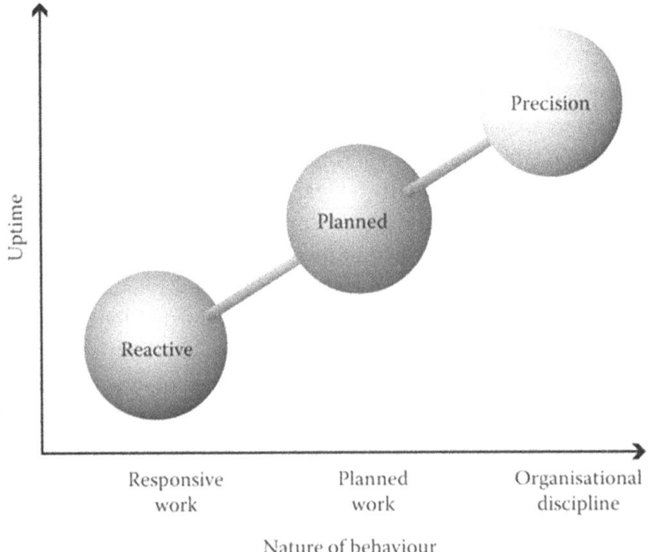

FIGURE 8.4 Uptime versus various maintenance strategies.

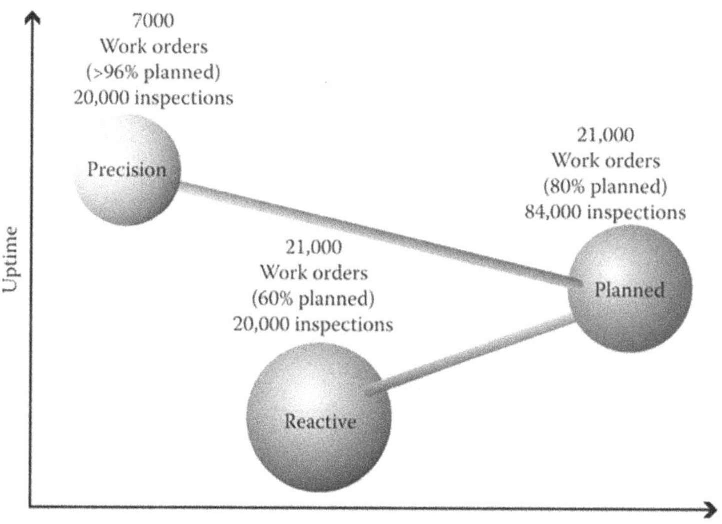

FIGURE 8.5 Uptime versus cost for the three strategies.

TABLE 8.4
Uptimes and Costs for the Three Strategies

Maintenance Strategy	Uptime	Cost
Reactive	85%	Medium
Planned	95%	Highest
Precision	97.5%	Lowest

in cost and work would be needed to get to the precision domain. Fortunately for us, we had seen the data from the DuPont benchmark plants in Japan that had won the TPM awards. In these plants, they showed us that the amount of work, and therefore the cost of maintaining a highly reliable plant, was in fact even less than the work and cost of remaining in the reactive domain. In Table 8.4, these points have been combined to show that the precision domain has both the highest uptime and the lowest costs.[3]

NOTES

1. Note the similarity to Deming here.
2. Reprinted with permission of Realiabilityweb.com®. Reliabilityweb.com is a registered trademark of NetexpressUSA Inc. d/b/a Reliabilityweb.com. Copyright © 2016. All rights reserved. For more articles like this one, visit online, www.reliabilityweb.com/.
3. Reprinted with permission of Reliabilityweb.com®. Reliabilityweb.com is a registered trademark of NetexpressUSA Inc. d/b/a Reliabilityweb.com. Copyright © 2016. All rights reserved. For more articles like this one, visit online, www.reliabilityweb.com/.

9 Design Issues in Reliability Engineering and Maintenance

Poor quality control adversely affects the reliability of a product, but good quality control cannot improve it. Either reliability is designed into the product in the first place or it will not be there. In one sense every engineer should be a reliability engineer and design reliability into everything he turns out.

William H. Roadstrum, *Excellence in Engineering*, Wiley, 1967

INTRODUCTION

This chapter will cover areas of design for reliability, testing for reliability, and other related issues.

FACTORS OF SAFETY AND PROBABILISTIC DESIGN

A reliability engineering approach to factors of safety will show that there is a rational approach to such matters. In many cases, the strength of a component and the load applied to it are not constants but vary according to some or other statistical distribution. For our purposes, we will assume that these distributions are normal.

We will now define two variables, safety margin (SM) and loading roughness (LR):

$$SM = \frac{S - L}{\left(\delta_S^2 + \delta_L^2\right)^{0.5}} \tag{1.14}$$

$$LR = \frac{\delta_L}{\left(\delta_S^2 + \delta_L^2\right)^{0.5}} \tag{1.15}$$

Where S = the mean of the strength distribution.
L = the mean of the load distribution.
δ_S^2 = the standard deviation of the strength distribution.
δ_L^2 = the standard deviation of the load distribution.

The safety margin is the relative separation of the mean values of the load and strength distributions. The loading roughness is a measure of the variability of the

DOI: 10.1201/9781003326489-9

load distribution, relative to the combined standard deviation of both the load and strength distributions.

The safety margin can also be regarded as an area under the normal distribution. For any area less than unity, there is a probability of failure when the load is applied. (The proof of this is not given here.) Refer to Appendix 1 for a table of the normal distribution.

When the two distributions have no variability, or when it is assumed that whatever variability there may be will not lead to an overlap of the distributions, then we have *intrinsic reliability*, as shown in Figure 9.1. This is the tacit assumption of traditional design practice.

We will also achieve intrinsic reliability as long as the load and strength distributions do not intersect. Normal distributions go from –infinity to +infinity, but in practice, they effectively are at zero after three standard deviations. If the distributions do intersect, then we will have failures in the shaded areas as shown in Figure 9.2.

Notice that in Figure 9.2, the loading roughness, as indicated by the standard deviation of the load distribution, is more of a problem than a large standard deviation for the strength distribution, as it will result in more failures.

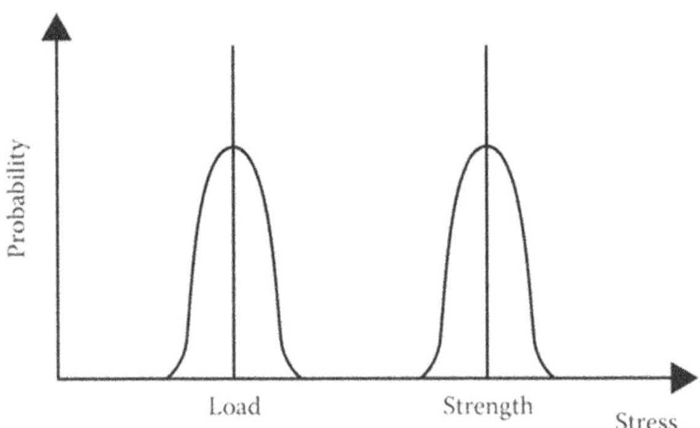

FIGURE 9.1 Intrinsic reliability.

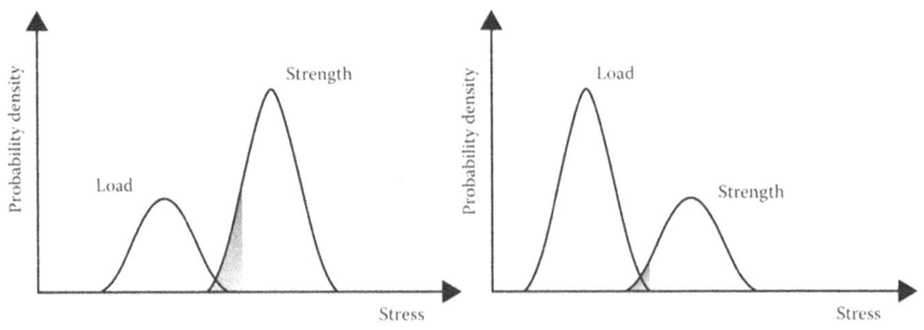

FIGURE 9.2 Intersection of load and strength distributions.

TWO CASES WITH THE SAME SAFETY MARGIN

One technique to eliminate in-service failures is to apply a proof load to a population of new items as shown in Figure 9.3. The value of the proof load should be the maximum value of the load distribution. This eliminates weak items from the inventory and hopefully ensures that in-service reliability will not be compromised. Figure 9.4 shows the resultant strength distribution after proof loading.

So in designing for reliability, we have four options:

- Increase the mean of the strength distribution.
- Decrease the mean of the load distribution.
- Decrease the standard deviation of the strength distribution.
- Decrease the standard deviation of the load distribution.

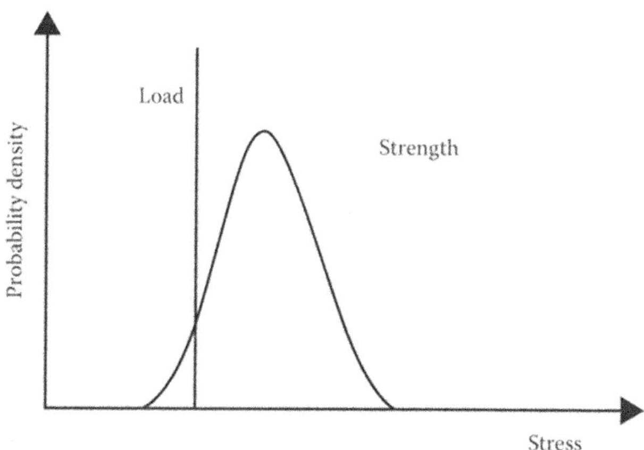

FIGURE 9.3 Proof loading.

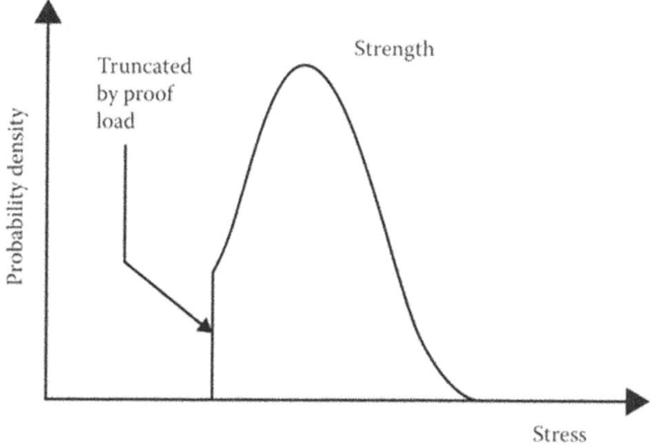

FIGURE 9.4 Resultant strength distribution after proof loading.

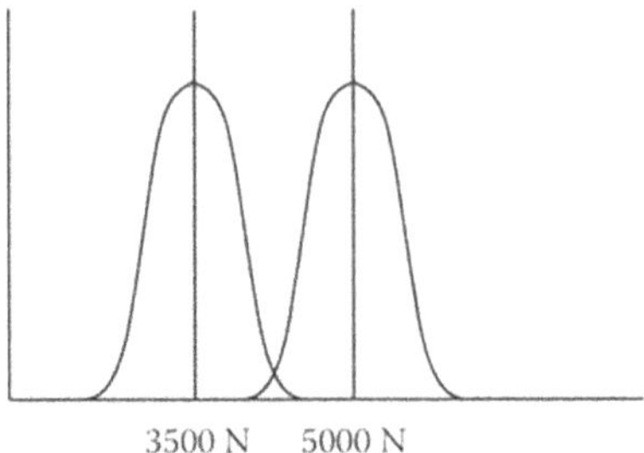

FIGURE 9.5 Strength and load distribution.

This Concept Is Shown in Figure 9.5

It must also be noted that we have not considered changes over time for the two distributions. In practice, both the mean and the standard deviation of the strength distribution might change over time. One reason for this might be corrosion and the effect that corrosion will have on fatigue life and also, in extreme cases, on failure due to wall thinning, in the case of a pressure vessel.

Why use probabilistic design instead of relying on determinate methods with fixed safety factors, derived from experience? The one advantage might be a cheaper, lighter design, by reducing the safety margin. The caution that must be exercised is in the determination of the distributions and particularly in the establishment of their standard deviations. Secondly, it might be possible to accurately estimate the strength distribution, taking into account manufacturing variability, but the determination of the load distribution will be more difficult, as it is usually dependent on persons other than the designer.

Example

1. A component has a strength which is normally distributed with a mean value of 5,000 N and a standard deviation of 400 N. The load is also normally distributed with a mean of 3,500 N and a standard deviation of 400 N. What is the reliability per load application?

Answer:

The safety margin is $SM = \dfrac{5000 - 3500}{\sqrt{(\delta_S^2 + \delta_L^2)}} = 2.65$

Consulting the normal distribution table in Appendix 1, we get for 2.65 the value 0.4960. As the table only covers half the distribution, this figure must be multiplied by 2 to give 0.992. Hence, the answer is that the reliability per load application is 0.992. In other words, in 0.008 cases, that is, in 0.8% of the cases, failure will occur.

2. For the aforementioned standard deviations, what should the safety margin be to have no more than a 0.2% chance of in-service failure, given that the

mean of the strength distribution is to remain at 5,000 N? In other words, what must the mean of the load distribution be reduced to?

Consulting the normal distribution table, we see that for 3.1 standard deviations, the area under the distribution is 0.499 × 2 = 0.998. The complement of this is 0.002, or 0.2%, as required.

$$\sqrt{\left(\delta_S^2+\delta_L^2\right)}=565.68$$

So SM = 3.1 = (5000 – x)/565.68. So x = the mean of the load distribution = 3248.5 N.

CASE 9.1 THE WOODRUFF KEY PROBLEM

Author's Note: This is a simple design for reliability problem.

The basic principles of design for reliability are, (1) "Use reliable parts" and (2) "Use as few parts as possible."

Shown in Figure 9.6 is a Gear G, which is positively located on shaft S by Woodruff key N.

The right-hand end of the shaft is threaded for a nut, which will positively locate the gear on the shaft. Unfortunately, the design is inadequate. With intermittent starts and stops, there are peak start-up torques which the Woodruff key cannot cope with. The groove in the shaft bell mouths and the key bends.

Fortunately, the torque is unidirectional. Redesign the assembly so it is both cheaper and more reliable. (This is an actual case, relating to a make of lawnmower.)

For those unfamiliar with the technology, please refer to the internet. Wikipedia is helpful in describing different forms of keyway, their purpose, etc.

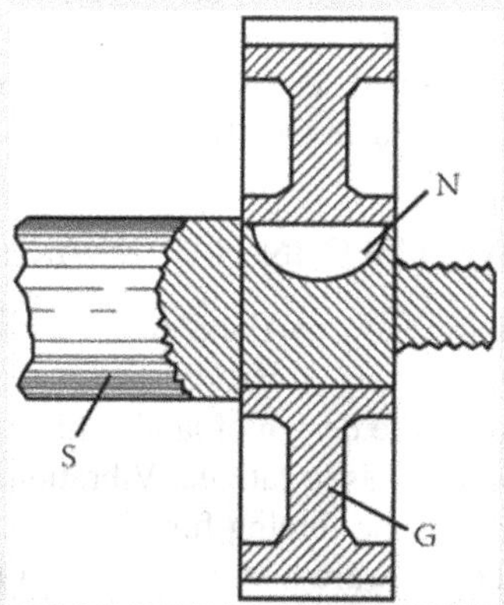

TAKEOUT: FIGURE 9.6 Assembly in Cross Section

COMPUTERS IN RELIABILITY ENGINEERING AND MAINTENANCE

No engineer works without computers in this century. He might not be directly involved, but computers and programs are all around him. It is undeniable that computers and their software have increased the average engineer's productivity by orders of magnitude. (The average engineer has not seen a concomitant increase in his remuneration to match the productivity increase, however!)

It must also be remembered that much can be accomplished in the areas detailed following with nothing more than an Excel spreadsheet. Having said that, the various types of software described next have large customer bases and many satisfied customers.

We will divide our discussion into three logical areas:

1. Computer programs used in hardware design, in the design-for-reliability phase. Such programs will be used to analyse failures, so as to pre-empt them in new designs. Or they will be design-for-reliability programs, applying statistical techniques to the probability of a failure due to extreme conditions—the "perfect storm" type of scenario. Or they will be programs that stress designs to failure in the computer, through vibration and fatigue, so that the design can be perfected in the software, and therefore, as stated earlier in this text, the prototypes are proof-of-concept devices rather than testbeds.
2. Computer programs used during prototype testing.
3. Computer programs used in the systems engineering phase of design, such as programmes for optimising system layouts for reliability. Such programs deal with levels of redundancy, levels of logistic support, numbers of critical spares, etc.
4. Computer programs used in the middle of the life cycle to assist with maintenance planning, costing, and scheduling, as well as failure reporting.

It is the very nature of this chapter that proprietary names must be mentioned as the details of the program are explained. The author does not recommend the use of any program over any other but uses them simply as examples of what is available.

COMPUTER PROGRAMS USED IN HARDWARE DESIGN

The LMS company in Belgium, now part of Siemens, is famous for its programmes for analysing and predicting noise, vibration, and harshness (NVH). Such programmes are indispensable for the design of land vehicles, aircraft, and space vehicles. And vibration, of course, leads to fatigue. Vibration levels can be modified by changes in mass, damping, and the forcing function itself. The structure, whatever it is, can be modified in the software and optimised so that fatigue-driven unreliability is eliminated in the hardware. As an example using the durability module of the LMS's Virtual Lab suite, a designer may take the finite element design from any Finite Element package and subject it to various vibration modes, inducing fatigue in the design. The design may then be modified—stiffened, made more or less massive, or having damping incorporated. The modified design is then rerun to see if fatigue problems have been solved, in an iterative manner. As recently as the 1970s, this was not possible, and systems were unwittingly designed and built to operate at or near resonance, thereby shaking themselves to pieces. The writer has had experience of at least two such systems, one a building housing vibratory mineral processing equipment at first-floor level and one a coal mill whose foundation and running speed had to be changed before it could be accepted.

As further examples, the De Havilland Comet jet airliner of the 1950s failed in service due to fatigue at the passenger windows. With present-day software, this would have been pre-empted, as the software would have indicated that a failure would have occurred in the existing design.

As a final example, the MGA Twin Cam sports car suffered serious reliability problems with melted pistons and had to be withdrawn from production. It was subsequently found that the rubber-mounted carburators vibrated so badly that the fuel foamed. This caused the engine to run very lean, which led to higher combustion temperatures and melted pistons. A vibration analysis using NVH software would have warned the designers of the foaming, and the carburator setup could have been changed

Another type of program already described early in this text is that which uses Weibull analysis to analyse failures so that they can be eliminated in future designs. The program used in this text is Weibull-Ease, but there are several others on the market.

COMPUTER PROGRAMS USED IN THE TESTING PHASE

Even though design software has become so sophisticated, it is still sometimes necessary to test products and components. The LMS division of Siemens has programs such as Test Lab for this purpose. The program, in conjunction with a controller (a piece of hardware also supplied by LMS), sends signals to a shaker, an electromagnetic table which will vibrate the test piece according to any desired regime. The component may be tested to destruction, or accelerometers might be mounted on it to observe its response to the imposed vibration. In this way, resonant phenomena may be observed. The design may be changed according to the test results and the design algorithms modified to improve future designs.

Test Lab software is also made in portable versions, for mounting in moving vehicles so that the NVH characteristics of the vehicle can be determined.

COMPUTER PROGRAMS USED IN THE SYSTEMS ENGINEERING PHASE OF A SYSTEM DESIGN

Such programs cover much of the ground covered in Chapters 1, 2, 3, and 5 of this text. They offer menus and purchase options for various modules including RBD/ABD, fault trees, RCA, RCM, Weibull, FRACAS, and others. Life tends to be short in this area of business, so some of the companies referred to here might only exist at the time of writing. ARM, Windchill, Visual Spar, Isograph, and Reliasoft are some of the more well-known.

COMPUTER PROGRAMS USED IN THE MIDDLE OF THE LIFE CYCLE

Such programs are Enterprise Resource Planning (ERP) suites. These programs assist with maintenance planning, contract management, inventory management, and historical recording of data, among other things. There were, until recently, many such programs, but the market has shaken down until only a few well-known suppliers are left, including:

- SAP
- J D Edwards (part of Oracle)
- Maximo (part of IBM)
- Microsoft Dynamics

Such software solutions are hugely expensive, but there are very large installed bases for each of them. Most installations are successful, but some definitely are not. Several companies have sued the suppliers for unsuccessful installations.

ASSIGNMENT 9.1

Conduct "website research" on all the aforementioned classifications. That is, discover programs that assist in designing for reliability in hardware, programs that facilitate fatigue testing, programs that assist with design for reliability in the systems engineering phases, and programs that assist with maintenance and reliability in the operational phase. In each case, at least two programs from each phase should be evaluated.

Write a short report on each of the phases. In the report, compare the programs you have chosen for characteristics like scope, ease of use, support, etc. As a guide, each of the reports should be about five pages in length.

ASSIGNMENT 9.2

Consider how one achieves *intrinsic reliability* for an engineering component or system. In particular, consider the ways of ensuring intrinsic reliability for a pressure vessel, considering the four options given prior, viz:

- Increase the mean of the strength distribution.
- Decrease the mean of the load distribution.
- Decrease the standard deviation of the strength distribution.
- Decrease the standard deviation of the load distribution.

If you are not familiar with that technology, consider any component or system that you are familiar with. Note that a qualitative answer is all that is required, but formulae and calculations may be used to illustrate one's argument. Your report should include, but is not limited to, the following:

- Title and author page
- Statement of originality of the work
- Summary
- Table of contents
- Table of illustrations
- Table of tables
- Introduction
- History of the technology
- Present status (basis of present factors of safety and results of recent research)
- Possible future developments
- Recommendations
- Appendices
    1. References
    2. Others as may be required

Photographs and drawings are to be included if possible, including photographs of failures. Reports are to be typewritten, single spacing, in Times New Roman, 12.

Quality of presentation and depth of thinking are required, rather than a set number of pages. But as a guide, a report to fulfill the previously mentioned requirements will probably be not less than 10 pages long, including illustrations, tables, photographs, etc.

CASE 9.1 THE BENT TRAILER

Alan Riley's first job, after graduating from university and completing military service, was to work in the trailer manufacturing subsidiary of a large truck manufacturer. One morning, he was called into the manager's office and told to investigate a problem:

"A new trailer which we have just supplied to an important customer has arrived back at our depot, bent. Here are the drawings. See what you can make of it. The customer assures us that the trailer was not overloaded."

The trailer's chassis consisted of two fabricated I-beams running the length of the trailer, connected with cross bracing. The axles were secured at the one end and the connection for the fifth wheel at the other. The I-beams are not visible in Figure 9.7, hidden behind the decorative, perforated side-strips shown in the figure.

Alan studied the drawing of the cross section of one of the I-beams as shown in Figure 9.7a and as modified for use in the trailer in Figure 9.7b. Getting out his Timoshenko, his textbook on strength of materials as used in his university days, he calculated that the I-beams had failed in compression of the upper flange.

FIGURE 9.7 The horse and trailer before it bent.

Presenting his findings to the manager, he was told that steel cannot fail in compression. He tried to explain that the additional lower strip, welded to the lower flange of the I-beam, had lowered the centroid of the moment of inertia, leading to an asymmetrical stress pattern as shown in Figure 9.7b. This led to failure of the upper flange, experiencing plastic deformation, while the lower flange remained sprung in elastic deformation.

The manager countered by saying that the additional strip on the lower flange was often used to strengthen the I-beam, not weaken it. When the matter was discussed with the trucking company representative at a subsequent meeting, he agreed with the manager.

Alan was perplexed. Could he be wrong? He phoned his friend Mark Francis, who worked at another company. "Mark, can a steel beam fail in compression?" Referring to a figure in Timoshenko, Mark confirmed that it could.

The service engineer for the company inspected the trailer and confirmed Alan's prognosis. "The top surface of the I-beam is rippled and definitely has failed in compression," he said. The manager apologised to Alan.

Years later, working at a company producing pressure vessels for the cryogenic industry, Alan encountered a similar problem in the saddle supports for a horizontal liquid oxygen vessel, where the supports had failed by plastic deformation during a hydrostatic test.

Alan summarised his philosophy as follows: "*Always be careful with the design and fabrication of built-up sections.* They do not always perform as designed. Consider Figure 9.7b: If the welds clamp the lower strip securely to the flange so that the frictional force generated allows the two sections of the bottom flange to work as one, then the result will be as we have discussed. But if the additional strip is not firmly held against the bottom flange, then all the stress must past through the welds only, which may or may not yield, but in any event, the lower strip then adds nothing to the strength or configuration of the I-beam. The de facto configuration is then as for Figure 9.7a."

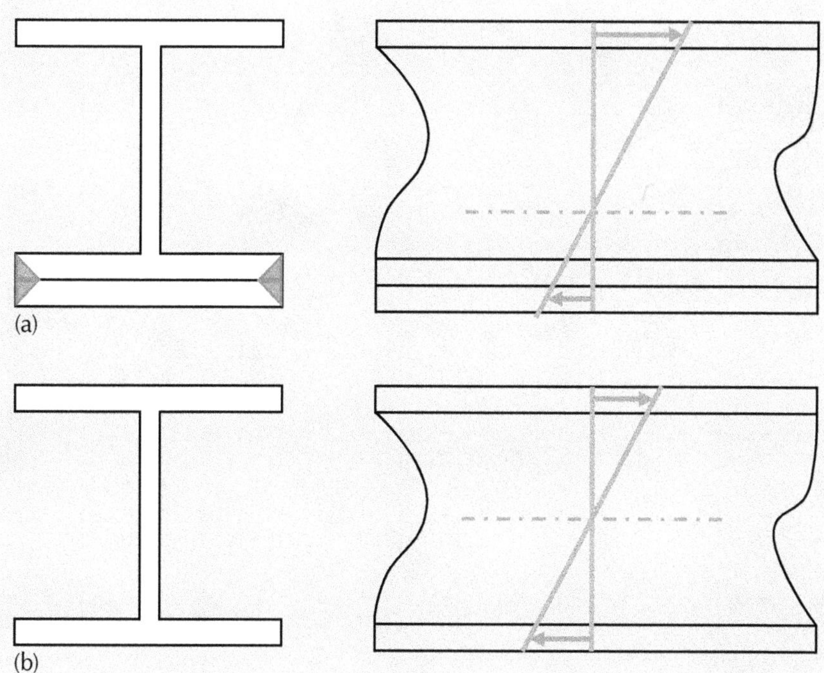

FIGURE 9.7 (a) Beam Cross-section with Extra Plate added. (b) Original Beam Cross-section.

10 Maintenance Planning and Scheduling

> The real people who hold our civilization together are the maintenance people. If it weren't for them—pumping water out of subways, painting bridges to keep from rusting, fixing a steam pipe that is 70 years old—we'd be sunk. If we got rid of all the politicians and the policymakers in the world, the world would keep going. If you get rid of maintenance people, the whole thing breaks down.
>
> **Alan Weisman (2007)**

MAINTENANCE PLANNING AND SCHEDULING

This chapter will cover areas of vital importance but often neglected. Some of the material has been adapted, with permission from Maintenance Planning Scheduling and Work Control by Roland Bergh of Roland Bergh Consulting.

THE HISTORY OF MAINTENANCE

The following Table 10.1 describes maintenance as performed in the various stages of the Industrial Revolution. This is the author's conception of the process—other references will differ on the details. The table describes an ongoing increase in knowledge and in complexity. Any column builds on the columns before, not replacing them, but including them. Maintenance planning in the twenty-first century will still include techniques from previous eras, when their use is appropriate.

It will be seen from Table 10.1 that maintenance practice is tied up with phases of the Industrial Revolution. As systems became more complicated, better maintenance practices were developed and introduced. We also see from the table that the generally agreed four types of maintenance are included, viz run-to-failure, planned or history-based maintenance, predictive or condition-based maintenance, and maintenance elimination. The period in history when the two main maintenance optimisation techniques, RCM and TPM, were introduced is also shown. A rough estimate of overall equipment effectiveness or OEE, as developed in the TPM technique, is also shown for each period.

No Easy Task

It will be seen by what follows that maintenance planning and scheduling, if done properly, is a lot of work. Many systems must be in place before any realistic planning can be attempted. What follows is a bare-bones description of the process. A company without the necessary will and resources will not be able to succeed in this endeavour.

DOI: 10.1201/9781003326489-10

TABLE 10.1
A History of Maintenance through Time

Phase of the Industrial Revolution	Pre–Industrial Revolution	Phase 1	Phase 2	Phase 3	Phase 4
Type of Production	Cottage Industry	Factories, One-off Production	Assembly Lines	Automation	Supply Chain
Typical Innovations	Cottage Industries, Hand Tools	Coal and Iron, Railways, Weaving of Cotton Cloth	Automobiles, Aircraft, Radio, Telegraph, Steel	Electronics, Computers	Internet, Smartphones
Power	Human, Animal, Watermills, and Windmills	Steam	Electricity	Nuclear	Renewables
Period	Pre-1760s	1760–1900	1900–1945	1945–2000	2000–
Type of Maintenance	Run-to-Failure	Run-to-Failure	Planned Maintenance	Predictive Maintenance RCM and TPM developed	Maintenance-Free Systems
Type of Surveillance	Visual Inspection	Visual Inspection	Statistical Analysis of Failures	Condition Monitoring, CMMS	Online and Remote Monitoring
OEE	< 50%	50%	50–75%	75–90%	90% +
Maintenance Team Support	Skilled Artisans Alone	Inspectors	Statisticians, Industrial Engineers	Reliability Engineers	Reliability Engineers

Preparation for Maintenance Planning and Scheduling

Before installing a maintenance planning and scheduling system in a company, the following concepts should be in place:

10.1 Maintenance Key Performance Indicators

A typical list should include:

- Availability
- Laplace statistics to check whether the trend in availability is improving
- Overall equipment effectiveness
- Backlog in weeks of work
- Overtime percentage
- Wrench time

These indicators have been discussed and defined elsewhere in this text.

10.2 Maintenance Costing System

This should be in place for all maintenance significant items and should cover:

- Maintenance labour cost
- Maintenance material cost
- Downtime cost

10.3 Wrench Time

It is essential that an accurate figure for wrench time is available for the various trades in the plant. This should have been assessed by an industrial engineering study, using the work sampling technique.

Maintenance Policy Document

This document should be in place as well so that the reasons for maintenance planning and scheduling are clearly laid out. If there is ant contention as to the procedures, this policy document may be referred to.

CMMS

It goes without saying that a CMMS should be in place and that it is the vehicle used to implement the maintenance instructions coming from the planning and scheduling.

Maintenance Plan

The annual maintenance plan should be generated from whatever method of maintenance optimisation was used, such as RCM. Some companies do not have such a method in place and so must rely on existing plans, past experience, etc. The point is that a plan must be generated and put in place, perhaps refined as time goes on.

Maintenance Budget

A maintenance budget should also be in place so that finances are in place for the work that has to be done. This budget might have been generated from the annual maintenance plan.

The Importance of the Planners and Schedulers

In many companies, the planner and the scheduler roles are combined. In any event, planning and scheduling is an important job, and the persons placed in these positions must have the right qualifications and experience. Often, the best planners and schedulers are former senior artisans who have also demonstrated an aptitude for planning work. The positions must be at such a level in the organisation that the incumbents are treated with respect by the other persons in the planning process, for example, the operations manager, the maintenance manager, etc.

The Maintenance Planning Process (As Undertaken by the Planning Department)

1. Tasks and frequencies are aggregated into a yearly plan.
2. Weekly plans are then derived from the yearly plan.
3. Job times are derived, incorporating wrench time percentages.
4. Allowances must be made for leave, sick leave, and other times to be subtracted from the man-hours available.
5. An allowance must also be made breakdown work, derived from past experience.
6. There is also an amount of work in the backlog.

Maintenance Scheduling (As Undertaken by the Maintenance Planner and Scheduler)

1. The weekly maintenance plan is presented by the planner at the weekly meeting (usually on Thursday, for maintenance planned for the following week).
2. The plan is firmed up, agreed by maintenance and production, and issued for the following week (usually on Friday).

Case Studies in Maintenance Planning and Scheduling

In this text, the cases "Trouble at the Smelter" and "The Embraer" case deal directly with maintenance planning and scheduling. The case "Maintenance at Automotive Pressings" is also valuable in demonstrating as facility where there was no maintenance planning at all.

Outage Management

In addition to the weekly maintenance plan, there is sometimes the need to shut down sections of the plant or the entire plant on a regular basis (perhaps once in five years, for example). In such cases, the outage is handled using project management techniques as discussed elsewhere in this book. The outage becomes a project, just as the construction of a new facility is a project. Outage management might be handled by the planning and scheduling department or by a separate outage management department.

Requirements Planning

Requirements planning is the process whereby the maintenance department determines its requirement for maintenance resources. These resources include the skills, facilities, and special tools that are essential for the execution of the maintenance programme. The following illustrates what items must be included in the requirements planning process:

- **Spares**. Spare parts are a key element in the planning process. The right spare is required at the right time at the right place. On the other hand, spares holdings represent capital tied up that cannot be used for other purposes in the firm, for example, for marketing campaigns, establishing of overseas facilities, etc. So spares holdings have to be optimised, probably using probabilistic criteria.
- **Skills.** Artisans and technicians with the appropriate experience and skill sets are required to perform the maintenance tasks in the organisation. Changes in workload due to seasonal effects or market forces dictate that there may have to be resorting to overtime and temporary contract labour.
- **Facilities.** The use of specialised equipment such as mobile cranes has to be optimised—whether they should be purchased or rented as and when required. Certain items of plant may not be maintained in-house due to the specialised skills required, and so such items may be subject to supplier/maintainer contracts, for example.
- **Special tools** may be required for dismantling equipment, such as hydraulic wrenches, electrical test equipment, etc. Once again, decisions have to be made concerning in-house or contracted equipment.
- **Downtime requirements.** It is the responsibility of the maintenance function to explain to the operations department the times and frequencies for maintenance on all the plant.
- **Budget.** A maintenance budget must be prepared and agreed by all stakeholders. In fact, all the items mentioned in the bullet points mentioned previously must be considered in the budget. The best way of developing such a budget is to use the RCM process, discussed elsewhere in this volume. Once all the tasks and frequencies for all maintenance actions have been determined, then the necessary manpower, tooling, and services required can be justified.

Man-Hours Required

The man-hours required in the maintenance section must be determined in two ways: man-hours per skill available and the jobs required to be done per skill. The method is described in Tables 10.2 and 10.3 that follow, which are only partially filled in for this example.

From the aforementioned two tables, it will be seen that the company has 20,855 hours available in the year, and a requirement of 11,638 hours to perform the work.

The Master Schedule

This document is generated weekly from the yearly maintenance plan. The diagram that follows illustrates its development.

TABLE 10.2
Man-Hours per Skill per Month

	Total Hrs	Jan	Feb	Mar	Apr	May	Jun	Jul	Aug	Sep	Oct	Nov	Dec
Fitter	8,450	600	220										
Elect.	5,550	350	550										
Instr.	2,885	220											
Welder	3,970	350											
Total	20,855	1,520											

TABLE 10.3
Maintenance Tasks in Hours per Skill per Month

	Total Hrs	Jan	Feb	Mar	Apr	May	Jun	Jul	Aug	Sep	Oct	Nov	Dec
Fitter	5,105	400											
Elect.	4,175	300											
Instr.	1,485	110											
Welder	873	50											
Total	11,638	860											

A typical page from a weekly master schedule is given in the following Table 10.4.

The following are important points to note in the compilation of the weekly master schedule:

- All planning work for a job must have been completed before it may appear in the weekly master schedule.
- All high-priority jobs must be included in the schedule even if the available plannable hours are insufficient.
- The backlog must be kept between tw.o and four weeks.

A certain amount of time must be available for low-priority jobs.

Plannable Work Hours

The plannable hours are the quantity of man-hours available for the allocation of planned work in the next planning week. The planner must know what the plannable hours are for each maintenance section so that the master schedule can be loaded with an appropriate amount of work.

The planner must obtain an estimate of the plannable hours from each maintenance supervisor prior to the development of the master schedule. Various limiting factors must be considered, as demonstrated in the example that follows.

Step 1: Establish the manpower strength in the section = 10 mechanics
Step 2: Number of normal work hours per week = 45
Therefore, potential work hours available = $10 \times 45 = 450$

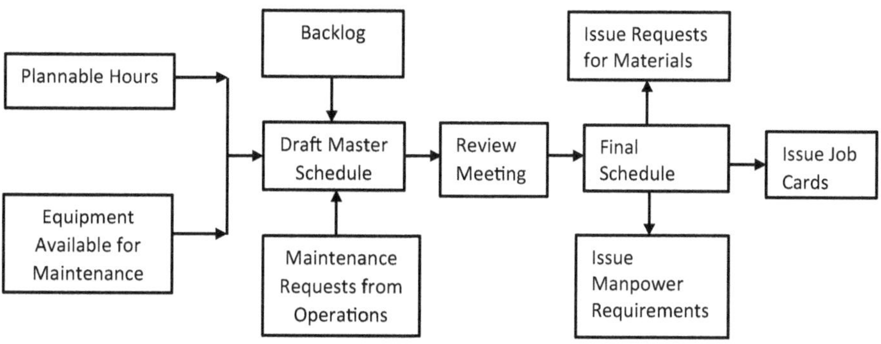

FIGURE 10.1 Weekly master schedule development.

TABLE 10.4
Sample Page from the Weekly Master Schedule

Planner	**A N Other**	**Maintenance Supervisor**	**T Boss**	**Week No.**	**20**	**Page No.**	**7**
Plant	**Ammonia No 1**	**Equipment**	**Coal Mill**	**Craft**	**Fitter (Mechanic)**		
WO ID	**Description**	**Status**	**Offline Date**	**Earliest Start**	**Priority**	**Hours**	**Preceding WO**
1/47	Replace all grinding elements		May 8	May 9	1	40	
1/07	Replace bellow no 4 as damaged		May 8	May 9	1	4	1/06
Total Hours						44	
Notes:			Allow 24 hours for cooling				Leak detected May 7
Signatures	**Planning**	**Maintenance**	**Operations**				
	xxx	yyy	zzz				

Step 3: Deduct hours for persons on leave, etc: 2 persons on leave for 1 day
Therefore, available work hours = 450 – (2 × 9) = 432

Step 4: Productive available work hours = 432 × (wrench time percentage)
= 432 × 60% = 260 (approx.)

Step 5: Plannable hours = 260 – (historic unplanned percentage)
= 260 – (260 × 0.2) = 208 hours

We see that this figure can be considerably lower than the potential work hours.

The aforementioned calculations must be repeated for each trade in each section of the plant.

Equipment Utilisation

Apart from the manpower calculations, the planner must also compute the hours for which each type of equipment is available for maintenance in the week considered.

Step 1: Identify any regular periods of time when the plant in question is not being used. Such periods could include cleaning, product changeover, etc.
Step 2: Identify periods in the week when the equipment is scheduled to be idle.
Step 3: Confirm the times in the previous two steps with the operations department.
Step 4: Record the start dates and times of all stoppages. Record the durations of all these stoppages.

The Draft Master Schedule

The draft master schedule must be compiled by Thursday, with the artisan hours and equipment utilisation recorded. According to the shutdowns available and the manpower available, a proposed work programme is compiled for each plant area and for the trades allocated to that area.

Step 1: List all high-priority jobs for the plant area for the week ahead. Check that the available hours are sufficient for the jobs to be done. Work still in progress from the previous work, if any, must be allowed for.
Step 2: If there are still hours available, see if any other jobs which are not high-priority can be included in the workweek.
Step 3: Retain the list of jobs for the draft master schedule.

Delay Report

If there are high-priority jobs that cannot be allocated to this week's master schedule, this must be highlighted in a report. The reasons for the delays must be stated, for example, lack of spares, insufficient labour, etc. The planner must make suggestions at the weekly planning meeting on possible corrective actions.

The Review Meeting

The objective of the review meeting is to achieve agreement between the planner, the maintenance supervisor, and the operations representative as to what work must be performed in the coming week. The operations representative must confirm what equipment will be available for maintenance.

An attempt must also be made at the meeting to ascertain whether there are any ways to improve the schedule for the week. It must also be confirmed that there are no clashes in the schedule; for example, mechanics and electricians are able to access the plant as required.

Supporting documentation for the meeting should include:

- A copy of the previous week's master schedule
- Draft master schedules for each plant section at the meeting
- High-priority jobs that are delayed
- Plannable work hours per supervisor
- All equipment downtime

Guidelines for the meeting are as follows:

1. The planner must present a list of delayed priority jobs. Discussion must ensue on possible ways to resolve delays causing these jobs to be postponed.
    - If the delay can be resolved, record how the delay is to be resolved and mark the job for inclusion in the week's programme. Set a planned start and end date.
    - If the delay cannot be resolved, return the job to the "not allocated" list.
2. Discuss each on the draft master schedule with the maintenance supervisor and the operations representative to achieve agreement in principle. Then discuss the jobs in detail and determine:
    - If the job can be done with the plant in operation. If so, mark the job as accepted and update or add a plant start date and duration.
    - If the job must be done with the plant offline, check that the job time is less than or equal to the downtime of the plant.
    - If all parties agree to going ahead with the job, mark the job as accepted and confirm a start time and duration.
    - If the job time is longer than the downtime window, discuss with operations and maintenance and attempt to resolve the issue with extra manpower, off-site assistance (e.g., hire of a mobile crane), etc.
    - If a solution is reached, confirm the job and add a start time and duration. Edit the downtime window.
    - If a solution cannot be reached, mark the job as *not accepted*

Note: The process must ensure that all available planned hours are used up. On the other hand, if the available planned hours are insufficient, the meeting must agree on the removal of any low-priority jobs.

THE OFFICIAL MASTER SCHEDULE

The planner must now construct the official master schedule for the week. To create this, the planner must use:

- The draft master schedule with comments as added at the review meeting
- The equipment downtime list, as modified in the review meeting

- A list of delayed high-priority jobs, with comments from the meeting
- Minutes of the review meeting

The planner now interrogates the outstanding work database of the CMMS for all jobs accepted at the planning meeting and updates the following fields:

- Planned start date
- Planned start time
- Job Status: "Allocated"
- Sort the report by planned start date and time
- The planner, maintenance supervisor, and operations representative must sign off the CMMS document and the final master schedule

JOB CARDS

The planner now implements the master schedule by printing the job cards that represent each job on the master schedule. These cards are then issued to the maintenance supervisor.

11 ISO 55000

Asset Management is the optimal management of physical assets over their life cycle.

Asset Management gives assurance that assets will fulfil their required purpose. Assurance applies to the assets themselves, to the process of asset management and to the asset management system.

International Standards Organisation,
ISO 55000 Committee (Wikipedia, 2014)

PRELUDE: STANDARDISATION

ISO 55000 is all about standardisation—about a standard way to maintain assets. So to begin at the beginning, we shall first talk about standardisation, what it is, and why it is a good thing.

STANDARDISATION: A SHORT HISTORY

> Standardization started thousands of years ago. For example, from around 1,000 BC, the Bible demanded a standardized life of justice and order: 'Do not use dishonest standards when measuring length, weight or quantity. Use honest scales and honest weights, an honest ephah and an honest hin.'
>
> (Leviticus, Chapter 19, Verses 35–36)

This is a quote from the DKE, a German standards organisation.

The first national standards organisation was the British Standards Institute, or BSI, established in 1901 in London. Furthermore, the BSI oversaw the first ever Commonwealth Standards Conference, which led in turn to the establishment of the International Organization for Standardization (ISO).

The organisation, which today is known as ISO, began in 1926 as the International Federation of the National Standardizing Associations (ISA). This organisation focused heavily on mechanical engineering. It was disbanded in 1942 during the Second World War but was reorganised under the current name, ISO, in 1946.

For the first 40 years of its existence, ISO focused on developing *technical standards* for products and technologies. The big turning point came in the 1980s, when ISO began developing *process standards*, the first of which became known as the ISO 9000 Quality Management system standards. ISO 14000, a standard concerned with the environment followed, as did others.

The ISO 55000 standard on asset management, ISO 55000 is an international standard covering management of assets of any kind. Before it, a Publicly Available Specification (PAS 55) was published by the British Standards Institution in 2004 for

DOI: 10.1201/9781003326489-11

physical assets. The ISO 55000 series of asset management standards was launched in January 2014.

30 countries are currently (2022) full participants in the ISO 55000 programme: Argentina, Australia, Belgium, Brazil, Canada, Chile, China, Colombia, Czech Republic, Finland, France, Germany, India, Ireland, Italy, Japan, Republic of Korea, Mexico, Netherlands, Norway, Peru, Portugal, Russian Federation, South Africa, Spain, Sweden, Switzerland, United Arab Emirates, United Kingdom, and the United States. 14 other countries have Observer status.

It is seen from the aforementioned that ISO 55000 is widely accepted as the process standard for asset management. In other words, it prescribes an optimal way to procure, operate, maintain, and divest in physical assets like plant and equipment.

ISO 55000 is divided into three main sections:

1. ISO 55000, entitled Asset Management: Overview, Principles, and Terminology
2. ISO 55001, entitled Asset Management: Management Systems: Requirements
3. ISO 50002, entitled Asset Management: Management Systems: Guidelines for the Application of ISO 55001

In practice, the section which is mainly used is ISO 55001. The contents of which is broken down as follows:

1. The organisation
2. Leadership
3. Planning
4. Support
5. Operations
6. Performance
7. Improvements

The full description of the aforementioned is nothing more or less than good management practice as taught at business schools. But a note of caution must be introduced here. The ISO 55000 system of management is very comprehensive and requires that various procedures and systems be put in place, which may tax the resources of the organisation to an impossible extent. To quote just one example, a full configuration management system must be in place. This requirement might be beyond the ability of the organisation to implement. (Configuration management is discussed elsewhere in this textbook.)

Nevertheless, many organisations across the world have adopted ISO 55000 successfully and will attest to its efficacy in optimising their assets, particularly in areas such as plant maintenance.

Various other organisations have preceded ISO 55000, but with a narrower focus. The Japan Institute of Plant Maintenance, for example, have for many years awarded their PM prize to organisations that have exhibited excellence in plant maintenance. It could be said that the implementation of an ISO 55000 system could result in an organisation being audited for the PM prize and achieving it.

ISO 55000 is closely allied to other ISO standards like ISO 9000 for wuality and ISO 14000 for environmental management, as well as standards for safety and for risk management.

To return now to the specifics of ISO 55001. We will now attempt to unravel the legalese of the document and thereby provide a summary thereof. The numbering scheme next follows the sections so numbered in the standard.

The Organisation

This section speaks about the necessity of a Strategic Asset Management Plan (SAMP) and the need to get all the stakeholders in the organisation on board. Also emphasised is the need to tailor the ISO 55000 requirements to the size of the organisation and its specific needs, especially as regards reporting of financial and non-financial factors.

Leadership

- **Leadership and commitment.** Here the standard states the obvious fact that to succeed, the ISO 55000 initiative must be driven by top management and become a "way of life" feature of the organisation. All such initiatives, such a TPM and RCM, that precede ISO 55000 and are less ambitious in scope, have a history of success or failure dependent on top management's involvement. The standard describes ten specific actions that top management must continuously perform. These are in general terms, ensuring that the ISO 55000 system works, that everyone concerned understands the system and its purpose, and that the system is integrated with the other aspects of the company's business.
- **Policy.** The top management must also establish an asset management policy. Policy can be best described by "this is the way we do things around here." The policy must be *appropriate* to the organisation's size and needs. It is not a case of "one size fits all."
- **Roles and responsibilities.** These must be clearly spelled out.

Planning

- **Risks and opportunities.** This is the first heading in the planning section. Risks and opportunities must be identified and dealt with.
- **Asset management objectives.** Here follows a description of the characteristics of the objectives, that they must be consistent with overall company objectives, that they must be appropriate, measurable, recordable, monitored, and capable of being modified as necessary
- **Planning to achieve the asset management objectives.** This will be done using the Strategic Asset Management Plan (SAMP), which is the second document mentioned after the **policy**, mentioned earlier. The plan (or sub-plans) go into details according to the "Five Ws and the H," also known as Kipling's serving-men, as discussed elsewhere in this textbook. For example, for any activity, *who* will do it, *when* will it be done, *why* must it be done, *where* will it be done, and *how* will it be done? Such is the nature of plans.

Support

This section covers three areas as given next:

1. **Resources.** The resources needed for the asset management system are emphasised here,
2. **Competence.** The competence of everyone in the organisation is to be defined, measured, and assured.
3. **Awareness.** Everyone in the organisation must be aware of their role in the asset management policy and the implications of their possible non-conformance.

Operations

This section also covers three areas: operations p;lanning and control, management of change, and outsourcing.

Performance Evaluation

In this section, the need for internal audits is stressed, as is the need for management reviews, which are not the same. The standard spells out the differences.

Improvement

Sections covered here are the need for identification and correction of non-conformances (a quality function), preventive actions, and continuous improvements.

CASE 11.1

Refer back in the textbook to Case 5.6, "Trouble at the Smelter." Re-analyse this case using the principles of ISO 55000. Lay out your answer using the headings given prior (organisation, leadership, etc.). Notice how your appreciation of the situation is clarified by the application of the ISO 55000 principles, as described next. Further insights will be obtained by full reference to ISO 55000, rather than just the summary given previously.

ISO 55000 PRE-AUDIT

Next is an audit process developed over several years and used successfully by the author in approximately 30 companies. The audit is not strictly aligned to the ISO 55000 system but can be used to indicate the relative strength of an organisation before attempting an ISO 55000 system.

MODEL AUDIT BOOK

Notes to auditors:

1. Award points as shown if answer is given. Documentary proof required for at least one answer per subsection, that is, for 1.1, 1.2, etc.
2. A few questions may be repeated in different sections of the audit—this is deliberate (different persons supply the answer).
3. One point scored for each question, unless otherwise stated.

SECTION 1

Overall Indices

Respondent: General manager, managing director

1.1 Industry Benchmarks

1.1.1 Availability Index (Describe)
1.1.2 Reliability Index (Describe)
1.1.3 Maintenance Cost as a Percentage of Turnover
1.1.4 Cost of Downtime per Hour
1.1.5 Maintenance Cost Ratio: Labour: Material: Downtime
1.1.7 Backlog of Maintenance Work
1.1.8 Other Relevant Indices Used by Your Organisation (List Them)
1.1.9 Source of Industry Benchmarks

1.2 Target Figures

1.2.1 Availability Index
1.2.2 Reliability Index
1.2.3 Maintenance Cost as a Percentage of Turnover
1.2.4 Cost of Downtime per Hour
1.2.5 Maintenance Cost Ratio: Labour: Material: Downtime
1.2.6 Works Power Cost
1.2.7 Backlog of Maintenance Work
1.2.8 Other Indices Used by Your Organisation (List Them)
1.2.9 How Were the Targets Determined from the Benchmarks?

1.3 Current Figures (For Last Complete Year, as Compared to Targets)

1.3.1 Availability Index
1.3.2 Reliability Index
1.3.3 Maintenance Cost as a Percentage of Turnover
1.3.4 Cost of Downtime per Hour

1.3.5 Maintenance Cost Ratio: Labour: Material: Downtime
1.3.7 Backlog of Maintenance Work
1.3.8 Other Indices Used by Your Organisation (List Them)

Total out of 24+ ☐

SECTION 2

Maintenance strategy

Respondent: Works manager or equivalent and general manager/managing director

2.1 Does your company have definitions for terms such as *strategy*, *policy*, *vision*, *mission*, etc.?
2.2 Are these terms understood throughout the company?
2.3 Does a documented maintenance strategy exist, with objectives, goals, and targets?
2.4 Does the maintenance strategy form part of overall business plan, and is it part of company's overall strategy?
2.5 Are plant additions and improvements properly coordinated and handled separately from regular maintenance?
2.6 Describe how the maintenance strategy was devised (e.g. through an RCM [reliability-centred maintenance] programme).
2.7 Is the strategy reviewed regularly and updated if necessary (e.g. yearly)?
2.8 Does an asset register exist?
2.9 Has an equipment criticality list been defined according to a standardised model or procedure?
2.10 Are internal maintenance audit procedures in existence, and are they applied regularly?
2.11 Is there statistical support for the maintenance function (e.g. trending of data, etc., with statistical verification of significance, statistical analysis of failures, etc.—may be part of a reliability engineering function)?
2.12 Is there operations research or industrial engineering support for the maintenance function (e.g. crew size optimisation by queuing theory or simulation studies, value engineering studies, etc.)?
2.13 Is a life cycle asset management process in place (e.g. are calculations in place for optimal equipment replacement periods, etc.) (examples to be shown)?
2.14 Are any contact maintenance agreements in place ?
2.15 If so, how is the decision made to self-maintain or use a contractor?

2.16 Is there a loss control function in place which covers the maintenance function?
2.17 Who has overall responsibility for the maintenance strategy, its design, implementation, and revision?
2.18 Is there reliability engineering support for the maintenance function (failure analysis, statistical and metallurgical, equipment redesign to minimise or eliminate maintenance, etc.)?

Total out of 18 ☐

SECTION 3

Planned Maintenance (Value ~ 20%)

Respondent: Works manager or equivalent

3.1 Maintenance Organisation

3.1.1 Maintenance Organisational Chart in place and current.
3.1.2 Job descriptions available for all maintenance positions, with a clear definition of functions and a clear definition of each person's or team's key performance area and accountability.
3.1.3 Ratio of hourly paid maintenance personnel to supervisors (answer should lie in range 8–12:1).
3.1.4 Ratio of hourly paid maintenance personnel to planners (answer should lie in the range 15–20:1).
3.1.5 Does the company have an incentive scheme for workers, based on plant/business performance? If so, does the maintenance department participate?
3.1.6 Percentage of skilled personnel, that is, qualified artisans in hourly paid group? (Answer should be > 80%.)

3.2 Training

3.2.1 All supervisors trained in maintenance management fundamentals.
3.2.2 All planners trained in modern maintenance planning techniques.
3.2.3 All artisans trained in plant-specific technologies.
3.2.4 Skills assessment and skills development programmes in place and operative.

3.3 Planning and Scheduling

3.3.1 Percentage of planned work orders experiencing delay due to inadequate planning? (Should be < 5% of artisan hours booked to jobs.)

3.3.2 Outstanding work reports available by craft, department requesting, and date required.
3.3.4 Actual job times are compared to estimates.
3.3.5 Plant register in place and up to date—all equipment numbered in register and physically.
3.3.6 Job reference packs, giving job instructions, tools required, materials required, and estimated times per craft are available for critical jobs/equipment.

3.4 Work Orders

3.4.1 All maintenance actual man-hours are recorded and costed against specific work orders.
3.4.2 All maintenance materials are costed against specific work orders.
3.4.3 All significant maintenance jobs (i.e. jobs worthy of a history) are covered by work orders.
3.4.4 All work orders are cross-referenced to specific equipment numbers or addresses.
3.4.5 Work is checked by supervisor for work quality and completeness.
3.4.6 Work orders are checked by supervisor for correctness and completeness before issue and for completeness and feedback after job is complete.
3.4.7 Work orders include estimates of and are completed with actuals of downtime, craft hours for each craft, materials, requestor's name.

3.5 Maintenance Optimisation (e.g. RCM—Reliability-Centred Maintenance)

3.5.1 RCM programme or similar in place (10 points).
3.5.2 RCM programme or similar regularly updated.
3.5.3 Manning levels, spares levels, condition monitoring on programme, etc. determined from RCM programme or similar?
3.5.4 All maintenance tasks and frequencies determined by RCM or similar process?

3.6 Configuration Management (Modification Control)

3.6.1 Configuration management programme in place and being used to ensure that plant changes are documented in all relevant places, for example, manuals, parts lists, training programmes, etc. (10 points).
3.6.2 All equipment manuals in place and up to date.

3.7 Outage Management (If Applicable)

3.7.1 Outages managed according to project management principles, using critical path analysis for project duration, resource allocation, and costing.
3.7.2 Outages managed by a specifically designated outage manager.
3.7.3 Training in outage management given to all involved in the process—line production managers, maintenance management, and planning.

3.8 Condition Monitoring or Predictive Maintenance

3.8.1 Condition monitoring programme in place and continuously managed.
3.8.2 Standard work orders available for lubrication, detailed inspections, and condition monitoring readings.
3.8.3 The following condition monitoring techniques are in use: e.g. Vibration analysis, oil analysis, thermography, etc. List the techniques used.

3.8.4 How successful has the ConMon programme been in eliminating forced outages?
3.8.5 Has an audit been done on ConMon programme effectiveness?
3.8.6 What has been the benefit/cost ratio of the ConMon programme, and is this tracked?
3.8.7 How are ConMon warnings integrated into the work order programme?

3.9 Incident Investigations (Reactive Analysis of Failures)

3.9.1 Does a formal procedure exist for the investigation of serious incidents? (Reports of recent investigations to be shown.)
3.9.2 Is a procedure in place to ensure the implementation of any recommendations made as a result of an investigation?

3.10 Work History

3.10.1 How is work history recorded?
3.10.1 How is work history used?

Total Out of 64 ☐

SECTION 4

Inventory Management (Spares and Materials Management) (Value ~ 8%)

Respondent: Materials manager, stores manager or equivalent, together with the maintenance manager

4.1 Stores

4.1.1 Maintenance material stock-out % (answer should be < 5%).
4.1.2 Aisle/bin locations specified for all in-store items.
4.1.3 Maintenance stores catalogue available in hard copy or on computer and is up to date.
4.1.4 Register of company-owned tools in place and up to date.
4.1.5 Shafts in storage are stored horizontally and are adequately supported along their entire length.
4.1.6 Mounted bearings (particularly in motors) are "index-turned" on a regular basis.
4.1.7 Procedures are in place to adequately cope with stores items with a shelf life problem, for example, paints and rubber components (e.g. inspect at regular intervals/discard at regular intervals, etc.).
4.1.8 Rubber products are properly stored to prevent mechanical stress points and oxidation (e.g. v-belts are not hung on nails, or in sunlight).

4.2 PURCHASING

4.2.1 Vendor selection programme in place.
4.2.2 Tracking system in place to ensure timeous delivery from suppliers.

4.3 Planning

4.3.1 Stock reorder programme in place and linked to risk of downtime as a reorder criterion, at least for strategic spares.
4.3.2 Annual review of stock for obsolescence.
4.3.3 Reservation system in place for spares and materials against works orders.
4.3.4 Return to stores procedures in place.
4.3.5 Audit process in place to identify and eliminate substores.
4.3.6 Do you have a programme involving both maintenance and stores to investigate standardisation of suppliers and commonality of parts, for example, bearings?
4.3.7 Is there a method in place for the inspection of incoming stores, to ensure the correctness and quality of the stockholding?

4.4 Costing.

4.4.1 How are spares costed (annual holding cost, etc.).
4.4.2 Is the true cost of spares holdings known?

4.5 Discussion

4.5.1 Question for most senior person in charge of the inventory function: How does inventory management fit in with the overall strategy of the company (5 points)?

Evaluators' Assessment of Question 4.4.1:

....

....

....

....

....

Total Out of 24 ☐

SECTION 5

Maintenance Reporting and the CMMS (Computerised Maintenance Management System) or MMMS (Manual Maintenance Management System)

Respondents: Works manager (IT manager—for CMMS), chief planner

5.1 CMMS

5.1.1 Is a CMMS in place?
5.1.2 Is there data integrity?
5.1.3 Is there system integrity?
5.1.4 Percentage of functions actually used.
5.1.5 If there is no CMMS in place but a manual system is used, has the benefit/cost of a CMMS been investigated?
5.1.6 Reason for not using a CMMS, if a manual system is used.

5.2 Reports (MMMS or CMMS)

5.2.1 Maintenance reports delivered timeously.

5.2.2 Maintenance exception report in place, indicating work orders incomplete, equipment awaiting spares, and any other cause of non-compliance.

5.3 Support for the Management Function

5.3.1 Data collection system in place and used to collect failure frequency data.
5.3.2 Data collection system in place and used to collect cost of failure data.
5.3.3 MMMS/CMMS integrated with condition monitoring system.
5.3.4 MMMS/CMMS integrated with inventory programme.
5.3.5 Inspections and planned maintenance interventions are all generated by the MMMS/CMMS.

5.4 Discussion

5.4.1 Question for most senior person in charge of the maintenance. Reporting function: How does maintenance reporting and/or the MMMS/CMMS fit in with the overall strategy of the company (5 points)?

Evaluators' Assessment of Question 5.4.1.

....

....

Total out of 18 ☐

SECTION 6

Maintenance Excellence and Continuous Improvement (e.g. TPM)

Respondents: General manager, managing director

6.1 Policies and goals in place.
6.2 Formal programme structure in place.
6.3 What stage of the programme has been reached, for example: Stage 1: Operator cleaning of equipment, etc.?
6.4 Equipment improvement programme in place.
6.5 Autonomous maintenance programme in place.
6.6 Operator training in plant fundamentals in place (for all other training, see Section 3).

6.7 Integration programme in place, that is, all other company departments have been sensitised to the importance of the maintenance function and support it accordingly.
6.8 Maintenance prevention programme in place (equipment re-engineering).
6.9 Trends in improvement criteria being produced regularly.

Total out of 9 ☐

SECTION 7

Safety, Health, Environment, and Quality (SHEQ)

Respondent: Works manager, safety officer, environmental manager, quality manager

7.1 Safety and Health

7.1.1 ISO 18000 or similar safety system in place and applied.
7.1.2 Safety system accredited and audited by recognised authority (state authority).
7.1.3 Discussion: How does the safety system fit in with the overall company objectives, and what is the interface between safety and maintenance (5 points)?

7.2 Environment

7.2.1 ISO 14000 or similar system in place and applied.
7.2.1 System accredited and audited by recognised authority (state which authority).
7.2.3 Discussion: How does the environmental system fit in with the overall company objectives, and what is the interface between environmental management and maintenance (5 points)?

7.3 Quality

7.3.1 ISO 9000 series or similar system in place and applied.
7.3.2 ISO 9000 accreditation and audit (state accrediting body).
7.3.3 Discussion: How does the quality system fit in with the overall company objectives, and what is the interface between quality and maintenance (5 points)?

Total out of 21 ☐

SECTION 8

Artisan interview

Respondent: Artisans

8.1 Artisan Selected by Management (Tenure at the firm > 12 months)

8.1.1 Have you been on a craft upgrade programme, or are you due to go on one in the next twelve months (e.g. a course on a specific machine at the factory where you work, or a general course on bearings, seals, etc.)?
8.1.2 Have you been exposed to an RCM or similar programme in any way?
8.1.3 Have you been given a course in the work order procedure, and do you understand the importance thereof?
8.1.4 Do you book waiting time on jobs?
8.1.5 On how many jobs do you have to wait? (Answer must be 20% or less.)
8.1.6 Are your work instructions clear and complete?
8.1.7 Do you receive feedback on your section's/department's performance on a regular basis?
8.1.8 Do you ever make suggestions to your superiors on improvements to plant or procedures or methods of working?
8.1.9 If your answer to 8.1.8 is yes, are your suggestions well received by management?
8.1.10 Discussion: What is your role in helping the company achieve its objectives?

Evaluator's assessment of Section 8.1.10 (5 points)?

....

....

....

....

8.2 Artisan Selected by the Audit Team; Other Details as for 8.1

8.2.1 Have you been on a craft upgrade programme, or are you due to go on one in the next 12 months (e.g. a course on a specific machine at the factory where you work or a general course on bearings, seals, etc.)?

8.2.2 Have you been exposed to an RCM or similar programme in any way?
8.2.3 Have you been given a course in the work order procedure, and do you understand the importance thereof?
8.2.4 Do you book waiting time on jobs?
8.2.5 On how many jobs do you have to wait? (Answer must be 20% or less.)
8.2.6 Are your work instructions clear and complete?
8.2.7 Do you receive feedback on your section's/department's performance on a regular basis?
8.2.8 Do you ever make suggestions to your superiors on improvements to plant or procedures or methods of working?
8.2.9 If your answer to 8.2.8 is yes, are your suggestions well received by management?
8.2.10 Discussion: What is your role in helping the company achieve its objectives?

Evaluator's assessment of Section 8.2.10 (5 points)?

....

....

....

....

Total out of 30 ☐

SECTION 9

Interview Assessment (Between Auditors Only)

9.1 Employee attitudes: subjective assessment of attitudes from the interviews. Score each item as follows:

Excellent = 5
Very Good = 4
Good = 2
Fair or Less = 0

9.1.1 Management positive, optimistic, and realistic. .... (5, 4, 2, 0)
9.1.2 Artisans positive and enthusiastic and do not use the interview as a lobbying platform for a wage increase................. (5, 4, 2, 0)

Total Out of 15 ☐

SECTION 10

Site Visit

10.1 State of plant

Score each item as follows: TPM standard ("eat off the floor") = 10

Very Good	= 8
Good	= 6
Fair	= 4
Unacceptable	= 2
Poor	= 0

10.1.1 Condition of plant.. (10, 8, 6, 4, 2, 0)
10.1.2 Condition of maintenance shops. .... (10, 8, 6, 4, 2, 0)
10.1.3 Condition of stores. .... (10, 8, 6, 4, 2, 0)

Total out of 30 ☐

Auditor's notes on site visit:

....

....

....

....

....

SECTION 11

Maintenance Costing

Respondent: Chief cost accountant

11.1. What is the current level of maintenance expenditure as a percentage of income?
11.2. What is the current split between maintenance and investment expenditure?
11.3. How is maintenance expenditure prioritised at present?
11.4. What is the current level of capital investment?
11.5. What is the current method of prioritising investments?
11.6. What indices of maintenance performance are currently in use (e.g. availability/planned downtime/forced downtime)?
11.7. Are costs split up according to the "basic five"?

Maintenance costs—labour
Maintenance costs—materials
Operating costs—labour

Operating costs—other (e.g. power)
Cost of unavailability (e.g. lost revenue)

11.8. What is the cost structure in the maintenance department?

Fixed
Variable

11.9. What is the out-of-service cost (i.e. unavailability cost) for:

- An item of equipment?
- A production line?
- The plant as a whole?

11.10. How is this cost determined?

Total out of 10 ☐

SECTION 12

Labour

Respondent: Maintenance manager + HR manager

12.1. What are the restrictions, if any, of transferring labour from one depot to another (e.g. union rules, technical expertise, etc.)?
12.2 How is the culture and work ethic of the maintenance personnel measured?
12.3 How are manning levels determined?
12.4 Discussion on labour issues as they affect maintenance (10 points).

....

....

....

....

Total out of 13 ☐

SECTION 13

Employee Benchmarking

Respondent: Works manager and HR manager and possibly CEO

13.1 In any part of corporate literature, on your website, etc., do you make the claim that people are the most important asset of the company?

13.2 If you do not, please comment on the prior claim, made by many ....1 companies?

13.2 Do you (a) select, (b) promote, and (c) categorise employees according to the following four factors ....(12 points)?

13.2.1 Qualifications ....(a, b, c)
13.2.2 Experience....................(a, b, c)
13.2.3 Health ....(a, b, c)
13.2.4 Intelligence..................(a, b, c)

13.3 Do you have inventories of your overall employee ratings according to the aforementioned categories ....(10 points)?

For example, experience level per trade, individual + average
For example, level per trade
For example, level per engineer, etc.

13.4 Do you have comparisons of your average ratings with those of other companies (benchmarking) ....(6 points)?

13.5 Discussion with company representatives on the aforementioned aspects....................(20 points).

Total out of 50 ☐

SECTION 14

Alignment with ISO 55 000

Respondents: Works manager, CEO
For each, answer either yes or no; reasons must be given. Full marks may be awarded even if there is no alignment, provided cogent reasons are presented. 1 point per answer.

14.1 Is your asset management system aligned with ISO 55000?....................(y/n)

14.2 If the answer to 14.1 is negative, is it your intention to attain alignment to ISO 55000? ....(y/n)

14.3 Does your terminology align with ISO 55000 terminology? ....(y/n)

14.4 Do you have a life cycle management system in place?.............(y/n)

Totals

POSSIBLE GRAND TOTAL (SECTIONS 1–14)

Note: Discretionary bonus/penalty points may be added/deducted for some items not included in the audit sheets but which the evaluators feel require special consideration in this specific case (may add/deduct 0–20 points). Specify why bonus/ penalty awarded, if any:

....

....

....

....

....

....

Grand total ☐

Possible total before bonus points:

Appendix 1
The Standard Normal Distribution

Area Under the Curve from the Mean to a Distance z Standard Deviations from the Mean

z	**0.00**	**0.01**	**0.02**	**0.03**	**0.04**	**0.05**	**0.06**	**0.07**	**0.08**	**0.09**
0.0	0.0000	0.0040	0.0080	0.0120	0.0160	0.0199	0.0239	0.0279	0.0319	0.0359
0.1	0.0398	0.0438	0.0478	0.0517	0.0557	0.0596	0.0636	0.0675	0.0714	0.0753
0.2	0.0793	0.0832	0.0871	0.0910	0.0948	0.0987	0.1026	0.1064	0.1103	0.1141
0.3	0.1179	0.1217	0.1255	0.1293	0.1331	0.1368	0.1406	0.1443	0.1480	0.1517
0.4	0.1554	0.1591	0.1628	0.1664	0.1700	0.1736	0.1772	0.1808	0.1844	0.1879
0.5	0.1915	0.1950	0.1985	0.2019	0.2054	0.2088	0.2123	0.2157	0.2190	0.2224
0.6	0.2257	0.2291	0.2324	0.2357	0.2389	0.2422	0.2454	0.2486	0.2517	0.2549
0.7	0.2580	0.2611	0.2642	0.2673	0.2704	0.2734	0.2764	0.2794	0.2823	0.2852
0.8	0.2881	0.2910	0.2939	0.2967	0.2995	0.3023	0.3051	0.3078	0.3106	0.3133
0.9	0.3159	0.3186	0.3213	0.3239	0.3264	0.3289	0.3315	0.3340	0.3365	0.3389
1.0	0.3413	0.3438	0.3416	0.3485	0.3508	0.3531	0.3554	0.3577	0.3599	0.3621
1.1	0.3643	0.3665	0.3686	0.3708	0.3729	0.3749	0.3770	0.3790	0.3810	0.3830
1.2	0.3849	0.3869	0.3888	0.3907	0.3925	0.3944	0.3962	0.3980	0.3997	0.4015
1.3	0.4032	0.4049	0.4066	0.4082	0.4099	0.4115	0.4131	0.4147	0.4162	0.4177
1.4	0.4192	0.4207	0.4222	0.4236	0.4251	0.4625	0.4279	0.4292	0.4306	0.4319
1.5	0.4332	0.4345	0.4357	0.4370	0.4382	0.4394	0.4406	0.4418	0.4429	0.4441
1.6	0.4452	0.4463	0.4474	0.4484	0.4495	0.4505	0.4515	0.4525	0.4535	0.4545
1.7	0.4554	0.4564	0.4573	0.4582	0.4591	0.4599	0.4608	0.4616	0.4625	0.4633
1.8	0.4641	0.4649	0.4656	0.4664	0.4671	0.4678	0.4686	0.4693	0.4699	0.4706
1.9	0.4713	0.4719	0.4726	0.4732	0.4738	0.4744	0.4750	0.4756	0.4761	0.4737
2.0	0.4772	0.4778	0.4783	0.4788	0.4793	0.4798	0.4803	0.4808	0.4812	0.4817
2.1	0.4821	0.4826	0.4830	0.4834	0.4838	0.4842	0.4846	0.4850	0.4854	0.4857
2.2	0.4861	0.4864	0.4868	0.4871	0.4875	0.4878	0.4881	0.4884	0.4887	0.4890
2.3	0.4893	0.4896	0.4898	0.4901	0.4904	0.4906	0.4909	0.4911	0.4913	0.4916
2.4	0.4918	0.4920	0.4922	0.4925	0.4927	0.4929	0.4931	0.4932	0.4934	0.4936
2.5	0.4938	0.4940	0.4941	0.4943	0.4945	0.4946	0.4948	0.4949	0.4951	0.4952
2.6	0.4953	0.4955	0.4956	0.4957	0.4959	0.4960	0.4961	0.4962	0.4963	0.4964
2.7	0.4965	0.4966	0.4967	0.4968	0.4969	0.4970	0.4971	0.4972	0.4973	0.4974
2.8	0.4974	0.4975	0.4976	0.4977	0.4977	0.4978	0.4979	0.4979	0.4980	0.4981
2.9	0.4981	0.4982	0.4982	0.4983	0.4984	0.4984	0.4985	0.4985	0.4986	0.4986
3.0	0.4987	0.4987	0.4987	0.4988	0.4989	0.4989	0.4989	0.4989	0.4990	0.4990
3.1	0.4990	0.4991	0.4991	0.4991	0.4992	0.4992	0.4992	0.4992	0.4993	0.4993

Appendix 2
A Perspective on Robert Lusser[1]

Reid Collins

He knew all along that the odds of safe space travel were not on our side. Will the *Columbia* accident investigators acknowledge his insights?

The *Columbia* Accident Investigation Board is on the verge of declaring the probable cause of the February 1 shuttle disaster to be the chunk of foam that struck the leading edge of the left wing during the first moments of lift-off. What it will not say is that the tragedy is the latest fateful application of "Lusser's law."

The irony is that the late Dr Robert Lusser, regarded as the "father of reliability," had always harboured a suspicion that the propulsion systems for which he was chiefly responsible could not be made reliable enough. He was chief of the reliability section at Redstone Arsenal, Wernher von Braun's development centre in Huntsville, Alabama, in the mid-1950s. In his native Germany, Lusser had an early-established reputation for genius. He developed the Messerschmitt Me-109, once the world's fastest aircraft and a mainstay fighter of the German Luftwaffe. Lusser also engineered the Heinkel He-219, a Luftwaffe nightfighter. Cantankerous and confrontational, he fought with both Willy Messerschmitt and Heinkel and at one time or another quit both.

Most famously, or infamously, Lusser engineered the first cruise missile, the ramjet-powered V-1, the "buzz-bomb." Thousands were launched against Britain after the Royal Air Force had defeated the German-manned bombers. And thousands of Britons died.

A few weeks before the official end of World War II in Europe, on March 13, 1945, one plane left a flight of Allied bombers flying high over Upper Bavaria, peeled off, and sped for a farmhouse in the middle of nowhere. It was the sanctuary to which Lusser had sent his wife and four children. A stick of seven bombs was dropped, one hitting the house, killing Mrs Hildegarde Lusser. French prisoners employed on the farm said the plane bore British markings, odd for the fact that the others in the formation appeared to be American on the traditionally American daylight raids. The surviving Lussers have always wondered. But Robert Lusser was away at the V-1 works at that time.

Unlike von Braun, who arrived in the West at war's end, Robert Lusser didn't get to the United States until 1948, first with the US Navy at Point Mugu, then the Pasadena Jet Propulsion Labs, and then at Redstone with von Braun in 1953. Lusser would apply his law of probability to the rockets being developed with which whose

[1] Article used with the permission of *The American Spectator:* Wladyslaw Pleszcznski, Editorial Director, *The American Spectator*, Arlington, Virginia.

formula, $R_S = R_1 \times R_2 \times \ldots R_n$, presaged systems theory thinking, never caught the von Braun fever for manned space exploration.

In an interview with this reporter in the mid-1950s, Lusser declared that "man can never go to the moon, let alone to the Mars." He explained in layman's terms that there was simply too much to go wrong; the probability odds with which he had wrestled a lifetime were simply too great for the risk. And this from a man charged with trying to make it happen! Impolitic? Yes, and especially so at a time when a Senate Majority Leader named Lyndon Johnson was leading the budget battle for a nascent NASA.

Lusser's daughter, Traute Grether, recalls her father was "completely convinced re-entry would not work." But an America alarmed by Soviet space achievements was not to be deterred by mathematical formulae. (It became legendary among US engineers and programme managers at the Cape that if they waited for "the Germans" in Huntsville to declare their rocket ready, they would never leave the ground.)

In January of 1959, Lusser left Huntsville, returned to Germany, and joined the combined Heinkel–Messerschmitt–Boelkow concern. Later, he would make European headlines by accurately predicting the failures of the German-bought F-104 Lockheed Starfighters because he observed they were being converted into all-purpose craft that could not meet those requirements; 20 of the planes crashed, killing a dozen pilots. Lusser's law was not to be denied.

Lusser spent his last days and a lot of his personal fortune trying to market a ski binding he had designed to release at just the right time of stress. He was 69 when he died in January of 1969, seven months before men did what he thought to be "reliability forbidden"; they landed on the moon and came back.

However, a review suggests that not enough attention has been paid Lusser's law. The Soviets lost a single space pilot and then a three-man crew on failed re-entries. The United States lost three astronauts in a launch pad fire when pure oxygen was used in the early Apollo craft, necessitating a switch to an oxygen–nitrogen mix and elimination of flammable materials in the craft. Of five space shuttles built, two have been lost, the *Challenger* on launch and, most recently, the *Columbia* on re-entry, with total fatalities of the crews.

As in the past investigations, the *Columbia* Accident Investigation Board will assign a probable cause, and in hindsight, it will all look so preventable. There will ensue a debate over funding for safety. Heads have already rolled. More may. Beneath it all, a constant, a mathematical rule of reliability intuited by a German engineer whose name is unknown to most Americans. Unless they want to know the odds and then consult "Lusser's law."

References

Abernethy, R. B. *The New Weibull Handbook*, 5th Edition. Gulf Publishing Company, Houston, 2004. ISBN 0-9653062-3-2.

Ascher, H. and Feingold, H. *Repairable Systems Reliability*. CRC/Marcel Dekker, Inc, 1984. ISBN-13 978–0824772765.

Bompas-Smith, J. H. *Mechanical Survival: The Use of Reliability Data*. McGraw-Hill, Maidenhead, UK, 1973. 07–084411–9.

Cameron, K. *Top Dead Center*. Motorbooks, MBI Publishing Company, St. Paul, Minnesota, 2007. ISBN 13 978-0-7603-2727-2.

Caplen, R. H. *A Practical Approach to Reliability*. Business Books Ltd, London, UK 1972. ISBN 13 978–022066099.

Dean, J. *Capital Budgeting*. Columbia University Press ISBN-10: 0231018479.

Deming, J. E. *Out of the Crisis*. MIT Press, Boston 1986. ISBN 10- 0262541157.

Drucker, P. *The Concept of the Corporation*, Routledge, 1993 ISBN 9781560006251.

Drucker, P. *The Practice of Management*, 1955, Harper & Brothers, New York.

Fayol, H. *Administration Industrielle et Generalle*, 1916, Bulletin de la Sociétè de l'Industrieminerale, Paris, *Industrial and General Administration*, 1949, Pitman, London.

Galison Manufacturing Company, *Welkom South Africa. Brochure on Scraper Winches*. www.galison.co.za.

Gano, D. L. *Apollo Root Cause Analysis*, 2nd Edition. Apollonia Publications, Yakima, Washington, 2003. ISBN 1-883677-01-7.

Guess, V. C. *CMII for Business Process Infrastructure*. Holly Publishing, Scottsdale, Arizona. ISBN0-9720582-0-6

Hackman, J. R. and Oldham, G. R. *Work Redesign*. Addison-Wesley, Reading, MA, 1980.

Herzberg, F. *One More Time: How Do You Motivate Employees?* Harvard Business Review, January 2003.

Higgins, L. R. and Brautigam, D. P. *Maintenance Engineering Handbook*. McGraw-Hill, New York, 1994.

International Atomic Energy Agency. ASSET Guidelines Revised 1991 Edition. Reference material prepared by the International Atomic Energy Agency for Assessment of Safety Significant Events Teams, IAEA TECHDOC No. 632, IAEA, Vienna, 1991.

International Standards Organisation, ISO 55000 Committee, 2014. https://en.wikipedia.org/wiki/ISO_55000. Last edited 17 August, 2022.

Irving Fisher. 1930 no longer in print. Now available as a Kindle edition.

Jenkins, G. Piper Alpha—Lessons for life cycle safety management: Proceeding of a symposium organized by the Institution of Chemical Engineers, September 1990, *The Institution of Chemical Engineers*. Published in Utilities Policy, Vol. *1* (*4*), pp. 356–357, 1991.

Kepner, C. H. and Tregoe, B. B. *The New Rational Manager*. Princeton Research Press, Princeton, New Jersey, 2013. ISBN-13: 978–0971562714.

Kipling, R. *The Elephant's Child*, various publishers including Random House, 1902. ISBN 978–1939228246.

McDermott, R., Mikulak, R. J. and Beauregard, M. *The Basics of FMEA*, 2nd Edition. CRC Press, 2009.

McGregor, D. *The Human Side of Enterprise*, 1960, McGraw Hill, New York.

Narayan, V., Wardhaugh, J. W. and Das, M. *Case Studies in Maintenance and Reliability*. Industrial Press Incorporated, New York, 2012. ISBN 978-0-8311-3323-8.

O'Connor, P. D. K. and Kleyner, A. *Practical Reliability Engineering Handbook*, 5th Edition. John Wiley & Sons, Chichester, UK, 2012. ISBN-13: 978–0470979815.

Palmer, R. D. *Maintenance Planning and Scheduling Handbook*, 2nd Edition. McGraw-Hill, New York, 2006. ISBN 0-07-145766-6.

Porkess, R. *Collins Dictionary of Statistics*. Harper Collins, Glasgow, 2014. ISBN 0-00-714501-2.

Reason, J. and Hobbs, A. *Managing Maintenance Error*. Ashgate Publishing Company, Aldershot, Hampshire, UK, 2003. ISBN 0-7546-1590-1.

Roadstrum, W. H. *Excellence in Engineering*. Wiley, 1967. Library of Congress Catalog Card Number: 67–26524.

Ross, P. *The Night the Sea Caught Fire: Remembering Piper Alpha*. The Scotsman, 15 June 2008.

Rudd, T. *It Was Fun*. Haynes Publishing, Sparkford, Somerset, UK, 2000. ISBN 1-85960-666-0.

Shirose, Kunio. *TPM for Workshop Leaders*. Productivity Press, New York, 1922. ISBN 0-915299-92-5

Suzuki, T. *TPM in Process Industries*. CRC Press, Taylor and Francis Group, Boca Raton, Florida. ISBN 1-56327-036-6

Taylor, F.W., *Principles of Scientific Management*, 1911, Harper & Brothers, New York and London.

Trotti, J. *Phantom over Vietnam*. Berkley Books, New York, 1986. ISBN 0-425-10248-3.

Wang, W. *Reverse Engineering—Technology of Reinvention*. CRC Press, Florida, USA, 2011. ISBN 978-1-4398-0630-2.

Westerkamp, T. 'Maintaining Maintenance', *IIE Solutions*, Vol. *38* (*7*), pp. 37–48, 2006.

Weisman, A. Thomas Dunne Books, St Martins Press, New York, 2007. ISBN-13: 978-0-312-34729-1.

Williams, J. B. *The Theory of Investment Value*, Harvard University Press, 1934 now published by Frazer Publishing Company, Virginia 2002. ISBN 0-87034-126-X.

Index

Note: Page numbers ending in 'f' refer to figures. Page numbers ending in 't' refer to tables.

G

H

I

J

K

L

M

N

O

P

Q

R

S

T

For Product Safety Concerns and Information please contact our EU
representative GPSR@taylorandfrancis.com
Taylor & Francis Verlag GmbH, Kaufingerstraße 24, 80331 München, Germany

www.ingramcontent.com/pod-product-compliance
Lightning Source LLC
Chambersburg PA
CBHW060746220326
41598CB00022B/2341

* 9 7 8 1 0 3 2 3 5 3 4 9 4 *